FISHERIES

FISHERIES

Editor

Dietmar Wirnhier

scitus
academics

Fisheries

Edited by **Dietmar Wirnhier**

This Book is Printed on Acid-Free Paper. ♾

ISBN: 9781681179513

Library of Congress Control Number: 2017949085

SCITUS Academics LLC,

616, Corporate Way, Suite 2, 4766,

Valley Cottage, NY 10989

www.scitusacademics.com

Notice

Printed in United States of America.

Preface

Fish play an important role in the balance of life on which we depend. Fish are essential to life in the world's oceans. Fishery is an entity engaged in raising or harvesting fish which is determined by some authority to be a fishery. Fisheries have occupied an important place in the economic structure of many countries throughout history. Fisheries future explains the method in new advancements of fisheries science. The ocean plays a vital role in the global economy by providing food and a source of income for millions of people. More than 50 million people work in the fishing and aquaculture sector, many in small-scale fisheries that are critical to the economies of their communities. More than three billion people, many of whom live in the poorest and least developed countries, rely on food from the ocean as a significant source of protein, highlighting the role of marine species in food security worldwide.

This book covers the novel and current scientific studies taking place across universities and research centers in various parts of the world in the areas of fisheries science, fishing technology, fisheries management and relevant socio-economics. It highlights the main freshwater fishes and the major fisheries of the Nile River basin. The ecological status of the Nile fisheries and their economic importance to populations living in the countries of the Nile basin are also reviewed. It also describes some of the major Pan-Arctic fisheries, the stock assessment methods applied to assess them and how the fisheries might change with climate warming and further development of the northern regions. This book is intended for fisheries scientists, biological oceanographers, gear technologists, economists, managers, administrators, policy makers and legislator.

Table of Contents

CHAPTER 1

The Nile Fishes and Fisheries

Waleed Hamza[1]

[1]Biology Department, Faculty of science - United Arab Emirates University, Al-Ain, UAE

1. INTRODUCTION

The Nile basin extends from 4° S to 31° N and includes ten African countries: Burundi, Egypt, Eritrea, Ethiopia, Kenya, Rwanda, Sudan, Tanzania, Uganda and Democratic Republic of Congo (Fig. 1). The source of the Nile water is one of the upper branches of the Kagera River in Rwanda. The Kagera follows the boundary of Rwanda northward, and continues along the border of Tanzania before draining into Lake Kyoga between Lake Victoria and Lake Albert, the Nile rushes along its course for 483 km within rocky valleys and over rapids and cataracts. The section between the two lakes is called Lake Kyoga and is part of the Victorian Nile, which name is used for the river section till its confluence with the Blue Nile in Sudan. The Nile discharges from the northern end of Lake Albert and flows through northern Uganda to the Sudan border where it becomes the Bahr El-Jebel. At its conjunction with Bahr Al-Ghazal, the river becomes the White Nile. At Khartoum, the White Nile is joined by the Blue Nile.

FIGURE 1.The Nile River trajectory from source to outfall

The 1529 km long Blue Nile has its source in Lake Tana in the Ethiopian Highlands. From Khartoum the Nile flows 322 km north to its junction with the Atbara River. Downstream from its confluence with the Atbara River, the Nile traverses the Nubian Desert, and is marked by two profound bends in its course, the first north of Khartoum and the second near Aswan in the Egyptian territory. The Nile flows towards the Mediterranean Sea through the Nile delta that splits into the two main Nile delta branches, the Rosetta and Damietta branches. The Nile River flows generally north to the Mediterranean Sea for a distance of 5,584 km. From its remotest head-stream, the Ruvyronza River in Burundi, the river is 6,671 km long, and it has a drainage area of more than 2,590,000 km^2 [1].

With a length of 6,695 km from source to Mediterranean outfall, the Nile River is the longest river in the world. The uniqueness of the Nile is not only related to morphometric features of its basin, but to other many facts. No other river traverses such a variety of landscapes, and such a spectrum of cultures and peoples as the Nile. No other river has historically had such a profound material effect upon those who dwell along its banks, representing the difference between plenty and famine, between life and death, for multitudes since the earliest stages of human history [2]. The course of the Nile flows from highland regions with abundant moisture to lowland plains with semi-arid to arid conditions [1].

For Nile basin countries with no access to the coast for economic activities, the Nile River is their artery of life and economic sustenance. Its fertile sediments supply agriculture soil with necessary elements for plant growth, and the fish stock in its large natural and manmade lakes are crucial to the food and freshwater security of the Nile basin countries. Fish industry and fishing activities on the major water bodies of the Nile basin are not only a source of affordable animal protein, especially in poor countries, but also provide employment, income and export earnings to the riparian communities.

The principal objective of the present review is to highlight the main freshwater fishes and the major fisheries of the Nile River basin. The ecological status of the Nile fisheries and their economic importance to populations living in the countries of the Nile basin are also reviewed. The main lakes and river systems of the Nile basin will be compared on the basis of their fish species, fisheries potentiality and economic importance to the countries that host them.

2. MAJOR LAKES OF THE NILE BASIN

The lengthy course of the Nile River is interrupted by several important lakes. Some of them are natural, while others were engineered by dam constructions and artificial reservoirs projects. Within the Nile basin, there are five major natural lakes with a surface area totaling more than 100,000 km^2(Victoria, Edward, Albert, Kyoga and Tana). These are vast areas of permanent wetlands and seasonal flooding (the Sudd, Bahr Al-Ghazal, and Machar marshes). There are also five major reservoir dams (Aswan High Dam, Roseires, Khasham El-Girba, Sennar and Jebel Aulia) in the basin. Before discharging its water into the Mediterranean Sea, the Nile fills four coastal lakes within its Delta (Lake Mariut, Lake Edku, Lake Burullus and Lake Manzala) with a total area of 1100 km^2 [2].

The share of the Nile basin area for each of the ten countries hosting the river may represent a major or minor fraction of the total area of the country (Table 1). However, the economic significance of the basin may be quite disproportionate to its area, such as it is for Kenya and Tanzania [3].

TABLE 1. National areas within the Nile Basin "Source [14]"

Country	Area in basin (km^3)	Percentage of total country area in Nile Basin
Burundi	13 000	46%
D R Congo	22 300	1%
Eritrea	25 700	21%
Ethiopia	366 000	32%
Egypt	307 900	33%
Kenya	52 100	9%
Rwanda	20 400	83%
Sudan	1943 100	78%
Tanzania	118 400	13%
Uganda	238 700	98%

Nile waters are the mainstay for freshwater supply for agriculture irrigation, navigation, water for human use ("drinking, industrial, domestic uses"), hydroelectric power generation, and of course for exploitation of natural fish stocks and aquaculture fish farming projects.

3. FISH AND FISHERIES OF TROPICAL LAKES (L. VICTORIA AND L. TANA)

Fishes and fisheries of Lake Victoria and Lake Tana will be reported here, as natural examples of tropical large water bodies of the Nile basin. The combined surface area of these lakes is 75% of the total of the five tropical lakes of the Nile basin. They are also important because they lie at the source of both the White Nile and the Blue Nile, respectively. Their main fishes are representative of the fish species living in the other lakes (Edward, Albert and Kyoga). Details on the other tropical lakes (Edward, Albert, Kyoga, and George) fish fauna and its fisheries are found elsewhere [4], [5].

3.1. Lake Victoria (Kenya, Tanzania and Uganda)

Lake Victoria is the largest natural lake in Africa (Fig. 1) and the second largest lake in the world after Lake Superior (USA-Canada). It has a surface area of 68,800 km^2 and a very large catchment area of 193,000 km^2. The shoreline measures 3450 km in length and the lake has a mean depth of 40 m and a maximum depth of 80 m. The lake is shared by Kenya (6%), Uganda (43%), and Tanzania (51%). The lake basin is estimated to have a population of 30 million individuals which is growing at a rate of >3% per annum. Three major cities (Kampala, Kisumu and Mwanza) with a combined population of about six million inhabitants depend on the lake for domestic and municipal water supply and waste disposal. Recent data has shown it is becoming eutrophic [6].

Two introduced species, Nile perch (*Lates niloticus*) and Nile tilapia (*Oreochromus niloticus*), and one native cyprinid (*Rastrineobola argentea*) dominate the lake's fisheries. The introduction of the Nile perch (*L. niloticus*), in the early 1960's had major ecological consequences. It is believed that some 200 endemic halochromine species (previously comprising 90% of the lake fish biomass) have become extinct in the lake due to predation by the Nile perch [7]. Most of these species had no commercial value, but are now considered relevant to the sustainability of the Lake Victoria ecosystem [8]. Lake Victoria fisheries production has increased markedly since the introduction of the Nile perch. Despite the general unreliability of published data, there is agreement that between the late 1980's and early 1990's there was an increase in the quantity of fish landed in Lake Victoria fisheries reaching a total of 500,000 tons per year. This figure was in decline by the late 1990's and early 2000's when it reached a total of 340,000 ton per year (Fig. 2). This reduction was mainly due to the overexploitation of the fisheries resources, and to the increase in the number of motorized boats, and to the Nile perch processing industries. Reportedly the number of fishing boats in the lake increased from 12,000 boats during 1982 to 22,700 boats during 1990 and more than 52,000 boats in 2002 [6]. In recent years, as fishing pressure on the Nile perch has intensified, there are signs of recovery in at least some of the former haplochromine species that were thought to be extinct from the lake environment [9], [10], [5].

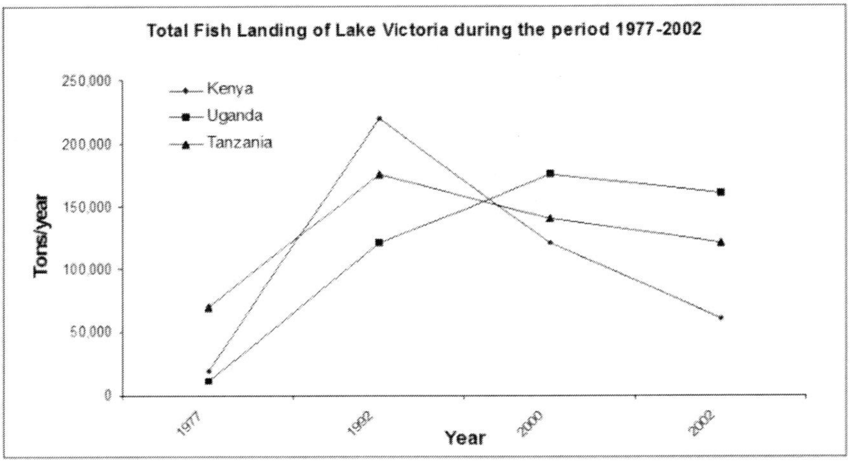

FIGURE 2. Variation of the total fish landing of Lake Victoria during the period 1977-2002 [8]

Extracted data from the results obtained by [6], from the ECOPATH prediction model have revealed that the stock biomass (tons km^{-2}) of the fishes (Nile perch, tilapia and sardine) are different in the waters of Lake Victoria shared by different countries. The same is also true for the ratios of production per biomass except for the sardine species (Fig. 3). The data also shows that the Ugandan lake areas have lowest stock of Nile perch biomass, while the highest stocks are found in the Kenyan lake areas. Interestingly, the Ugandan stock biomass of tilapia is the highest while the Tanzanian stocks are the lowest. The sardine stock biomass is higher in the Kenyan lake areas followed by the Tanzanian, with the lowest recorded in the Ugandan lake areas (Fig. 3). An overall estimate shows that the average stock biomass for the whole of Lake Victoria is 4.59 tons km^{-2} for the Nile perch, 3.27 tons km^{-2} for the sardine and 2.36 tons km^{-2} for the tilapia. The same trend is observed for the estimate of production/biomass $year^{-1}$, giving the ratios 3.75, 3.0 and 1.75 for the three fishes, respectively.

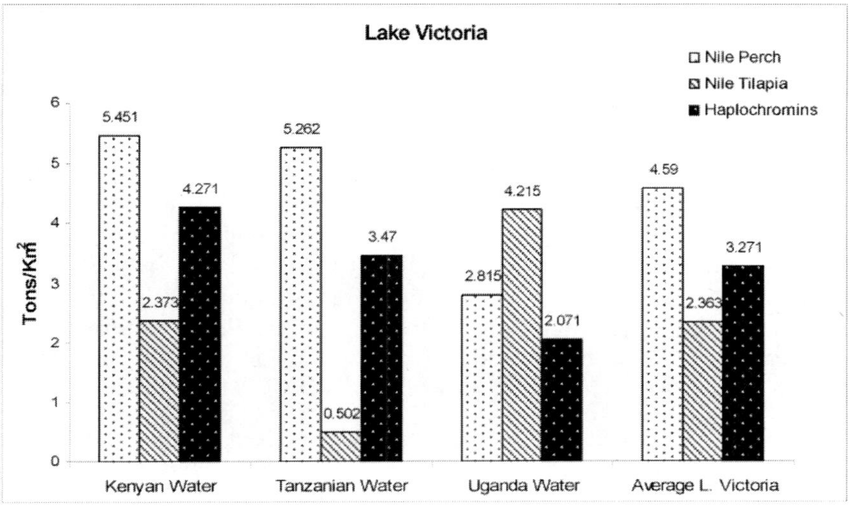

FIGURE 3. Biomass potentials of different water sectors of Lake Victoria for the main three fish species (tons km^{-2}) [8]

The artisanal fishing gear was in use on traditional fishing boats in Lake Victoria fisheries until the industrial exploitation of the Nile perch for international commercialization. At that time, motorized boats and improved fish gear "even illegal" was used. The most common gear used in Lake Victoria fisheries are gill nets and long-line hooks. During the period of 2000 and 2004, the numbers of gill nets in use increased by 50% reaching 980,000 gill nets, while the numbers of long-line hooks is estimated at 8 million in 2001. This number was adjusted to only 6 million during 2006 [6]. The current haul is estimated at 45 kg boat^{-1} day^{-1} compared with the 80 kg boat^{-1} day^{-1} figure during 1990's. This indicates over-intensive exploitation of the fisheries of Lake Victoria fish resources and argues for sustainable management intervention.

3.2. Lake Tana *(Ethiopia)*

Lake Tana is the largest lake in Ethiopia (surface area 3,050 km^2). It lies at an altitude of approximately 1800 m a.s.l. in the north-western highlands of Ethiopia, 500 km north of the capital Addis Ababa (Fig. 1). The lake is shallow (average depth 8 m, maximum depth 14 m), and its trophic status is described as oligo-mesotrophic [11]. The lake's fishery is 30 km downstream from the Blue Nile outflows, and is isolated from the lower

Nile basin by 40m high waterfalls. Physical and hydrological features of the lake water follow a seasonal pattern. Rainfall peaks in July-August are followed by a rise in the lake water level by 1.5 m, reaching highest levels in September-October [12].

In their study, DeGraaf et al. [12] identified the three endemic fish species groups Barbs or yellow fish (*Labeobarbus* spp.), African catfish (*Clarias gariepinus*), and Nile tilapia (*Oriochromus niloticus*), as the main representatives of Lake Tana fisheries. They noted that before 1986 Lake Tana was a subsistence fishery exploited predominantly by reed boat fishing vessels. This type of fishery was limited to the shore areas and targeted the native Nile tilapia using locally made fish traps and small gill nets.

In 1992-1993 about 113 reed boats were counted on the lake and about 374 gill nets were in use, with an overall daily haul averaging 12.3 kg (7.8 kg Nile tilapia, 4.3 kg *Labeobarbus* spp. and 0.2 kg African catfish). Here it is important to indicate that *Labeobarbus* spp. in this lake is represented by nearly 15 species. It was also reported at that time that the fishermen in the lake numbered about 400 individuals. This meant that each reed boat had only one fisherman [12]. The introduction of motorboats to Lake Tana fisheries was a consequence of the increasing demand for fish from the capital Addis Ababa. This introduced an additional 130 professional fishermen, which markedly negatively influenced the fish stock in the lake. In fact, the annual total catch fell to 255 metric tons in 2001 from a value of 360 metric tons in 1997. Not only was there a reduction in total catch but also the percentage composition of the individual fish species yield varied between 1993 and 2001, so that the catch of *Labeobarbus* spp. declined by almost third in 2001 compared to 1993. Notably, the catch of Nile tilapia increased 50% in 2001 compared to 1993 (Fig. 4). This may reflect the intensive exploitation of *Labeobarbus* spp. and the superior ability of the Nile tilapia to multiply in the lake.

4. FISH AND FISHERIES OF MID NILE (SUDAN FRESHWATER FISH AND FISHERIES)

Although Sudan is a country with a very large surface area (2,505,825 km^2), its inland water bodies occupy only 114,000 km^2 during periods of high water level. The inland fisheries of Sudan are based on the Nile river tributaries, contributing over 90% of the estimated production potential of the country. The Sudd swamps in the south (Fig. 1), and the man-made lakes on the White Nile, the Blue Nile, Atbara river and the main Nile river

(Lake Nasser portion in Sudan "Nubia") count as the major fishing localities with respect to fish resource magnitude and exploitation thrust [13]. The total fish production of the various inland water bodies and wetlands are estimated at 50,000 tons year^{-1}. This figure is not representative of the full potential of the various fisheries (Table 2), as in some areas civil war disturbances and dense cover of aquatic macrophytes, and inadequate fishing gear have had a negative effect on local fish production.

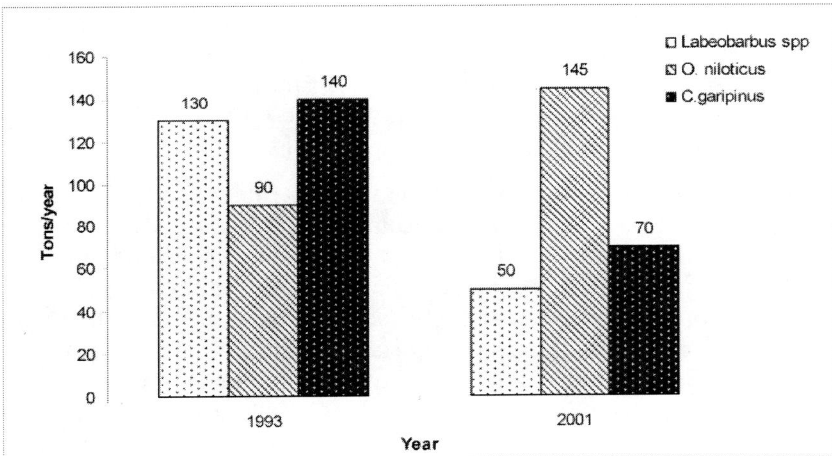

FIGURE 4. Total catch of the three main species from Lake Tana during the years 1993 and 2001 [13]

TABLE 2. Calculated Potential and Effective Fish Catch from Different Inland Water Bodies of Sudan

Area	Fish Catch Potential Tons year^{-1}	Actual Yields Tons year^{-1}	Percentages of Exploitations
Sudd Region	75,000	30,00	43%
White Nile Reservoir	15,000	13,000	86%
Blue Nile Reservoir	1,700	1,500	88%
Senner Reservoir	1,100	1,100	91%
Lake Nubia	5,100	1,000	19.6%

Portion			
Others	4,000	4,000	100%
Total	101,900	50,500	≈ 50%

The predominant fishing gear includes active and passive gill nets, seine nets, trammel nets, long line hooks, cast nets and baskets. The commercially important fish are Nile perch (*Lates niloticus),* Bagrid catfish *(Bagrus bayad),* Silver catfish *(Bagruc docmac),* Nile tilapia *(Oreochromus niloticus),* Carp fish *(Labeo* spp.*),* Barbs fish *(Barbus binny),* Mormyrus fish *(Mormyrus* spp.*),* Nile Distichodus *(Distichodus* spp.*),* Tiger fish *(Hydrocyon* spp.*)* and Nile robber *(Alestes* spp.*).* There are many other species, though they have no commercial value [13].

It is important to state that, apart from the FAO reports, there is almost no scientific literature available on the ecological features and dynamics of fish populations in the inland waters of Sudan. This underlines the importance of organizing a research program on inland fisheries in this area in order to fill the gap of knowledge about the fisheries status in this important part of the Nile basin. This is especially important for the wet lands which are not limited to Sudan but also exist in the ten countries of the Nile basin, and account for an area of almost 200,000 km^2. A characteristic of wetlands fisheries is the variety of traps that are used to catch fish in dominantly submerged or emergent vegetation habitats. Many of these traps have traditional designs adapted to local conditions and most are made from local plant materials - often from wetlands themselves. One type of fish requiring a special fishing technique is the African lungfish (*Protopterus aethiopicus*), which is native to Ethiopian wetlands and inhabits the seasonal wetlands where it aestivates in the dried soil and is "hunted" during the dry season.

5. DOWNSTREAM FISHES AND FISHERIES (EGYPTIAN FRESHWATER FISHERIES)

Egypt is the downstream country hosting the last 1530 km of the Nile river channel. The Nile enters Egypt through the Nubian Lake which continues as Lake Nasser behind the Aswan High Dam (AHD). The water allowed to pass through the AHD is the Egyptian quota of Nile water (55.5 Km^3 $year^{-1}$) as outlined in an agreement with Sudan in 1959. This quantity satisfies the freshwater demands of the Egyptian population for use in agricultural irrigation, industry, domestic and navigation purposes. On the way to its Mediterranean destination, the Nile water fills many canals within the irrigation network in the western desert and the Nile delta. It also contributes to the water budget of the Northern Delta lakes [2]. Thus the freshwater inland water fisheries in Egypt include Lake Nasser (the second largest man made reservoir in Africa, after Lake Volta-Ghana), the Nile main river channel, irrigation channels, some water bodies in the Western Desert, the Nile branches (Rosetta and Damietta), and the Northern Delta Lakes (Manzala, Burullus, Idko and Mariut). The excess water is then discharged into the Mediterranean Sea through the lake connections with the sea via direct and indirect inlet openings [2]

The annual fish yield from freshwater fisheries of the Egyptian Nile basin has increased annually from 157,888 metric tons in 1990 to 224,940 metric tons in 2000 [14]. The catch increased for a number of reasons, such as the nutritive effect of sewage and fertilizer discharge in drainage channels and lakes, the intensified fishing activities and adjustment of incensement statistical techniques. Apart from the intensive aquaculture that has developed during the last twenty years in Egypt, the major contributors to the freshwater fish catch could be listed as Lake Nasser, Nile branches and irrigation channels as well as the northern Delta lakes. Their contribution may be summarized as follows:

5.1. Lake Nasser Fishes and Fisheries

Lake Nasser lies behind the AHD and its extension in the northern Sudan is referred to as the Nubia reservoir. This lake has a depth reaching 180m and covers an area of 6216 km^2, 5248 km^2 of which are in Egypt and the rest in Sudan (Fig. 1). The total fish catch from Lake Nasser has been estimated as 28,153 tons $year^{-1}$ [15]. Tilapia species, mainly the Nile tilapia represent 90% of the total fish catch, while Nile perch and *Barbus* spp. cover the remaining 10%. Despite the large area of the lake, it contributes only about 10% of the total freshwater fish landing in the Egyptian Nile basin. This is probably due to its deep waters and the low number of motorboats with suitable fishing gear adapted to its bathymetry.

5.2. Nile Branches and Irrigation Channels Fishes and Fisheries

There are more than 100 sites recorded along the Nile branches and major irrigation channels at which freshwater fishes are collected for marketing. Most fishermen in these areas are not registered. The fishing fleets at these sites comprise hundreds of small wooden boats (4-6 m in length). The fishing gear used is mostly primitive, though trammel nets are in use in some areas. The common fish species caught from these channels are tilapia species (*O. niloticus, O. aureus, Sarotherodon galilaeus* and *Tilapia zilli*) as well as *Clarias* spp.

Although these fishing sites are sparsely distributed and reliable statistics are in short supply, it has been estimated that 34% of the total Egyptian freshwater fish catch is slated to the two main Nile branches (Rosetta and Damietta) and the major irrigation channels [16].

5.3. Northern Delta Lakes (NDL)

As the final reservoirs of Nile river water before it flows into the Mediterranean, the four lakes (Mariut, Edku, Burullus and Manzalah) are the last opportunity for Egyptians to use the Nile water within the Delta area (Fig. 5). The four lakes occupy an area of about 1100 km^2. They are commonly shallow (average depth 1.10m), and their water salinity is known to change from fresh to brackish in the seaward direction. They are connected to the Mediterranean Sea either directly or indirectly.

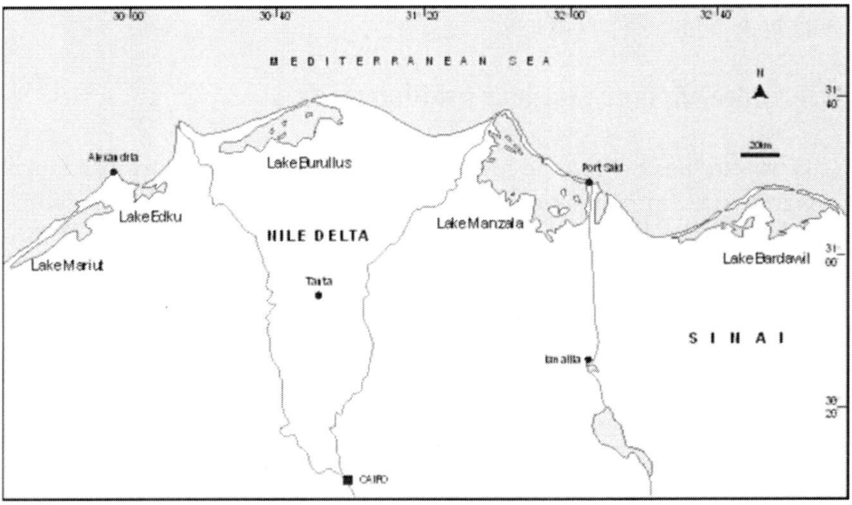

FIGURE 5. The Northern Nile Delta lakes

During the last 50 years, the surface area of the four lakes has been reduced to as little as 50% of the area they covered during the 1950's (Table 3). The lake areas lost have been used for highway construction, reclamation of new lands for agriculture, and in some cases, for residential constructions. Such unnatural use of the lake basins has certainly modified their ecosystems as a whole, beginning with hydrological changes and resulting in the reduction of their economic and natural value. Due to their location along the coastal area of the Egyptian Mediterranean, the lakes are surrounded by or included within the highly populated coastal cities of the country. This has become more pronounced, particularly during the last 20 years, in the various forms of human impact on these water bodies, such as industrialization and the construction of new infrastructures in response to population density increases. These impacts are manifested in the deterioration of the water quality and reduction in lake surface area [17]. The consequent negative effect of human impact on the NDL ecosystems became evident in the decline in fish production. Up to 1985 the annual fish production of these lakes accounted for >50% of the total annual yield for the country.

TABLE 3. Gradual changes in the surface areas (Km^2), of the Northern Delta Lakes during the last 50 years "After, [17]"

Year's	L. Mariut	Edku L.	L. Burullus	L. Mazalah
1950's	136.9	150.1	571.8	1274.2
1970's	68.4	129.2	534.4	997.3
1980's	66.1	115.3	481.6	904.1
1990's	62.3	109.7	350.8	650.3

According to the year 2001 records, the lakes fisheries had a total fish production of about 140,000 tons year $^{-1}$, which represented about 60% of the freshwater fish catch and about 30% of the total annual yield of Egypt [16]. The main freshwater fish species living in the NDL are Nile tilapia *(Oreochromis niloticus)*, Blue tilapia *(O. aureus), Mango tilapia (Sartherodon galilaeus)* and *Tilapia zilli* also known as *Tilapia melanopleura,* Nile perch *(Lates niloticus),* Bagrid catfish *(Bagrus bayad),* Nile carp *(Labeo niloticus),* Yellow fish *(Barbus prince),* Barbus fish *(B. bynni bynni), and* African catfish *(Calrias gariepinus).* Other species, such as Grey mullet *(Mugil cephalus),*Thinlip grey mullet *(Liza ramada),*Saline mullet *(Liza salinus) and Eel fish Anguilla anguilla,* belonging to the brackish-marine sectors of the lakes, represent a minor part of the total catch of these lakes. The fishing fleets and gear used in the lakes are generally primitive and they are adapted to the shallow water depth and the presence of abundant emergent and submerged macrophytes. Thus wooden boats varying in size from 4 to >20 m in length and a number of net types and sizes are used.

6. WATER QUALITY IN MAJOR NILE BASIN FISHERIES AND ITS IMPACT ON FISH PRODUCTION

6.1. Water Quality in The Upstream Basins

Studies of water quality of the Nile basin and its channels and lakes have demonstrated a trend of lower water quality in the downstream basins compared with those located upstream. This is mainly due to the following:

The all-year round continuous flow of the river from south to north which dilutes existing pollutants and contaminants and carries them downstream.

Less industrialization of upstream countries compared to the downstream ones.

Neither water multi-use nor recycling processes are common practice in upstream countries compared with the downstream ones.

Low population densities and less modernized sewage systems at upstream countries compared with the downstream ones.

In addition to the above, the main impact of human activities in the areas surrounding the upstream Nile river water bodies may result from the uncontrolled drainage and discharge of agriculture fertilizers, insecticides, herbicides and untreated sewage into the Nile basin. This is probably true for Lake Victoria and Lake Tana, since industrial pollution and discharge of inorganic pollutants are more limited. However, in these upper parts of the Nile basin, the existing civil wars and the absence of hygiene have resulted in periodic devastating epidemics such as cholera and other infectious diseases that may also negatively influence fisheries production either directly or indirectly.

Studies of Lake Victoria water quality have indicated that notable changes in physical, chemical and biological features of the lake ecosystem have occurred during the last three decades. These changes have resulted in increasing eutrophication and decreasing water transparency and decreasing dissolved oxygen concentrations at the hypolimnion layer during periods of stratification [18], [19]. These studies indicated also that water eutrophication in Lake Victoria is a direct result of nutrient enrichment from human activities in the catchments area as well as industrial and domestic sewage discharge from combustion processes. That is in addition to soil erosion and fertilizers' washout through drainage since the introduction of tea, cotton and maize farming.

Recent studies of pesticide residues in Nile tilapia and Nile perch from southern Lake Victoria has implications of recent exposure of these two species to DDT and endosulfan isomers. The study concluded that most of the analyzed samples contained residue levels higher than the method detection limits, though below or within the Accepted Daily Intake (ADI) limits [20]. Other studies have indicated that spraying of endosulfan in cotton fields near Tana River (Kenya) result in levels of fish contamination [21], [22].

The most surprising finding of these studies is the detection of these compounds in fish but not in local sediments and water. It has been found that the high demand for fish has pushed some Lake Victoria fishermen to use DDT as a mean to catch fish in certain areas of Lake Victoria. The reason of the absence or the inability to detect such compounds in the lake sediments could be due to dilution effect of the running water in these basins which made their residuals below the detection limits.

6.2. Water Quality in The Mid-River Basins

In Sudan the water quality of the southern wetlands (Sudd area) depends on how much water is supplied annually to its basins. The continuous alteration of wetlands by human activities and their utilization for other purposes have negatively influenced both the quality and the quantity of water reaching this area. The redirection of water away from the Sudd for human activities has resulted in the formation of isolated ponds, promoting water stagnation and growth of macrophytes, and weeds. These practices have negatively influenced the capacity of fisheries in this part of the Nile basin.

At Lake Nasser, where both the White Nile and the Blue Nile water mix in different portions, the residues of Nile water contaminants from the upstream countries via Sudan reach this huge man-made reservoir shared by Sudan and Egypt. Lake Nasser water quality is affected also by the continuous fluctuation of its water level. In this regard, most of the studies focused on detecting contaminant levels in the lake water and its effect through bioassay experiments of living biota. Unfortunately, these types of studies did not show the full effect of these pollutants on fish and the degree of pollution of the lake water. Recently, a monitoring study of a range of heavy metals (Co, Cr, Cu, Fe, Mn, Ni, Sr, and Zn) in Nile tilapia fish of different ages (1, 1.5, 2, 2.5 and 3 years) from Lake Nasser was undertaken by [23]. This study examined the trace element concentrations in water, sediments, and aquatic plants, as well as different edible and non-edible fish organs at different ages. The study showed that both water samples and aquatic plants have higher levels of Fe compared with concentrations detected in fish edible parts (muscles). On the other hand, the concentrations of the various metals in sediments were higher than their concentration in fish muscles. The author concluded that the concentrations of Co, Cr, Cu, Fe, Mn, Ni, Sr, and Zn in the edible parts of the examined Tilapia fish were within the recommended permissible levels for human consumption (Table 4).

TABLE 4. Heavy metal concentrations in fish (*T. nilotica*), aquatic plant (*N. armeta*), sediment and water samples from Lake Nasser. "Modified from, [23]". DW, dry weight; (±), Standard deviation

Elements	Zn	Sr	Ni	Mn	Fe	Cu	Cr	Co
Aquatic plantMean (μg/g DW)	175 (±6)	240 (±9)	19(±1.2)	740 (±8)	1720 (±14)	68 (±6)	29 (±1.1)	35(±1. 6)
SedimentMean (μg/g DW)	143 (±2.9)	455 (±12)	122(±3. 2)	1000 (±19)	51500 (±24)	109 (±2.6)	79 (±2.0)	89.5 (±2.6)
WaterMean (μg/g DW)	230 (±0.84)	1852 (±18)	145 (±1.02)	186 (±0.9)	1420 (±8)	220 (±1.1)	240 (±0.58)	185 (±0.8)
FishMean (μg/g DW)	1.55 (±0.08	1.92 (±0.99)	0.19 (±0.25)	0.5 (±0.066)	6.45 (±1.1)	0.27 (±0.08)	0.29(±0.22)	0.25 (±0.05)

6.3. Water Quality in The Downstream Basins

In their location farthest downstream in Nile basin, the Nile Delta branches (Damietta and Rosetta) and the Northern Delta Lakes (NDL) are, not surprisingly, the most polluted and have the lowest water quality. The trophic status in these water bodies has been described in the scientific literature as varying between eutrophic and hypertrophic conditions [17], [2]. The continuous aggressive human impacts on the water bodies of the NDL since the 1950's until now have resulted in various negative consequences, beginning with the reduction of their surface area (Table, 3), and increase of nutrients discharge through agricultural, industrial drainages and domestic sewage; and leading to the dramatic decline of their fisheries [24]. Both domestic and industrial sewage represent a major source of nutrient enrichment of water bodies of the Nile delta. These contribute significantly to the development of the eutrophication phenomenon and the consequent degradation of water quality. In a number of cases, municipal and rural domestic wastewater is discharged directly into waterways. The constituents of domestic and industrial input to water resources are pathogens, nutrients, trace metals, suspended solids, salts and oxygen demanding materials [25]. Siegel [26] reported that the nutrient base for aquaculture in Lake Manzalah is sewage carried by drains from as far away as Cairo, 140 km to the south. He added that untreated or poorly treated industrial wastes, heavy metals and other pollutants have been released into the Nile Delta drainage network and have been discharged, along with sewage and agriculture wastes, into the northern delta lakes and their associated wetlands. Studies of sedimentary deposits in the southeastern sector of Lake Manzalah have detected different trace metal concentration. They have detected Hg (up to 822 ppb), Pb (up to 110 ppm), Zn (up to 635 ppm), Cu (up to 275 ppm), Cr (up to 215 ppm), Sn (up to 14 ppm) and Ag (up to 4.7 ppm). The results suggest that high concentrations of heavy metals in lake sediments may cause contamination in fish, especially bottom feeders in such environments [26]. This may be the case in the northern lakes, since all of them receive the sewage discharges of major cities.

Although both agricultural drainages and sewage discharges into the lake environments may be considered as sources of nutrient that can compensate the nutrients reduction in the Nile water after the construction of Aswan High Dam (AHD), these discharges have different pollutants, such as organochlorine compounds, which are not only harmful to the fisheries but also to the consumers of fish. In his study, [27] noted that the use of organochlorine insecticides in Egypt began in the 1950's and was extensively used until 1981 to protect crops from insects, disease fungi and weeds, to remove undesired vegetation and for domestic household use in the control of insects. The reported active ingredients of major organochlorine pesticides during a 30-year period include: toxaphene 45,000 Mt (1955-1961), endrin 10,500 Mt (1961-1981), DDT 13,500 Mt (1952-1971) and lindane 21,000 Mt (1952-1978) [28]. DDT is still in limited use in the country as a rodenticide and termiticide. These pollutants are released into drainage canals and reach the delta branches and NDL. According to the data provided by the Ministry of Agriculture, the application of pesticides decreased in the country, from 20,500 tons in 1980 to 16,435 tons in 1995. Since chlorinated pesticides have been banned, the majority of the pesticides currently in use are organophosphorus compounds [27]. Table 5 shows concentrations of PCB's, DDT's, HCB and Lindane present in the water bodies located in the Delta region.

TABLE 5.Concentrations (ng/l) of HCB, DDT's, PCB's and lindane in freshwaters from different locations of the Nile Delta- Egypt "Modified from [27]" <DL: below detection limit

Location	Sampling	HCB	DDTs	PCBs	Γ-HCBs
River Nile					
Cairo	1993	<0.001	0.08-0.12	3-640	0.05-0.07
Rosetta Branch	1998	1-77	0.2-99	5-161	
Damietta Branch	1988	<DL-93	90-102	25-53 7-21	<DL-126
Nile Estuaries					
Rosetta	95-97	197-217	83-97	166-181	286-310
				390-430	
Damietta	95-97	195-240	109-128	270-330	312-352
				581-700	
Coastal Lakes					
Lake Manzala	1993	0.03-0.18	0.10-0.56	18-48	0.37-1.57
	92-93	0.84-2.28	26-55	<DL-20	8.6-12.1
			3-26		1-11

7. FRESHWATER AQUACULTURE IN THE NILE BASIN

Freshwater aquaculture activities in the Nile basin countries depend on the following main factors:

1. The status of fish as a main food source according to the customs of the local populations

The existence of a market demand for specific fish living in the local river and lake waters

The economic value of these activities for the private sector compared with other projects

This is clearly demonstrated by the estimation of aquaculture production within the Nile basin countries (Table 6). Although Lake Victoria produces almost 25% of the freshwater fish catch in Africa, the aquaculture productivity of countries sharing the basin of Lake Victoria is very limited. This could be due to the lack of interest of the private sector in developing aquaculture activities, since the Lake production easily satisfies local and international market demand. In Uganda the number of ponds totals 6200, covering a total area of 124.6 ha. In some of these ponds yields of about 10,000 kg ha^{-1} y^{-1} were recorded. The most widely distributed species are mirror carp and various tilapia species. However, since the Nile perch is the most popular fish for export, there is little interest in developing aquaculture. It is expected that future expansion in freshwater farming in this part of the Nile basin is promising, especially when the fish stock at Lake Victoria become overexploited.

TABLE 6.National aquaculture production in the countries of the Nile Basin [29], [30]

Country	1987	1988	1989	1990	1991	1992	1993	1994	1995	1996 -	2007
Burundi	25	24	25	30	50	50	55	55	50	10	200
DR Congo	723	759	760	700	700	730	700	750	750	750	2970
Eritrea	-	-	-	-	-	-	-	-	-	-	-
Ethiopia	-	1	33	36	36	22	28	33	55	85	-
Egypt	45000	50000	55000	55916	877	54195	45380	46887	45473	53302	635516
Kenya	126	251	603	698	722	749	763	813	893	500	4240
Rwanda	64	38	44	164	58	53	53	50	49	50	4038
Sudan	43	45	100	234	203	200	200	200	1000	1000	1950
Tanzania	35	37	375	375	400	350	200	150	200	200	10
Uganda	38	34	42	52	63	77	87	179	194	210	51110

Although freshwater aquaculture was not commercially practiced in Sudan until 1990, the population density increase and local market demands have encouraged large scale aquaculture of freshwater fish, especially Tilapia species. The main demand comes from the capital Khartoum, where population density is rising, with growing interest in tilapia fish as an alternative source of animal protein [3].

In Egypt, where water conservation is becoming the highest priority, the reutilization of drainage water from other activities has assisted private sector investment at aquaculture activities in general and freshwater fish farms in particular. This has been spurred by the Egyptian tradition of consuming fish as an essential food item. The local market demand for fish has increased tremendously during the last three decades due to population growth. Consequently, the price of captured fish has risen, assisted further by the degradation of the natural fisheries. The result was that fish moved out of the affordable range for poor people. The high price of fish was remedied by freshwater aquaculture, which is a very well-known practice in Egypt since 1930. Although records were not available at that time on the size of production, the FAO statistical services sector was able to establish records of the Egyptian fisheries and aquaculture beginning in 1950. These records indicate that the total fish catch from aquaculture has linearly increased up to 539,748 tons year^{-1} in 2005 (Figure 6). This quantity represents not only aquaculture freshwater fish but also brackish and marine cultured ones. Data suggests that >60% of this quantity consists of freshwater fishes, mainly Tilapia species.

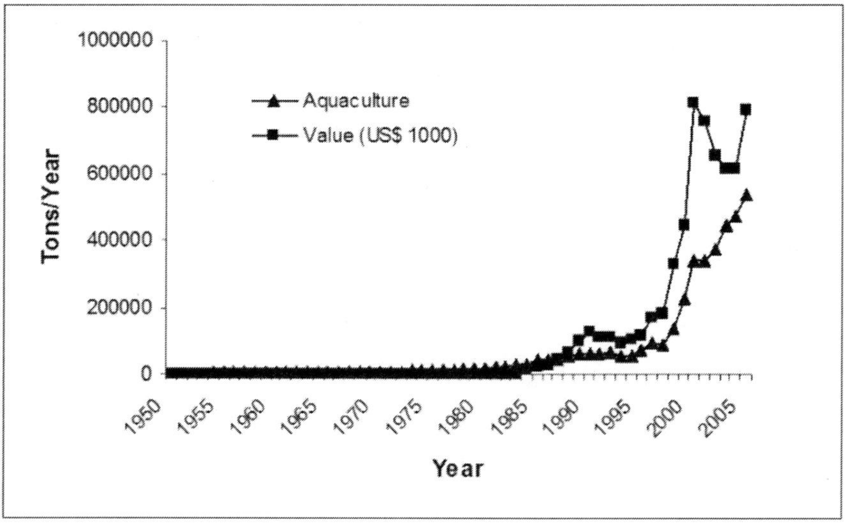

FIGURE 6. Aquaculture fish production in Egypt and its Economic value during the period 1950-2005 [14]

During the last two decades, and due to the reutilization of drainage water, the salinity in these water bodies has increased, leaving them brackish. These changes in water salinity are paralleled by changes in the cultured fish species. In fact, most of the fish farms responded to changes in water salinity by turning from Tilapia to carp fish farms and then to grey mullet. The total income of produced fish from aquaculture in Egypt is estimated at 971,846 * 10^3 US$ per year (Figure 6), according to the available estimate of 2005 [14]. Based on these figures it is estimated that the supply of fish per capita has increased from 7 kg year $^{-1}$ in 1990 to 15 kg year $^{-1}$ in 2003. In fact, Egypt is not only the leading country in the Nile River basin in freshwater aquaculture (Table 5), but also the first amongst the Mediterranean countries and the Arab world [31], [29], [3], [13], [14].

8. ECONOMIC VALUE OF FRESHWATER FISHERIES OF THE NILE BASIN

There is no doubt that freshwater fisheries provide an essential supply of animal protein to large parts of the developing world. At national, community and family levels these systems are critically important in sustaining food security. However, this is not the case for the upstream countries of the Nile River basin, where, although the great lakes have abundant fish production, there is a lot of concern about malnutrition of the populations surrounding the basin. An example is provided by Lake Victoria which supports Africa's largest fishery. Its most valuable product is the Nile perch, much of which is exported. This has given rise to arguments claiming a direct linear relationship between perch exports and disturbingly high rate of malnutrition along the lake's shore [32]. This state of affairs is mainly due to the foreign fish processing companies that control the fish exports, and the high demand on the Nile perch especially in the EU countries. With such high demands for Nile perch, the value of the fisheries has risen considerably. In 1983, there were 12,041 boats operating on the lake. By 2004, there were 51,712 boats and 153,066 fishermen. Of the 1433 landing sites identified in 2004 along the lake shores, just 20% of the population had communal lavatory facilities, 4% were served by electricity and 6% were served by potable water supply [32]. The authors have concluded that, Policy options can include targeting health care specifically at the lake's communities. They also added that local organizations such as LVFO (Lake Victoria Fishery Organization), should translate the largest inland fishery benefit into discernible positive impact on the ground. That means positive economic return has to be reflected on fishermen families.

The Ugandan portion of Lake Victoria fish is among the highest revenue foreign exchange earner for non-traditional exports. In fact, the fish exports registered unprecedented growth in earnings, coming second after coffee exports in 2001. Its proportional contribution to export earnings rose from 5% in 1994 to 17% in 2001. It is estimated that fish exports in 2001 were in the order of 28,000 metric tons, with a value of approximately US$ 78 million. That estimate is only for the Nile perch. Other estimates included Nile tilapia from other countries bordering the lake (Kenya and Tanzania). These account for a total catch of 110 metric tons. However, the difference in value is marked, since the Nile perch is exported to EU countries, while the other species and small size Nile perch are locally consumed or regionally exported at lower prices. Recent statistics on the annual total fish production of the three countries (Kenya, Tanzania and Uganda) from Lake Victoria is amounted to 960, 500 tones [30]

In Sudan freshwater fisheries contribute very little to the national economy, especially as the estimated fish consumption per capita is only 1.7 kg year $^{-1}$. In general, the fishing industry in 2003 is estimated to have exported 1,629 tons of fish with a value of US$ 533,000, while a quantity of 157 tons was imported for local consumption at a cost of US$ 324,000. This means that the total earnings from fish production did not exceed US$ 200,000 representing only 0.4% of the national GDP [13].

As previously highlighted, Egyptian fisheries, including both natural fisheries and aquaculture represent the largest producers of freshwater fishes within the Nile basin countries. This is mainly due to the rapid growth in freshwater aquaculture production, amounting to about 60% of the national total fish catch (Figure 7). Despite high fish production, the fishing industry plays a relatively minor direct role in the economy of Egypt. Nevertheless, domestic fish production makes a valuable contribution to the national food supply and to the traditional way of life, for which fish eating remains an important part. In addition, it is a significant source of food for the tourist industry.

Fishing industry is also important for the livelihoods of over 65,000 fishermen and other people employed full-time in related activities (estimated at some 300,000 men). The fish industry in Egypt also produces fish powder (from non-edible fish) of about 543 tons to be used as nutritive food additives in poultry farms [13].

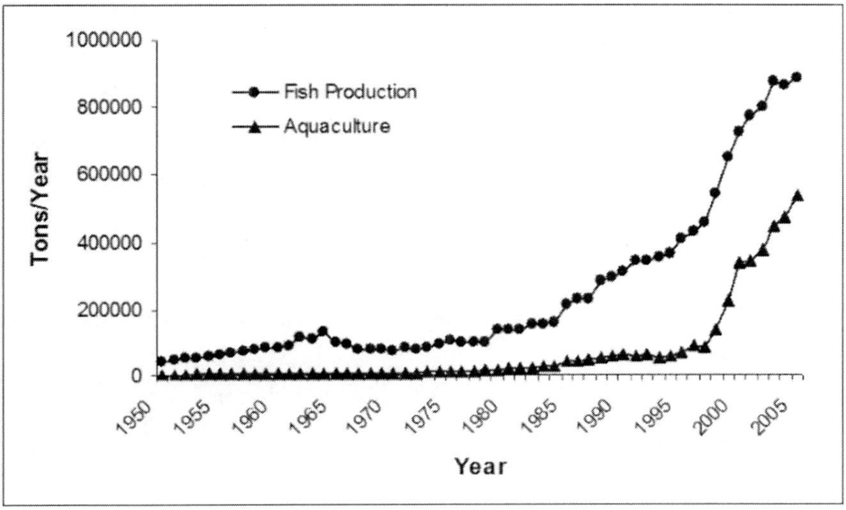

FIGURE 7. Growth trends in total fish production in Egypt (marine and freshwater catch) compared with aquaculture during the period 1950-2005 [14].

9. FUTURE OF FRESHWATER FISH PRODUCTION OF THE NILE BASIN

Without exception, the freshwater fisheries in the Nile basin countries are under severe and constantly stressful conditions. The Nile fisheries have continued to deteriorate despite its impressive freshwater resources and the many efforts by national and regional institutions with national and international support. This decline is mainly due to the rapid increase in population densities surrounding the water bodies of the Nile basin, which has negatively influenced the water quality of these basins and their fisheries. In addition, overexploitation of the fish resources has advantaged non-commercial and invasive species in their competition with native and economically valuable ones. The lack of regulatory fishing plans and management policy in most of the Nile basin fisheries has also added complexity to any future strategic plans to safeguard such valuable resources. In order to manage this important sector along the Nile basin there is a need to:

- Make appropriate information data available to guide decision makers

- Enforce laws and sustainable management of fisheries resources and the fish habitat.
- Control the use and dumping of Pesticides residuals in the Nile basin
- Promote sustainable practices with the help of scientific knowledge.
- Provide adequate financial resources and human capacity to implement fisheries program.
- Investigate and improve the socio-economic status of fishermen.
- Regulate fishing gear and practices that may otherwise damage fish stocks and habitat.

The poverty of many of the diverse populations living along the Nile basin remains a barrier to the implementation of concepts of natural resources conservation, despite the many efforts of governments and NGO's to disseminate this information.

Faced with a continuing large gap between global supply and demand for fish protein, with critical shortage in some regions, aquaculture is widely regarded as having a crucial role to play in meeting global and regional food requirements over the next 20 years. Aquaculture can be a water efficient means of food production, and also brings wider resources management benefits. To this end, it is advisable that the Nile basin countries take advantage of the existing water resources and encourage the private sector in implementing aquaculture activities as an actual and future alternative of animal protein and as a substitute for the degraded natural fisheries of the Nile basin.

REFERENCES

1. Roskar, J., 2000. Assessing the water resources potential of the Nile river based on data available at the Nile forcasting center in Cairo. Geografski zbornik. 53:32-65.
2. Hamza, W., 2006. Estuary of the Nile. In: Wangersky P. (Ed.), *Estuaries;* Hdb. Env. Chem. *Springer Verlag Publisher*, Vol. 5 H: 149-173 pp.
3. FAO, 2000. Nile basin initiative report to ICCON "Water and Agriculture in the Nile Basin", Land and water development division AGL/MISC/29/2000 Rome 1-68pp.
4. Witte, F., van Oijen and N. Sibbing, 2009a. Fish fauna of the Nile. In: Dumont H.J. (ed.), The Nile. Monographiae Biologicae, Vol. 89:647-675. Springer, Dordrecht.
5. Witte, F., M. deGraaf, O.Mkumbo, A.I. El-Moghraby and F. Sibbing, 2009b. Fish fauna of the Nile. In: Dumont H.J. (ed.), The Nile. Monographiae Biologicae, Vol. 89:647-675. Springer, Dordrecht.
6. Matsuishi, T., L. Muhoozi, O. Mkumbo, Y. Budeba, M. Najiru, A. Asila, A. Othina and I. G. Cowx, 2006. Are the exploitation pressure on the Nile Perch fisheries resources of Lake Victoria a cause of concern?. Fisheries Management and Ecology 13 (1) 53-71.

7. Witte, F., T. Goldscmidt, P. Goudswaard, W. Ligtvoet, M. van Oijen and J.H. Wanink, 1992. Species extinction and concomitant changes in Lake Victoria. Netherlands Journal of Zoology 42: 214- 232.
8. Balirwa, J.S., C.A Chapman, L.J. Chapman, I.G. Cowx, K. Geheb, L. Kaufman, R.H. Lowe-McConnell, O. Seehausen, J.H. Wanink, R.L. Welcomme and F. Witte, 2003. Biodiversity and fisheries sustainability in the Lake Victoria basin: an unexpected marriage? Bioscience 53: 703-716.
9. Cowx, I.G., M. van der Knaap, L. L Muhoozi and A. Othina, 2003. Improving fishery catch statistics for Lake Victoria. Aquatic Ecosystem Health & Management, 6(3): 299-310.
10. Njiru, M., J. Ojuok, A. Getabu, T. Jembe, M. Owili and C. Ngugi, 2008. Increasing dominance of Nile tilapia, *Oreochromis niloticus* (L) in Lake Victoria, Kenya: Consequences for the Nile perch *Lates niloticus* (L) fishery. Aquatic Ecosystem Health & Management, 11(1): 42-49.
11. Dejen, E., J. Vijverberg, and F.A. Sibbing, 2004. Temporal and special distribution of microcrustacean zooplankton in relation to turbidity and other environmental factors in a large tropical lake (L. Tana, Ethiopia), Hydrobiologia 513: 39-49.
12. DeGraaf, M, P.A.M. van-Zweiten, M.A.M. Machiels, E. Lemma, E. Dejen, and F.A. Sibbing, 2006. Vulnerability to a small scale commercial fishery of Lake Tana's (Ethiopia) endemic Labeobarbus compared with African catfish and Nile tilapia: an example of recruitment-overfishing? Fisheries Research 82: 304-318.
13. FAO, 2002. Fishery country profile, FID/CP/SUD. February, 2002.
14. FAO, 2008. FAO Fisheries and Aquaculture Information and Statistics Service (Egypt). http://www.fao.org/fishery/countrysector/FI-CP_EG/1
15. Earth Trend, 2003. Water resources and freshwater ecosystems-Egypt. Available online (www.earthtrends.wri.org).
16. GAFRD. 1995-2002. Annual fishery statistics reports (General Authority for Fish Resources Development), Cairo-Egypt.
17. Hamza, W., 1999. Differentiation of phytoplankton communities of Lake Mariut: A consequence of human impact. Bull. Fac. Sci. Alex. Uni. 39 (1-2): 159-168.
18. Mugidde, R., 1993. The increase in phytoplankton production and biomass in Lake Victoria (Uganda). Ver. Internat. Verein. Limnol. 25: 846-849.
19. Bootsma, H., & R.E. Hecky, 1993. Conservation of the African Great lakes: a limnological perspective. Conservation Biology 7: 644-656.
20. Henry, L., M.A. Kishimba, 2006. Pesticide residue in Nile tilapia (Oreochromis niloticus) and Nile Perch (Lates niloticus) from Southern Lake Victoria, Tanzania. Environmental Pollution 140: 348-354.
21. Munga, D., 1995. DDT and endosulfan in the fish from Holla irrigation scheme. M.Sc. Thesis. University of Nairobi, Tana River, Kenya.
22. Wadinga, S.o., 2001. Use and distribution of organochlorine pesticides; The future in Africa. Pure and Applied Chemistry 73:1147-1155.
23. Rashed, M.N., 2001. Monitoring of environmental heavy metals in fish from Nasser Lake. Environment International 27: 27-33.
24. Oczkowwski, A and S. Nixon, 2008. Increasing nutrient concentrations and the rise and fall of a coastal fishery; A review of data from the Nile Delta-Egypt. Estuarine, Coastal and Shelf Science. 23: 189-219.
25. Hamza, W. & Mason, S., 2004. Water availability and food security challenges in Egypt. Proceedings International forum on Food security under water scarcity in the Middle East: Problems and Solutions. Como- Italy 24-27 Nov., 2004.
26. Siegel, F., 1995. Environmental geochemistry in development planning: an example from the Nile delta, Egypt. Journal of Geochemical Exploration. 55: 265-273.
27. Barakat, A., 2004. Assessment of persistent toxic substances in the environment of Egypt. Environment International. 30: 309-322.
28. El-Sebae, A.H, Abou Zeid, M., & Saleh, M.A., 1993. Status and environmental impact of Toxaphine in the third world- a case study of African agriculture. Chemoshere. 27:2063-2072.

29. FAO, 1998. Aquaculture production statistics-1987-1996. FAO Fisheries Circular, No. 815, Revision 10 [FIDI/C815] Rome.
30. FAO, 2009. Fisheries and Aquaculture statistics. http://www.fao.org/statistics
31. FAO, 1994. Seminar on freshwater culture, Damascus, May 2-4, 1994; Edited by MEDRAP II Regional Center Tunis-Tunisia.
32. Geheb, K., S. Kalloch, M. Medard, A. Nyapendi, C. Lwenya and M. Kyangwa, 2007. Nile perch and the hungry of Lake Victorai: Gender, status and food in an East African fishery. Food Policy- Elsiver (in press).

CHAPTER 2

Pan-Arctic Fisheries and their Assessment

Ross Tallman , Muhammed Y. Janjua

[1] *Arctic Stock Assessment and Integrated Ecosystem Research, Central and Arctic Region,*

Fisheries and Oceans Canada, Winnipeg, Manitoba, Canada

[2] *Institute of Marine Research, Bergen, Norway*

[3] *Fisheries Joint Management Committee, Inuvik, NT, Canada*

[4] *Thüen Institute of Sea Fisheries, Braunschweig, Germany*

ABSTRACT

Pan-Arctic fisheries are highly diverse in their purpose, species biology, productivity, economic and strategic importance as well as in how they are prosecuted. They range from full industrial fisheries to community-based artisanal, sport and subsistence fisheries. The nature of Arctic ecosystems in the region varies from extremely productive to relatively barren in terms of fisheries production. Gear types vary, but offshore trawl fisheries and inshore and freshwater gillnet fisheries are the most common. Rights-based fisheries (e.g., for indigenous inhabitants) are more prominent in the Canadian and American Arctic than in European jurisdictions. The principal harvested species in freshwater environments tend to be from few taxa mainly *Salvelinus*spp. and from the family *Coregonidae,* while the marine taxa are more diverse. Compared to north temperate fisheries, Arctic fisheries have impressive variation across longitudes; some jurisdictions support only small-scale subsistence fisheries, whereas others contain some of the largest yields among industrial fisheries. Approaches to scientific assessment are also highly diverse with a range from catch-based indicators to sophisticated fully age-structured population models.

Keywords: arctic, fisheries, models

1. INTRODUCTION

This chapter describes some of the major Pan-Arctic fisheries, the stock assessment methods applied to assess them and how the fisheries might change with climate warming and further development of the northern regions. The chapter is a broad overview to introduce the reader to this topic which has not been included in most fisheries text books.

Pan-Arctic fisheries are highly diverse in their purpose, species biology, productivity, economic and strategic importance as well as in how they are conducted. They range from full industrial fisheries to community-based artisanal, sport and subsistence fisheries. Rights-based fisheries (e.g., for indigenous inhabitants) are more prominent in the Canadian and American Arctic than in European jurisdictions. The patchy nature of Arctic environments has a strong influence on species life cycles such that geographically extensive migrations between critical habitats for rearing and growth, spawning or calving and over-wintering are undertaken by many taxa. Species tend to be long-lived. The principal harvested species in freshwater environments tend to be from few taxa mainly *Salvelinus* spp. and from the family *Coregonidae*. While the marine taxa are more diverse, the dominance of marine mammals at the apex of the food chain and as a source of food for humans is important in driving fisheries policy for a large portion of the Arctic zone. Compared to north temperate fisheries, Arctic fisheries have impressive variation across longitudes; some jurisdictions support only small-scale subsistence fisheries, whereas others contain some of the largest yields among industrial fisheries.

The chapter is organized by geographic regions: Barents Sea, Arctic Atlantic–Norwegian Sea, Arctic Atlantic–Greenland Sea, Greenland–continental, Baffin Bay–Davis Strait, Hudson Bay, Canadian Archipelago, Canadian Arctic mainland, Alaska, Beaufort Sea, Siberia and Chukchi Sea (**Figure 1**).

2. BARENTS SEA

The Barents Sea (**Figure 2**) is on the continental shelf surrounding the Arctic Ocean. It connects with the Norwegian Sea to the west and the Arctic Ocean to the north and the Kara Sea to the east. Its contours are delineated by the continental slope between Norway and Spitsbergen to the west, the top of the continental slope towards the Arctic Ocean to the north, Novaya

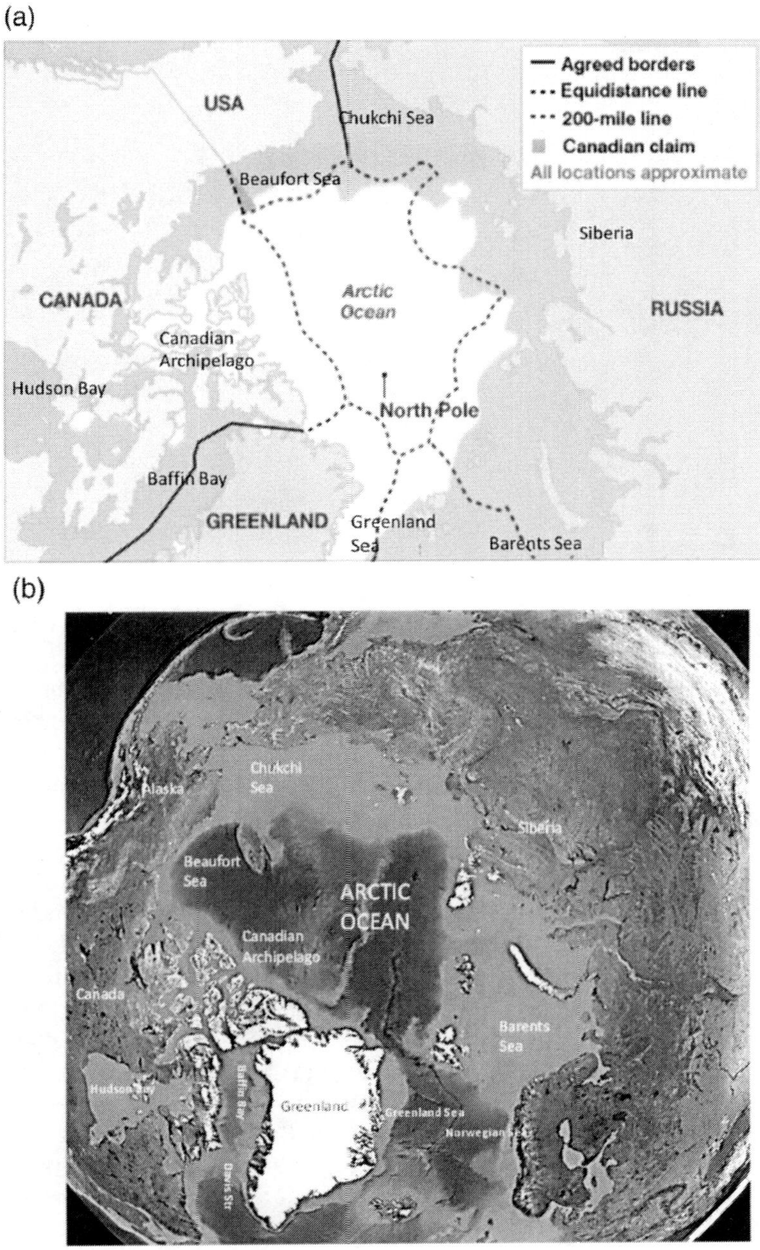

FIGURE 1. (a) Arctic Ocean and surrounding land masses showing approximate jurisdictional boundaries and (b) fishing areas discussed in the text.

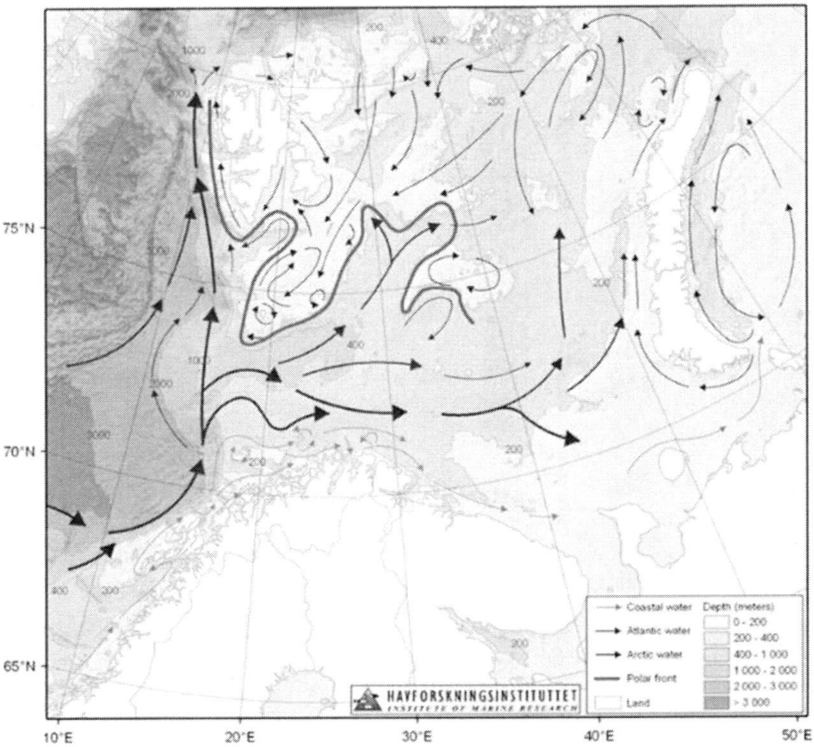

FIGURE 2. The main features of the circulation and bathymetry of the Barents Sea [2].

Zemlya archipelago to the east and the coasts of both Norway and Russia to the south. It covers an area of approximately 1.4 million km^2, has an average depth of approximately 230 m has and a maximum depth of about 500 m at the western end of Bear Island Trough. Its topography is characterized by troughs and basins (300–500 m deep), separated by shallow bank areas, with depths ranging from 100 to 200 m. The three largest banks are Central Bank, Great Bank and Spitsbergen Bank. Several troughs over 300 m deep run from central Barents Sea to the northern (e.g., Franz Victoria Trough) and western (e.g., Bear Island Trough) continental shelf break. These western troughs allow the influx of Atlantic waters to the central Barents Sea. The Barents Sea is shared between Russia and Norway, and there is long history of relatively successful cooperation in fisheries management, even during periods that were otherwise marked by political tensions [1].

The Barents Sea is home to the most productive commercial fisheries in the Pan-Arctic Region (**Figure 2**).

3. FISHERIES

3.1. Benthos and Shellfish

The sea floor is inhabited by a wide range of organisms. Some are buried in the sediments, others are attached to a substrate, some are slow and sluggish, and others are roving and rapid. More than 3050 species of benthic invertebrates inhabit the Barents Sea [3]. The benthic ecosystems in the Barents Sea have considerable value, both in direct economic terms and in their ecosystem functions. Scallop, shrimp and king crab are harvested in the region. Snow crab may be regarded as a potential commercial species in the Barents Sea. Many species of benthos, such as sea cucumber, snails and bivalves, are also of interest for bio-prospecting or as a potential food resource. Important fish species such as haddock, cod, catfish and most flatfishes primarily feed on benthos. Many benthic animals, primarily bivalves, filter particles from the ocean and effectively remove particulate matter from the water column. Others scavenge on dead organisms, returning valuable nutrients to the water column. Detritus feeders and other active diggers regularly move the bottom sediments around and therefore increase sediment oxygen content and overall productivity—much like earthworms on land.

The decline in the total biomass of benthos, from 1924–1935 to 1968–1970 [4], occurred throughout most of the Barents Sea and has been attributed to climate change by many investigators. The mechanism behind this biomass reduction is not clear, however.

The northern shrimp (*Pandalus borealis*) is distributed in most deep areas of the Barents Sea and Spitsbergen waters. The densest concentrations are found in depths between 200 and 350 meters. This species mainly feeds on detritus but will also scavenge for food. It is also important as a food item for many fish species and seals.

Red king crab (*Paralithodes camtschatica*) was introduced to the Barents Sea in the 1960s. Presently, it is an important commercial species. Adult red king crabs are opportunistic omnivores.

The snow crab (*Chionoecetes opilio*) is an invasive species deliberately introduced to this region. The first recordings of this species in the Barents Sea were in 1996. Since 2003, snow crab have been found in the stomachs of cod, haddock, wolffish and thorny skates, indicating that the crab abundance and settlement density have substantially increased.

The Iceland scallop (*Chlamys islandica*) is a slow-growing species common in all shallow areas (<150 m). It is usually associated with hard bottom substrate and most commonly in areas with strong currents [5]. The scallop is a filter feeder and is therefore highly dependent on seasonal phytoplankton production, which also has impacts on its growth [6]. The lifespan is 30 years and above.

There are eight species of squid inhabiting the Barents Sea [7]. The flying squid *Todarodes sagittatus*was a significant fishing resource in Norwegian waters during several periods up to1988 [8]. However, since then it has been almost absent from the waters and only sporadic catches have been recorded. *Gonatus fabricii* is another abundant squid species in the off shore waters of the Barents and the Norwegian Sea [9]. This species is important food for several bird and cetacean species, but could probably also be seen as a potential fishing resource.

3.2. Fish

More than 200 fish species are registered in trawl catches during surveys of the Barents Sea, of which nearly 100 occur regularly. The different water masses, together with bottom type and depth, are important factors determining the distribution of fish species. For pelagic species, the distribution and abundance of zooplankton are additionally important factors. The most important demersal fish species include Northeast Arctic cod (*Gadus morhua*), Northeast Arctic haddock (*Melanogrammus aeglefinus*), saithe (*Pollachius virens*), redfish (*Sebastes mentella* and *S. norvegicus*), Greenland halibut (*Reinhardtius hippoglossoides*), long rough dab (*Hippoglossoides platessoides*), wolffish (*Anarhichas lupus*, *A. minor* and *A. denticulatus*) and European plaice (*Pleuronectes platessa*), while the important pelagic species are Barents Sea capelin *(Mallotus villosus)*, polar cod (*Boreogadus saida)* and immature Norwegian spring-spawning herring *(Clupea harengus)*. In some warm years, increased numbers of young blue whiting (*Micromesistius poutassou*) have migrated into the Barents Sea. There have been large variations in abundance of most of these species. These variations are due to a combination of fishing pressure and environmental variability.

The recruitment of the Barents Sea fish species has shown a large year-to-year variability [2]. This variability in recruitment causes large variations in the biomass of pelagic forage fish, which are all either short-lived (capelin and polar cod) or spend only a short part of their lifespan in the Barents Sea (herring). The most important reasons for the recruitment variability are variations in the spawning biomass, hydrographic conditions, changes in circulation pattern, food availability and predator abundance and distribution. Recent work on larval drift has shown that even small changes in spawning locations can have a large impact on the drift pattern of fish larvae. Vikebø et al. [10] and Opdal et al. [11] investigated the drift of cod and herring eggs and larvae spawned at different locations along the Norwegian coast. Results showed that spawning further offshore and more to the south gave a much higher possibility for the larvae to end up west of Svalbard. By contrast, more northern spawning resulted in a higher proportion of larvae entering the Barents Sea. Also, retention of the larvae was affected by spawning site, and hence, the development stage for larvae when they reach the entrance to the Barents Sea. A report on the knowledge base in the Lofoten area showed that about 70% of the egg, larvae and juvenile stages of the total commercial stocks (measured in catch biomass) in the Norwegian and Barents Sea pass by the Lofoten--Vesterålen area. About 12% of the total stocks (measured in catch biomass) spawn in the Barents Sea. The Lofoten–Vesterålen area is therefore a vulnerable key area for the recruitment to the commercial stocks in the Norwegian and Barents Sea [2].

Cod is the most important predator among fish species in the Barents Sea. It feeds on a wide range of prey, including larger zooplankton, most available fish species, including their own juveniles, as well as shrimp [2]. Cod prefer capelin as prey, and fluctuations of the capelin stock may have a strong effect on growth, maturation and fecundity of cod, as well as on cod cannibalism and hence recruitment to the stock. The role of euphausiids in the cod diet increases in the years when capelin stock is at a low level [12]. Also, according to Ponomarenko [13], inter-annual changes in euphausiid abundance are important for the survival of cod during the first year of life.

Capelin is an important consumer of zooplankton biomass produced near the ice edge. Farther south, capelin is the most important prey species in the Barents Sea as it transports biomass from northern to southern regions [14]. The Barents Sea capelin stock underwent drastic changes in stock size during the last three decades. Three stock collapses occurred in 1985–1989, 1993–1997 and 2003–2006, and data from 2015 suggest that the capelin may be headed towards a fourth collapse. The collapses had effects both downwards and upwards in the food web [15]. The release in predation pressure from the capelin stock led to increased amounts of zooplankton during the two first collapse periods. When capelin biomass was drastically reduced, its predators were affected in various ways. Cod experienced increased cannibalism, growth was reduced, and maturation delayed during the first capelin collapse. Sea birds experienced increased rates of mortality and total recruitment failures, and breeding colonies were abandoned for several years. Harp seals experienced food shortage and increased mortality because they invaded the coastal areas and were caught in fishing gears and because of recruitment failures. There is evidence for differences in how the three capelin collapses affected the predators. The effects were most serious during the 1985–1989 collapse, but much less during the second and third collapse. This was probably related to increased availability of alternative food sources during the two last periods of collapse.

Herring is also a major predator on zooplankton. The herring spawns along the Norwegian western coast, and the larvae drift into the Barents Sea as well as into fjords along the coast. The juveniles of the Norwegian spring-spawning herring stock are distributed in the southern parts of the Barents Sea. They stay in this area for about three years before they migrate west and southwards along the Norwegian coast and mix with the adult part of the stock. The presence of young herring in this area has been described to have a profound effect on the survival of capelin larvae and therefore on the recruitment to the capelin stock. The three collapses during the last three decades were all caused by recruitment failures, and all three were associated with rich herring year classes inhabiting the Barents Sea. However, while the presence of herring is seemingly a necessary factor for total recruitment failures of the capelin stock, it is not the only factor, since in some years the capelin recruitment has been relatively good in spite of moderate to high amounts of young herring in the Barents Sea.

Haddock is also a common species and migrates partly out of the Barents Sea. The stock has large natural variations in stock size. Water temperature at the first years of the life cycle may be used as an indicator of year class strength. Food composition of haddock consists mainly of benthic organisms.

Saithe is found mainly along the Norwegian coast, but also occurs in the Norwegian Sea and in the southern Barents Sea. The 0-group saithe drifts from the spawning grounds to inshore waters. The smaller individuals feed on crustaceans, while larger saithe depend more on fish as prey [16]. The main fish preys are young herring, Norway pout, haddock, blue whiting and capelin, while the dominating crustacean prey is krill.

Polar cod is a cold-water species found particularly in the eastern Barents Sea and in the north. It seems to be an important forage fish for several marine mammals, but to some extent also for cod. There is little fishing of this stock, and relatively little fisheries data. However, it is clear that the stock abundance is in decline.

Deep-sea redfish and golden redfish are important elements in the fish fauna in the Barents Sea, but due to heavy over-fishing, these stocks declined strongly during the 1980s and have since then stayed at a low level. Young redfish are plankton eaters, but larger individuals take larger prey, including fish.

Greenland halibut is a large fish predator with the continental slope between the Barents Sea and the Norwegian Sea as its most important area. It is also found in the deeper parts of the Barents Sea and the continental shelf. Investigations in the period 1980–1990 showed that cephalopods (squids, octopuses) dominated in the Greenland halibut stomachs, as well as fish (mainly capelin and herring). Ontogenetic shift in prey preference was clear with decreasing proportion of small prey (shrimp and small capelin) and increasing proportion of larger fish with increasing predator length. The largest Greenland halibut (length more than 65–70 cm) had a rather large portion of cod and haddock in the diet. The stock was overfished leading to low (though uncertain) stock abundance in the 1980s, but a partial moratorium in the early 1990s led to stock recovery, and the stock is now assessed as above limit reference points.

The blue whiting has its main distribution area in the Norwegian Sea and Northeast Atlantic, and the marginal northern distribution is at the entrance to the Barents Sea. Usually, the blue whiting population in the Barents Sea is small. In some years, the blue whiting may enter the Barents Sea in large numbers and can be a dominant species in the western areas. This situation occurred from 2001 and during 2003–2007. Since then, the abundance has decreased strongly, but showed an increase in 2012. These fluctuations are probably due to a combination of variation in stock size and environmental conditions. In the diet of blue whiting, zooplankton (copepods, hyperiids and euphausiids) is dominant in the younger age groups, while fish is increasingly important as the blue whiting gets older [17].

3.3. Marine Mammals

Marine mammals, as top predators and keystone species, are significant components of the Barents Sea ecosystem. Twenty-five species of marine mammals regularly occur in the Barents Sea, including: seven pinnipeds (seals and walruses); 12 large cetaceans (large whales); five small cetaceans (porpoises and dolphins); and the polar bear (*Ursus maritimus*). Some of these species are not full-time residents in the Barents Sea and use temperate areas for mating, calving and feeding (e.g., minke whale *Balaenoptera acutorostrata*). Others reside in the Barents Sea all year round (e.g., white-beaked dolphin *Lagenorhynchus albirostris* and harbor porpoise *Phocoena phocoena*). Some marine mammals are naturally rare, such as the beluga whale *Delphinapterus leucas*. Others are rare due to historic high exploitation, such as bowhead whale *Balaena mysticetus* and blue whale *Balaenoptera musculus*.

Marine mammals are important predators on the commercial fish species in the Barents Sea. However, their consumption estimations are associated with high level of uncertainty. According to Folkow et al., [18] and Nilssen et al., [19], marine mammals may consume up to equally much or even more fish than those caught in fisheries. Minke whales and harp seals have the largest consumptions and may together consume around 5 million tons annually of crustaceans, capelin, herring, polar cod and other gadoid fish (cod, haddock, saithe) [2]. Functional relationships between marine mammals and their prey seem closely related to fluctuations in marine ecosystems. Both minke whales and harp seals are thought to switch between krill, capelin and herring depending on the availability of the different prey species [20, 21, 19].

The only marine mammal species commercially harvested in the Barents Sea are harp seals and minke whale. Harp seal pup production estimates are based on data collected during the traditional Russian multispectral aerial survey. Since 2004, the abundance of harp seal pup production in the White Sea has been sharply reduced, according to these surveys. One of the key factors which caused the reduction in the harp seal pup abundance in the last years is the diminished ice extent due to warming. The changed ice conditions were responsible for the redistribution of animals in the pup period. Abnormal ice conditions in the White Sea possibly also led to higher natural mortality of pups.

4. FISHERIES ASSESSMENT

The benchmark assessment for North East Atlantic (NEA) cod was done in 2015 [22]. The assessment advised continuation of use of the Extended Survivors Analysis (XSA) model as a main tool for NEA cod assessment. Some changes in model configuration were recommended.

The main model used for assessment of key stocks is the Extended Survivors Analysis (XSA) [23, 24]. The model is generally fitted to the catch at age and natural mortality data and works similarly to most typical Virtual Population Analysis (VPA) back-calculating models. The back-calculations in these implementations work the same way, but they differ in the statistical methods used for "tuning" to indices of population size.

For North East Atlantic cod, saithe, the XSA was used as the main assessment method. For Norwegian Coastal cod, the "separable VPA" (SVPA) model proposed by [25] was used. This model assumes that fishing mortality can be separated into an annual component common to all ages of a same year and into a component at age common to many years. According to this model, fishing mortality in year i and age j or f (i,j) is defined as follows: f(i,j) = F(i) x S(j) where F(i) represents fishing mortality for year i and S(j) the exploitation pattern or selectivity at age j.

Haddock and beaked redfish (*Sebastes mentella*) are both assessed using statistical catch-at-age (SCAA) models. In contrast to the VPA methods, SCAA is a forward calculation method and is becoming more widely used throughout ICES fisheries with the introduction of the SAM (REF) model, which is used for haddock in the Barents Sea.

Where age data are uncertain or absent, an alternative model, the age- and length-structured Globally Applicable Area Disaggregated General Ecosystem Toolbox model or GADGET, is used for assessment. The model is currently used for the golden redfish, *Sebastes norvegicus*, and Greenland halibut.

5. ARCTIC ATLANTIC—NORWEGIAN SEA, ICELAND AND GREENLAND SEA

5.1. Fisheries

5.1.1. Norwegian Sea

The Norwegian Sea is a deep open ocean system, covering around 1.1 million km^2 extending from the Norwegian coast and the continental slope west of the Barents Sea out into the open Atlantic, as far west as the Iceland Sea and Jan Mayen, and bounded to the south by the North Sea. The area covers two separate deep water basins, the Norwegian and Lofoten basins, with water depths of 3000–4000 m, while the shallower water is dominated by the Norwegian Atlantic current branch of the Gulf Stream [26]. As a result of this steady influx of warm water, the area, although mostly in the Arctic, is largely free of sea ice, and the zooplankton is dominated by *Calanus finmarchicus*, which in turn supports large pelagic stocks. In contrast with many of the other systems described here, there exists a sizable area outside national territorial waters, the so-called "banana hole," in the gap between the 200-nm zones extending east from Jan Mayen and west from the Norwegian coast. Fisheries in the Norwegian Sea are primarily targeted at three large wide-ranging pelagic stocks: Norwegian spring-spawning herring (*Clupea harengus*), mackerel (*Scomber scombrus*) and blue whiting (*Micromesistius poutassou*), with a limited amount of horse mackerel (*Trachurus trachurus*) in the south. The region also has a high biomass of squid. There are several abundant pelagic stocks that span both the Norwegian and Barents Seas, notably the Norwegian spring-spawning herring which use the Barents Sea as a nursery area before moving to the

Norwegian Sea at about age 3 or 4, and the beaked redfish which is caught in both the Norwegian and Barents Seas. The mackerel and blue whiting in the Norwegian Sea represent the northern extreme of stocks ranging down to the Iberian peninsula. There are thus significant connections between the Norwegian Sea and adjacent ecosystems, and consequently, much of the stock assessments are conducted on a wider scale than the Norwegian Sea alone. The seabed of the main basins is dominated by deep water benthos, while along the continental shelf closer to the Norwegian coast are a number of cold-water coral reefs. These corals are slow growing and support a high biodiversity and receive protection in the form of exclusion areas for bottom trawlers. The minke whale population described in the Barents Sea section passes through the Norwegian Sea on its migration route, and a fraction of the stock remains in the Norwegian Sea to feed in the summer, while the remainder continues to the Barents Sea, possibly based on the biomass of herring encountered en route [27]. Other main marine mammal predators in the region are larger whales, including humpbacks, blue whales and fin whales, as well as several species of dolphins, notably white-beaked (*Lagenorhynchus albirostris*) and white-sided (*Lagenorhynchus acutus*) dolphins, as well as harp and hooded seals. In addition to the top-down forcing imposed by the marine mammal predators, there is considerable bottom-up forcing arising from multiple large pelagic stocks feeding largely on the same plankton resource (Skjoldal et al. 2004). The recent large biomasses of mackerel and blue whiting are associated with a decline in the biomass of zooplankton.

The main fisheries target wide-ranging pelagic, and the assessment and management of these species are therefore conducted through an international ICES WG on widely dispersed species, WGWIDE [28], extending beyond the Norwegian Sea. In 2014, 13 countries reported catches of over 9000 tonnes of mackerel (with the Faroes, Greenland, Iceland, Ireland, Norway, Russia and the UK all reporting catches of over 75,000 tonnes) out of a catch of 1.4 million tonnes, 7 countries reported over 9000 tonnes of NSS herring (with the catch dominated by Norway and Russia) out of a total of 460,000 tonnes, and 11 countries reported over 9000 tonnes of blue whiting (dominated by the Faroes, Iceland, Norway and Russia) out of a catch of 1.15 million tonnes. Although discards are believed to exist, they are highly variable with either 0 or 100% discards when the entire haul is discarded. As a result, estimates of discards are difficult to compile and considered inadequate by the WG (although believed to be low for blue whiting).

Both mackerel and blue whiting are assessed with the SAM catch-at-age model [29]. Norwegian spring-spawning herring has been assessed with a VPA model, although at the time of writing, an ongoing benchmark process is reviewing the choice of assessment model.

The mackerel stock ranges from the Iberian coast to the northern Norwegian Sea and supports a highly valuable fishery. Although line fisheries exist in the south, the Norwegian Sea fishery is based on freezer trawlers, pelagic trawlers and purse seine vessels. The assessment is based on catch-at-age data, a recruitment index from the IBTS surveys in the North Sea, the Norwegian Sea ecosystem survey (IESSNS) and an SSB index derived from an egg survey, as well as tag recapture data. There is considerable uncertainty in the assessment of the stock, with the perception of the stock changing between assessments. In the absence of an agreed management plan, ICES gives advice on a MSY basis. Following the expansion of the mackerel stock into Icelandic and Faroese waters, no general agreement has existed between the main states catching the mackerel on distribution of the quota since 2009, and consequently, the sum of the declared quotas in 2015 was 1.24 million tonnes, which was 330 thousand tonnes above scientific advice.

The blue whiting stock also ranges widely in the East Atlantic, with the Norwegian Sea believed to be the key nursery area. The assessment showed a significant downward revision in 2015 in accordance with a reduction in the survey index. The assessment is considered to have high, but non-quantifiable uncertainties. The stock is managed with an ICES-approved precautionary HCR agreed between the EU, Norway, the Faroes and Iceland, and catch in 2014 was slightly below the advised quotas. The bulk of the catch comes from large pelagic trawlers (although demersal trawlers also target blue whiting in the Norwegian Sea), with 92% of the catch being taken in the first half of the year. The Norwegian Sea forms the main nursery area for the stock; however, most of the catches occur further south.

In contrast to the other main pelagic stocks, the NSS herring is largely confined to Norwegian Sea and adjacent Icelandic and Faroese waters, except for the nursery area in the Barents Sea (which is not subject to a commercial fishery). The management of the fishery is thus controlled by Norway and Russia, and a long-term management plan has been in place since 1999. The plan has been evaluated by ICES and found to be precautionary. Following a collapse to 0.1% of its previous SSB in the late 1960s, the Norwegian spring-spawning herring retreated to the Norwegian coast. However, an exceptionally large year class in 1983 led to the stock recovering and resuming a widely distributed habit, with spawning grounds near the Norwegian coast and feeding throughout the northern Norwegian Sea. A series of poor year classes led to a decline in the stock, to the extent that the 2015 assessment estimated the stock to be below the Bpa of 5 million tonnes, although predicted to stabilize above Blim. The fishery follows the annual migration of the stock. A fishery begins in January on the spawning aggregation near the Norwegian coast, no spring fishery, a summer fishery in Icelandic and Faroese waters and around Jan Mayen and Svalbard, and finally a fourth quarter fishery (taking the largest fraction of the catch) in the eastern Norwegian Sea. The Norwegian fleet is dominated by purse seiners (92%) and pelagic trawlers (8%), while the Russian fleet is a variety of trawl vessels. Within the Norwegian-administered Norwegian Sea, a minimum catch size of 25 cm restricts the fishery to largely targeting mature individuals.

In addition to the fisheries described above, a small developmental catch exists for *Calanus finmarchicus*. The Calanus is pressed to extract the oil, which is marketed as an omega-3 supplement, highlighting that as a species with low trophic level, the oil is unlikely to have accumulated pollutants. The remaining mass is sold as fishmeal. Although the fishery is profitable at the current small scale, and a harvest of 1% of the stock could produce in excess of 2 million tonnes of marine oils and protein [30], it is unclear whether the fishery could be commercially viable if it expanded beyond the relatively small market for omega-3 supplements.

5.1.2. Icelandic Waters

Although Iceland sits just below the Arctic Circle, the EEZ covers 758,000 km^2 and extends into the Arctic and adjoins Greenlandic waters to the west and the Norwegian Sea and Faroese waters to the north and east. Fisheries are of critical importance to the Icelandic economy. The country was one of the first to claim a 200-nm economic zone, and total landings in 2014 were 1017 tonnes—for a population of 323,000. Seafood exports represent 41% of total export value, and the fisheries and related industries comprise 25% of GDP in 2014. By far, the most economically important component of the fishery is on the cod stock, representing over a one-third of the value of the exported seafood, and the 2015–2016 quota of 239,000 tonnes equates to almost ¾ of a ton of cod per Icelander.

The extensive EEZ ensures that many stocks are largely or wholly within Icelandic waters. The major exception is a mixed pelagic fishery in the east, catching NSS herring, mackerel and blue whiting from the margins of the stocks in the Norwegian Sea (see above), as well as an Icelandic herring stock. The ocean redfish (principally *S. mentella* and *S. norvegicus*) extends into international waters and is regulated by NEAF. This is divided into several stocks as well as two species, although there is a dispute as to the number of biological stocks, and ICES advice is to avoid "disproportional exploitation rate of any one component." The other major fisheries are on cod, haddock and capelin, with Greenland halibut, ocean wolfish, plaice, shrimp and lobster also caught.

In addition to large vessels, the fishing sector is managed to encourage small boats (<15 m) to participate in the commercial fishery by allocating a fraction of the total quota to this sector, reflecting the importance of fishing as a source of employment. The main gears used for demersal fishing are bottom trawls, longlines, gillnets and Danish seines, while mid-water trawls and purse seines are the main pelagic gears.

A wide variety of stock assessment methods are used, depending on the stock biology and data availability, ranging from detailed analytic modeling to qualitative measures for some redfish stocks. For most stocks, a TAC is set based on either a MSY approach or an explicit HCR. For capelin, an escapement strategy is used. In addition to the stocks described in the Norwegian Sea section, the following assessment models are used for the major Icelandic stocks.

ADAPT-type models are used for the Icelandic herring stock and the haddock. Cod is assessed with a statistical catch-at-age model (implemented in AD model builder). Capelin uses a short-term forecast model to project from survey estimates to SSB, taking into account predation from cod, haddock and saithe, in order to set an escapement strategy TAC such that the final SSB in 95% likely to be above Blim. Greenland halibut uses a Baysean surplus production model. The GADGET age and length-structured model is used to provide assessment for golden redfish (*S. norvegicus*), tusk (*Brosme brosme*) and ling (*Molva molva*), where the ability to use length data directly is considered valuable. Other redfish (multiple *S. mentella* stocks) are assessed using qualitative or survey-based methods in the absence of more reliable data.

5.1.3. Greenland Sea

The Greenland Sea is bounded by Greenland, the Arctic Ocean, Svalbard, the Norwegian Sea, Iceland, and the Denmark Strait to the south (**Figure 3**). The cold and deep (>2000 m) Greenland Sea is separated from the deep and warmer southern Irminger Sea by a relatively shallow (maximum depth of 630 m), east-to-west-oriented submarine sill in the Denmark Strait. Along with the Norwegian Sea, it forms the Arctic Ocean's main outlet to the Atlantic. On the shelf area north of this sill, the hydrographical conditions are dominated by the cold southward-flowing East Greenland Current (EGC) which, to a large extend, is composed of Norwegian Deep Water with temperatures often below 0°C [31]. The EGC is formed in the Arctic by the cooling of warmer northerly flowing North Atlantic Water (NAW) that is taken into the Arctic by the Norwegian Atlantic Current. Warmer water enters the Greenland Sea Gyre where it undergoes cooling through contact with the Arctic Ocean and associated sea ice. Most part of the Greenland Sea falls under ICES area of North East Atlantic Fisheries Commission (NEAFC) XIVa and IIb2. The ICES provides scientific advice for fisheries in the area through its advisory committee which is used by the relevant management authorities, e.g., NEAFC.

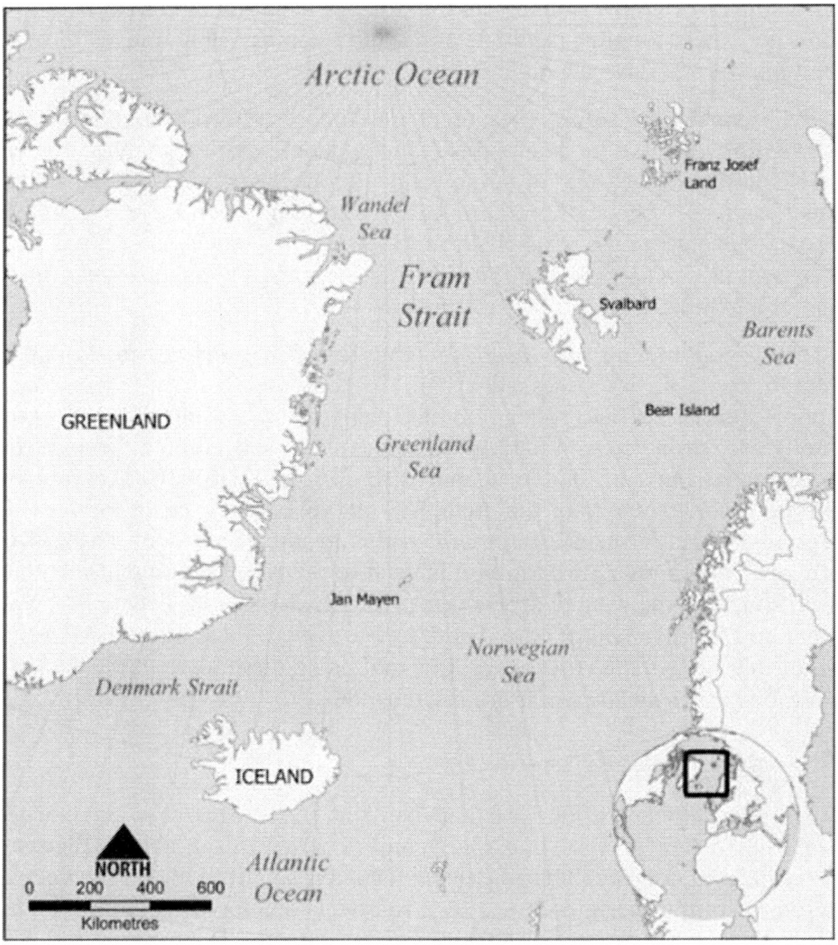

FIGURE 3. Arctic Atlantic: The Greenland Sea, Iceland and Norwegian Sea.

Arctic zooplankton produced in Greenland Sea migrates with the sea currents into the Iceland and Norwegian Seas and contributes to the feeding for the large stocks of pelagic fish [40]. In the past decades, the Greenland Sea has experienced warmer temperatures due to increased influx of North Atlantic water [32] which may have a significant influence on the anticipated yields of commercial species in this area. Cheung [33] has projected 15–45% increases in maximum catch potential in Greenland Sea between 2005 and 2050. Greenland halibut (*Reinhardtius hippoglossoides*), red fish (*Sebastes* spp.), Arctic cod (*Boreogadus saida*), herring (*Clupea harengus*), blue whiting (*Micromesistius poutassou*), tusk (*Brosme brosme*) and capelin (*Mallotus villosus*) are important commercial fish species in Greenland Sea and adjacent areas. Capelin was historically the largest fish stock in the adjacent Iceland Sea area. However, since the mid-2000s, there is a decline in its recruitment and stock size probably due to an increase in the inflow of Atlantic water [34].

5.2. Fish and Fisheries

The total number of fish species known from the Greenland exclusive economic zone (EEZ) is 269, whereas the lowest numbers of fish species with 47 are observed in the northeast of Greenland in the Greenland Sea [35].

Fisheries targeting marine resources off Greenland can be divided into inshore and offshore fleets. The majority of the Greenland fleet is comprised of approx 450 larger vessels and a big fleet of small boats. It is estimated that around 1700 small boats are dissipating in some sort of artisanal fishery mainly for private use or in the pound net fishery [36]. In East Greenland, fishing and other anthropogenic activities take place mainly in the southern parts.

The inshore fleet is constituted by a variety of different platforms from dog sledges used for ice fishing to small multipurpose boats engaged in whaling or deploying passive gears such as gillnets, pound nets, traps, dredges and longlines. The main targeted species are shrimp (*Pandalus borealis*), Atlantic cod (*Gadus morhua*), lumpfish (*Cyclopterus lumpus*), snow crabs (*Chionoecetes opilio*) and salmon (*Salmo salar*). The coastal fleets fishing for Atlantic cod, snow crab, scallops and shrimp are regulated by licenses, TAC and closed areas. Fishery for salmon and lumpfish is unregulated [36].

Apart from the Greenland fleet, the marine resources in Greenland waters are exploited by several nations, mainly EU, Iceland and Norway using bottom and pelagic trawls as well as longlines. The demersal offshore fishery is comprised of vessels primarily fishing Greenland halibut (*Reinhardtius hippoglossoides*), shrimp, redfish (*Sebastes mentella* and *S. norvegicus*) and Atlantic cod. Greenland halibut and redfish have been targeted since 1985 using demersal otter board trawls [36]. Longliners are operating on both the east and west coast with Greenland halibut and Atlantic cod as targeted species. Bycatches include roundnose grenadier (*Coryphaenoides rupestris*), roughhead grenadier (*Macrourus berglax*), tusk (*Brosme brosme*), Atlantic halibut (*Hippoglossus hippoglossus*) and Greenland shark (*Somniosus microcephalus*) (Gordon et al. 2003). The pelagic fishery in Greenland waters is conducted in East Greenland and currently targeted species are Atlantic mackerel (*Scomber scombrus*) and pelagic redfish (*Sebastes mentella*). A relatively small fishery for herring (*Clupea harengus*) is carried out in the border area between Greenland, Iceland and Jan Mayen. Additionally, Arctic cod (*Boreogadus, saida*), blue whiting (*Micromesistius poutassou*) and capelin (*Mallotus villosus*) are caught by the commercial fishery in the Greenland Sea and adjacent areas.

The main area of the Greenland Sea is part of divisions XIVa and IIb of the International Council for the Exploration of the Sea (ICES) and part of the convention area of the North East Atlantic Fisheries Commission (NEAFC). ICES provides scientific advice for fisheries in the area through its advisory committee which is used by the relevant management authorities, e.g., NEAFC. The demersal and pelagic offshore fishing together with longlines is managed by TAC, minimum landing sizes, gear specifications and irregularly closed areas.

Many species of cetaceans and pinnipeds feed in the Greenland Sea during the open water including walrus (*Odobenus rosmarus*), ringed seal (*Pusa hispida*), bearded seal (*Erignathus barbatus*), harp seal (*Phoca groenlandica*), hooded seal (*Cystophora cristata*) and bowhead whale (*Balaena mysticetus*). The Greenland Sea is an important whelping ground for harp seals and hooded seals. Hooded seal stocks in the area were quite large [37] and have been subject to commercial exploitation for centuries. However, giving an estimated total population of 84,020 hooded seals in 2013, this stock is below the conservation reference point in the precautionary harvest strategy developed by ICES and since 2007, the stock is protected from commercial hunt [38]. In contrast, harp seal are in abundance. The 2013 total abundance was estimated to be 627,410 with harvest level of only around 6000 per year [2].

Minke whales in the Norwegian Sea belong to the same stock as those feeding in the Barents Sea. The minke whales pass through on their way to summer feeding grounds in the Barents Sea, and a fraction remains in the Norwegian Sea.

There are few fisheries in the northern part of this area. In East Greenland, fishing and other anthropogenic activities take place in southern parts of east Greenland. Only small-scale subsistence hunting and fisheries take place near Ittoqqortoormiit. There are about seven marine fish species stocks which are harvested by commercial fisheries in Greenland Sea [39], mostly near Iceland. Arctic zooplankton produced in Greenland Sea migrates with the sea currents into the Iceland and Norwegian Seas and contributes to the feeding for the large stocks of pelagic fish [40]. The east Greenland drift ice forms a unique marine habitat under climate change that is not studied yet [41]. In the past decades, the Greenland Sea has experienced warmer temperatures due to increased influx of North Atlantic water [32]. Cheung [33] has projected 15–45% increases in maximum catch potential in the Greenland Sea from 2005 to 2050.

5.3. Fisheries Assessment

5.3.1. Greenland Halibut

The assessment model for Greenland halibut is a stochastic version of the logistic surplus production model using a new combined survey index and an Icelandic cpue index. Reference points as derived from this model are 30% BMSY as Blim, 1.7 × FMSY as Flim and an MSYBtrigger defined as 50% BMSY. BMSY and FMSY are inherited references in the model approach. An exploratory assessment on Greenland halibut using GADGET (Globally Applicable Area Disaggregated General Ecosystem Toolbox model) was presented in 2015.

5.3.2. Capelin

The assessment method is a stochastic projection of the stock starting from scientific acoustic measurements and finding the total allowable catch (TAC) that leads to the probability of the spawning stock biomass (SSB) < Blim being <5%. The initial quota is expected to be revised, based on in-season acoustic survey information in the autumn. The final TAC is expected to be set on the basis of survey information in the following winter.

5.3.3. Offshore Atlantic Cod

No stock assessment can be undertaken for this stock, due to the lack of significant rebuilding since the stock collapsed in the late 1960s. Two scientific surveys targeting cod are considered reliable indicators of the stock status. However, they are associated with large uncertainty due to single large hauls.

5.3.4. Golden and Beaked Redfish

GADGET is used for the assessment of golden redfish (*Sebastes norvegicus*) using commercial catch data and survey data from an Icelandic and a German survey. The demersal beaked redfish (*Sebastes mentella*) on east of Greenland and on the Icelandic slope is being assessed, based on trends in survey biomass indices from the Icelandic Autumn Survey in terms of the ICES "trends-based assessment" approach. Supplementary data used include relevant information from the fishery and length distributions from the commercial catch and the Autumn Survey.

5.3.5. Northeast Atlantic Mackerel

The model for Northeast Atlantic mackerel is SAM, the state–space assessment model. In SAM, the "states" (fishing mortalities and abundances at age) are constrained by the survival equation and follow a random walk process. The variances of the random-walk processes on abundances and fishing mortalities are parameters estimated by the model. SAM is a fully statistical model in which all data sources (including catches) are treated as observations, assuming a lognormal observation model. The corresponding variances, the so-called observation variances, are also parameters estimated by the model. Observational variances can be used to describe how well each data source is fitted in the model and effectively corresponds to the internal weight given by the model to the difference data sources. The other parameters estimated are the catchabilities of the surveys. Uncertainties (standard errors) are estimated for all parameters and for all states (fishing mortalities and abundances at age).

6. GREENLAND–CONTINENTAL

6.1. Fisheries

The fisheries of the Greenland continent (**Figure 4**) are mainly subsistence and rights based on species such as Arctic char and various marine mammals such as harp seal, hooded seal and ringed seal as well as walrus, narwhal, beluga whale and bowhead whale. Rivers with Arctic char are throughout the Greenland coast, and fishing is typically by gillnets set close to shore. The Arctic char is known to spawn in winter in river outlets in South East Greenland and utilizes the coastal areas, but no comprehensive reviews have been published [42]. Little information is available regarding the total harvest and management of Arctic char in Greenland. They are an important fish in Greenland providing a food resource for Greenlanders. They are also important in terms of socio-economic value as they are of interest to tourism with anglers and fly fishing enthusiasts traveling to Greenland.

6.2. Fisheries Assessment

As the Arctic char fisheries are prosecuted by local Greenlanders, there do not seem to be any published stock assessments. It is likely that assessments, if any, are carried out using catch-based methods.

FIGURE 4. Iceland, Greenland, the Norwegian and Greenland Seas.

7. BAFFIN BAY-DAVIS STRAIT

Baffin Bay and Davis Strait are two large basins between Baffin Island and Greenland (**Figure 5**). According to Hamilton and Wu [43], Baffin Bay is a semi-enclosed ocean basin between Baffin Island and Greenland that connects the Arctic Ocean and the Northwest Atlantic, providing an important pathway for exchange of heat, salt and other properties between these 2 oceans. To the south, the connection with the Atlantic is through Davis Strait, which is about 300 km wide and 1000 m deep. Baffin Bay's direct connection to the Arctic Ocean is far more restricted, being just 3 relatively small passages through the islands of the Canadian Arctic Archipelago (CAA). Arctic water also enters Baffin Bay–Davis Strait via the West Greenland Current which flows northwards along the western coast of Greenland. Melting ice sheets, changing sea ice conditions and changing weather also influence oceanographic conditions in Baffin Bay and Davis Strait. Trends and variability in the freshwater and heat input into the western North Atlantic via Baffin Bay–Davis Strait is of special interest because of the potential impact this input may have on global ocean circulation. Higher volumes of lighter, fresher water entering the Labrador Sea would increase stratification, with potential impact on the thermohaline circulation. The sinking of atmospherically cooled surface water in the Labrador Sea (the northwest arm of the North Atlantic) provides one of the driving forces for the "global ocean conveyor belt" which is vital in transporting heat and salt to northern latitudes. However, freshwater entering Baffin Bay is somewhat confined to the ocean's margins as part of a cyclonic circulation pattern, a principal component being the Baffin Island Current (BIC) that flows southwards along Canada's shelf and slope. Therefore, there is a potential for changes in this freshwater flux to impact the western North Atlantic ecosystem and fisheries by altering the physical properties of productive east coast banks and slope areas.

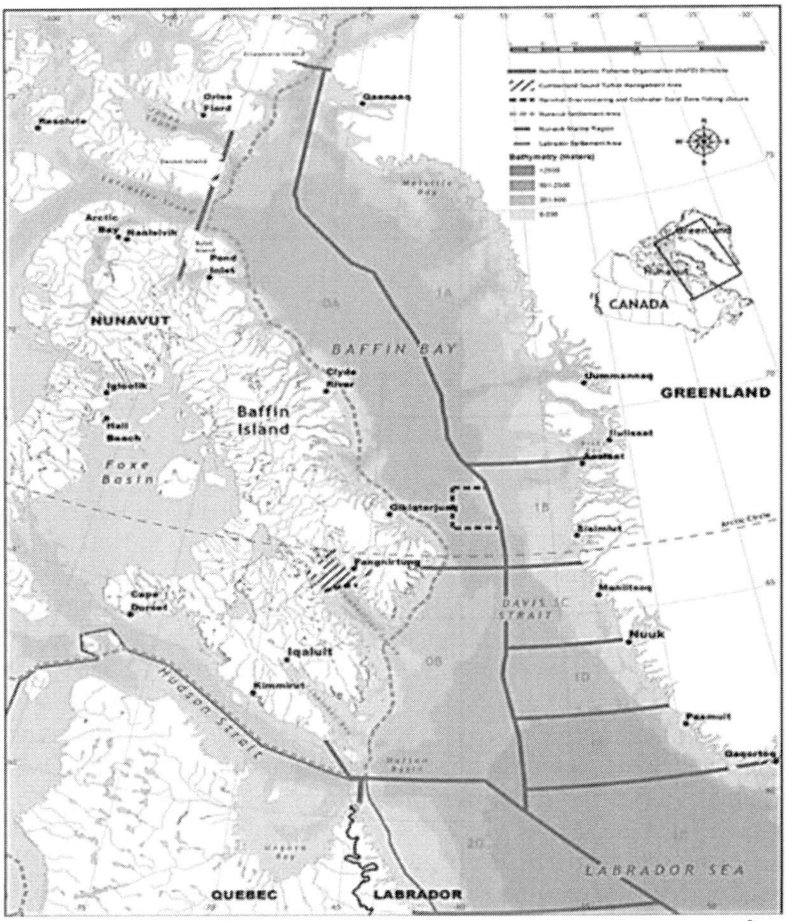

FIGURE 5. Map of Baffin Bay and Davis Strait.

7.1. Fisheries

A total of 183 species of marine fish have been recorded in Baffin Bay–Davis Strait near shore and offshore areas and Lancaster Sound region [44]. Jørgensen et al. [45] found seven assemblages of fishes in Davis Strait and the southern Baffin Bay by a standard type of cluster analysis. They found four of the assemblages in Baffin Bay, two in Davis Strait and one mainly in Davis Strait but scattered into Baffin Bay. The most important fish in the region is the Greenland halibut (*Reinhardtius hippoglossoides*). The area has many other species as noted by Jørgensen et al. [46] who collected 45 fish species from northern Baffin Bay between 72° 02' —76° 55' N at depth 150–1418 m. Their surveys found two species Greenland halibut and the sea snail very common and represented in large numbers in almost all trawl hauls. Greenland halibut from Greenland, Newfoundland and Labrador spawn in the deep waters of Baffin Bay. Arctic cod (*Boreogadus saida*) is important species and a dominant trophic link between zooplankton and higher predators such as seals and sea birds. The capelin and herring are important forge fishes in Davis Strait. Greenland halibut and shrimp are the main commercially important fish species in the region.

7.1.1. Greenland Halibut Fisheries

Greenland halibut is the only large-scale commercial fishery in Canada's Arctic. This fishery was begun in 1996 as a small exploratory fishery but has been expanding [47]. It takes place throughout the year. Greenland halibut is caught in both inshore and offshore areas. In fjords, fish are caught by longlines either from small vessels or from the winter ice. The offshore fishery for Greenland halibut takes place in summer and autumn on the shelf slopes. On the Canadian side, NAFO Area 0 is divided into Area 0A (North, Baffin Bay) and Area 0B (South, Davis Strait) (**Figure 6**). On the Greenland side, Area 1 is divided into 1A (offshore) and 1B (Baffin Bay), and 1C, 1D and 1 E (Davis Strait). Canada retains management authority for stocks in Subarea 0, while Greenland retains management authority in Subarea 1. NAFO Scientific Council conducts the stock assessment for the Subarea 0 and 1 for Canada and Greenland and recommends total allowable catch (TACs). Biomass, abundance, length frequency distribution and CPUE are the key metrics used in stock assessments and subsequent recommendations. The Greenland halibut fishery in western Baffin Bay is addressed by DFO's Integrated Fisheries Management Plan (IFMP) for NAFO Subarea 0 [47]. The Division 0A fishery quota is reserved exclusively for Nunavut, while the Division 0B quota is shared between Nunavut, Nunavik, Labrador, Newfoundland and Nova Scotia. Both mobile and fixed gears are used. In Area 0A, there is also a 100-t exploratory inshore quota. In 2015, the TAC is 16,000 tons in Div. 0A+ Div. 1AB, TAC for Div. 0B+ Div.

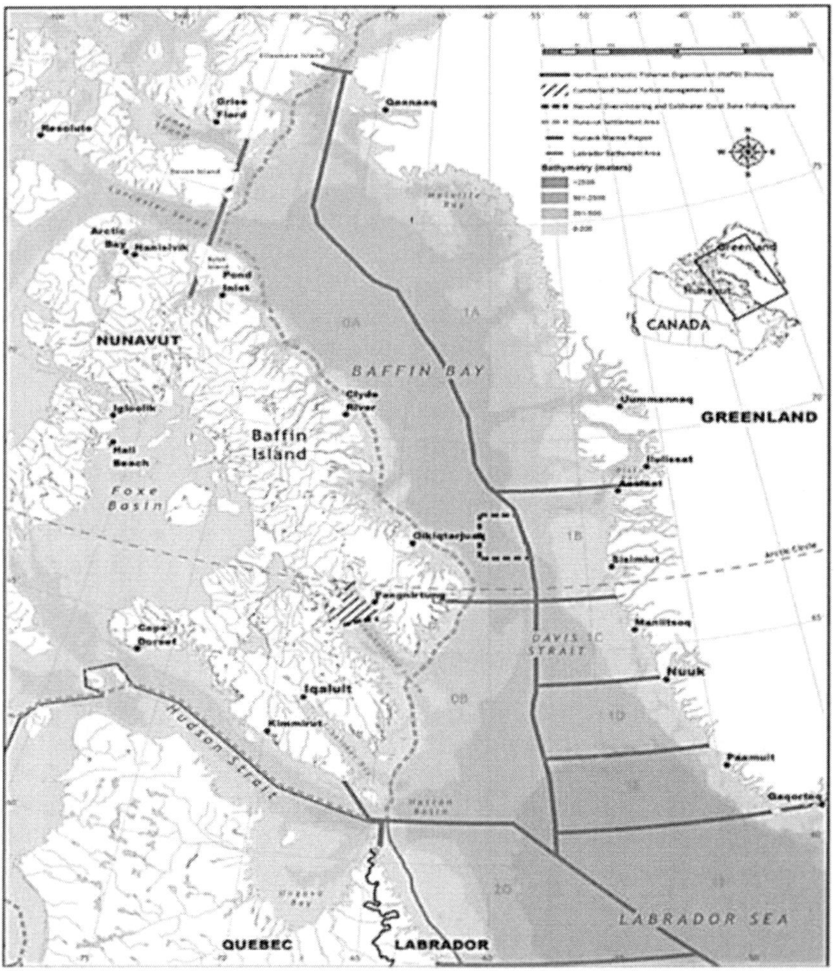

FIGURE 6. Northwest Atlantic Fisheries Organization Subareas and Divisions relevant to the Greenland Halibut fishery.

1CF is 14,000 tons, and the total TAC for the area (excluding inshore areas in Div. 1A) is 30,000 tons. In 2014, total catches were 31,083 tons [48]. In both areas of 0 Division, Greenland halibut catches are around 15,000 t in 2014, generating a landed value of more than $50 million/year. Inshore Cumberland Sound fishery is also an important winter fishery near Baffin Island. This fishery began in 1986 and has been operated during the winter months using longlines through holes in the ice. In 2005, a turbot management zone was established in Cumberland Sound with a TAC of 500 t. This quota is separate from NAFO Division 0B. In 1992, these catches peaked at 430 t. However, they were declined to levels below 100 t through the late 1990s. They peaked again to 245 t in 2003. However, harvest again declined significantly during the last decade because of poor ice conditions and reduced fishing effort (DFO 2008). Greenland halibut catches have averaged around 120 tonnes/year. Recently, there has been interest in a fishery in the open water season as well as exploring deeper areas (500–1000 m) in the center of Cumberland Sound, outside the winter fishing grounds and expanding turbot management areas up to the mouth of Cumberland sound. Greenland sharks are common bycatch. However, estimates of shark biomass, productivity and fishing mortality are unknown. Therefore, the ability to predict impact of Greenland halibut fisheries on the Greenland shark population is limited. In the eastern Baffin Bay, Greenland halibut can be found in all waters around Greenland both offshore and inshore, but the highest concentration is in NAFO division 1A inshore [49]. An inshore fishery for Greenland halibut developed in the beginning of the twentieth century, with the introduction of the longline in 1910. The majority of inshore fishing is concentrated near cities and settlements and in areas of iceberg producing glaciers having better fish catches. At present, this inshore fishery catches are around 20,000 tons [49]. The catches in Subarea 1 (offshore 1A + Div. 1B–1F) were 16,146 tons in 2014. Catches in these areas are taken by vessels from Greenland, Norway, EU-Germany Faeroe Islands and Russia mostly by trawl netting.

7.1.2. Northern Shrimp Fishery

In Baffin Bay, the Northern shrimp (*Pandalus borealis*) occurs on the continental shelf off West Greenland in NAFO Divisions 0A (Canada) and 1A–1F (Greenland) in depths between 150 and 600 m. For assessment and management purposes, the widely distributed northern shrimp population in East Canada, West Greenland and international waters in NAFO Subareas 0 and 1 is considered to be a single stock. Northern shrimp may represent a single meta-population within the Northwest Atlantic, but treating fishery management and ecological relationships at a smaller spatial scale appears consistent with precautionary fishery management [50]. Shrimp Fishing Area 1 (NAFO Division 0A) is a joint Canada–Greenland stock, and its management is bilateral, and annual assessments of this stock are done by NAFO Scientific Council. During 2005–2006, shrimp catches were recorded highest up to 157,000 tons. In 2012, a joint TAC of 105,000 tons was set by the Greenland Government for Subarea 1 (Greenland) and Div. 0 A (Canada) [51]. Canada has set a separate shrimp quota east of 60° 30'W. Canada has set a TAC of 16,921 for its fishery in SFA1 for 2012. Canadian shrimp catches have decreased in recent years, from about 7000 tons in 2003–2005 to 1300 tons in 2011. Northern shrimp fishery management is guided by the precautionary approach and ecosystem-based management approach [52]. The shrimp fishery in

Greenland is regulated by individual quotas. In in western Greenland, the fishery for northern shrimp has declined during the last decade. However, as a response to climatic changes with higher temperatures in southwest Greenland, the stock is moving further north and it is possible that the area could regain its importance for commercial shrimp fisheries [51]. Bycatch of ground fish species by the small meshed shrimp trawls is a concern for these fisheries. Two species of wolffish, *Anarhichus denticulatus* and *Anarhichus minor* listed as threatened under SARA, are a bycatch in the northern shrimp fishery. To reduce it, an exclusion device known as the Nordmore grate was introduced into the Canadian shrimp fishery. This device sorts out the larger fish, allows them to escape while retaining the smaller shrimp, and has decreased groundfish mortality remarkably. There are gear restrictions in Greenland that include a cod-end mesh size of at least 40-mm stretched, sorting grids with 22-mm bar spacing to reduce finfish bycatch, the use of rolling rockhopper ground gear and toggle chains to keep trawl netting off the bottom [51].

7.1.3. Arctic Char Fisheries

Arctic char is a diverse and abundant resource in Nunavut. Its anadromous (searun) form is common in many river systems, primarily targeted by exploitation in small-scale fisheries. Several rivers flow into the Baffin Bay, Davis Strait and Cumberland Sound which support anadromous char. They spawn and over-winter in freshwater habitats, and adults normally spend the summers feeding in the marine waters of Cumberland Sound in the vicinity of their natal stream [53]. They feed in near-shore areas along the coast during the summer and migrate into fresh water during the fall. Arctic char is an important fishery in Nunavut coastal areas both for subsistence and commercial purposes. Subsistence fisheries are managed by hunters and trappers organizations, whereas Fisheries and Oceans Canada (DFO) is responsible for the management of the commercial harvest. Commercial harvest of char has been ongoing in several communities since the 1960s [54]. Combined commercial harvests in Nunavut between 2001 and 2008 have ranged between 74,124 and 95,558 kg [54]. Commercial harvest quotas are usually assigned to different rivers separately. To minimize the chances of over-exploitation, a system of exploratory licenses was set in place where a conservative quota is assigned to a river, which is then fished for five consecutive years gathering biological information on the harvested catch. These data are then used to assess the sustainability of the current harvest level. In the absence of abundance estimates, quotas are set at a conservation level using DFO precautionary approach model on the basis of best available information. Some mark recapture, weir enumeration and modeling estimates of abundance have been done in certain locations. However, their complicated migratory behavior violates many assumptions of these methods [54], and high uncertainty is associated with these results.

The Cumberland Sound Arctic char commercial fisheries are operated on 17 stocks in the region [54]. All these fisheries take place in the fall during the upstream migration and utilize gillnets with 140-mm mesh size. Commercial quotas are set for each river, with none being greater than 5,000 kg [54]. Most watersheds sustaining Arctic char fisheries are fairly small, and these rivers probably support smaller populations However, no abundance estimates for these stocks have been done so far, and most of the fishery in the area is still at an exploratory phase. Although the distance between fish-bearing freshwater systems is comparatively larger, genetic data indicate that stock mixing does occur [55]. The community of Qikiqtarjuaq on western Baffin Bay traditionally harvests char from lake and river systems around Qikiqtarjuaq. Paddle Fiord, Nudluit and Confederation Fiord are important fishing areas. The quotas are usually for the areas, not for the water bodies. As compared to other Arctic char fisheries in Nunavut, very few studies have been conducted on Qikiqtarjuaq fisheries areas. Arctic char are fished from many waters close to Clyde River. There is no significant bycatch in Arctic char fisheries, and ecosystem impact of these fisheries is also negligible.

7.1.4. Other Fisheries

The commercial fishery for snow crab (*Chionoecetes opilio*) was also once an important fishery in western Greenland, but its stocks are decreasing, and landings have dropped from 15,000 tons in 2002–2000 tons in 2007, and it is unlikely that a fishery for snow crab will develop in the near future in this area [56]. Snow crab are fished for using traps, but tangle nets may also be deployed. There are other small subsistence fisheries in near shore and fjords mostly in spring and summer.

Atlantic cod (*Gadus morhua*) were targeted using gillnets and trawls in West Greenland. The fishery was started in the 1920s and reached up to 400,000 tonnes per annum in the 1960s. However, due to over-fishing, the stock size declined and the fishery collapsed completely in the early 1990s. A recent survey in 2014 in West Greenland estimated at 110 million individuals with a biomass at 84,900 tons showing an increase. In 2014, the catches were 116 t. This fishery occurred in spring from March to May [36].

7.2. Fisheries Assessment

A variety of methods are used to assess the major stocks in the Region. An Otter Trawl survey index (biomass) is the primary approach for assessment of Greenland halibut. Additionally, gillnet and longline surveys are conducted for the inshore stock of Greenland halibut in Cumberland Sound. For northern shrimp, the Northwest Atlantic Fisheries Organization uses a

Bayesian model which incorporates cod predation on shrimp. For northern shrimp within Canadian waters, a survey biomass is used.

Arctic char fisheries have been assessed using trend analysis of age structure and mortality [57]. A number of modeling approaches for the assessment of Arctic char have been used including basic catch-based models such as depletion-corrected average catch, status quo models, Bayesian and life history invariant-based surplus production models [58, 59], age-structured models and zero-inflated generalized linear models (Xinhua Zhu, pers. comm.).

8. HUDSON BAY

8.1. Fisheries

Cree and Inuit harvest fish during the open water season from James Bay and Hudson Bay estuarine or coastal waters (**Figure 7**). They do not have a tradition of offshore marine fishing. Fishing is a traditional social and cultural activity. Anadromous Arctic char are most sought after by Inuit in Nunavut and north of Kuujjuarapik in Nunavik. The reasons for its preference are predictable times and locations, growth and large size, and they are free of parasites. Further south, they harvest anadromous cisco, whitefish, longnose sucker and brook trout. Most fish are caught using gillnets set near the communities, either along the coasts or at river mouths. However, subsistence fisheries are not restricted in terms of the fishing area, season, or harvest. Capelins are also harvested when they spawn on the beaches. Subsistence harvests of cod and sculpin are much greater on the eastern side of Hudson Bay [60]. Commercially, marine fish species have not been found in sufficient quantity to support a marine fishery in Hudson Bay or James Bay. Small near-shore commercial fisheries for anadromous Arctic char have developed along the Kivalliq coasts and at Puvirnituq. There is an international standard meat-processing plant that processes fresh and frozen Arctic char for sale to domestic and international markets. This fish-processing operation has not received enough fish to consistently meet operating expenses. Transportation is a particular problem for these fisheries. Fishermen generally participate in the commercial fishery to subsidize subsistence harvests [60]. Commercial harvesting of coastal and estuarine fish especially Arctic char is also conducted by many communities along the Quebec coast during summer. Commercial harvest quotas at these locations are opened on request of the Hunters and Trappers Organizations. Commercial fisheries are closely regulated; however, over-harvesting can occur in areas with large subsistence fisheries. There is no marine trophy fishery in Hudson Bay or James Bay, and most sport angling is by local residents mostly for Arctic char.

FIGURE 7. Hudson Bay, James Bay and surrounding territories.

8.2. Fisheries Assessment

Fisheries assessment has mainly been using trend analysis in age structure with the exception of a virtual population analysis approach used for the last formal assessment of the Diana River stock [61] on the Kivalliq coast, but recently data-limited models have been applied to historic data [57].

9. CANADIAN ARCHIPELAGO AND ARCTIC MAINLAND

9.1. CANADIAN ARCTIC ARCHIPELAGO

Arctic char are distributed across the Canadian Arctic Ocean including around the islands of the Arctic Archipelago (**Figure 8**). While this species may also be found in many rivers and lakes located in Canada's far north, the sea-run Arctic char are the most sought after for food and commercial uses. Arctic char are an important cultural, subsistence and economic resource in the Arctic. A number of commercial fisheries take place in the ocean tidal waters and river waters, as well as many subsistence fisheries for Canada's Inuit. Arctic char are a highly priced delicacy, marketed mainly fresh and frozen as whole-dressed fish and steaks. A small quantity is also processed into value-added products including smoked char and jerky.

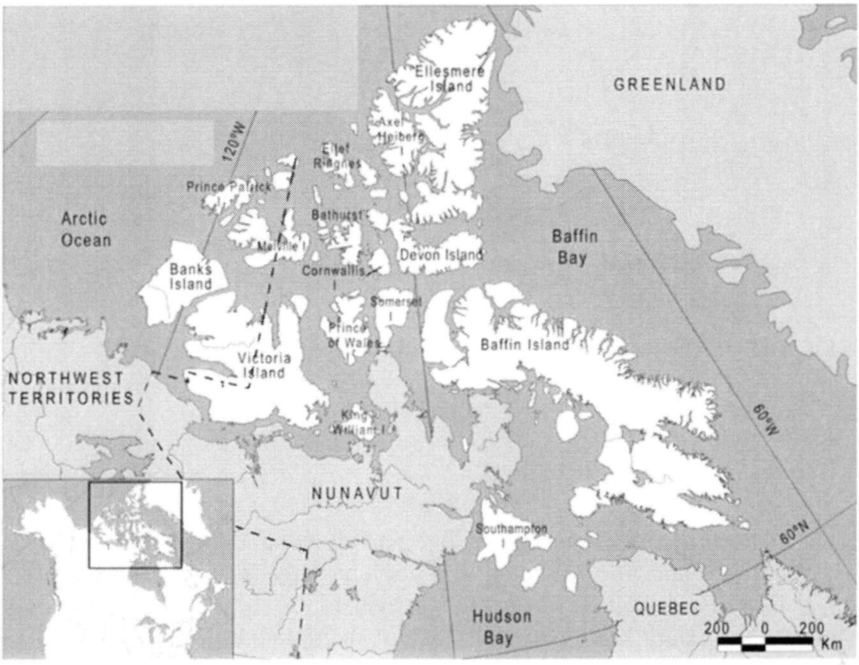

FIGURE 8. The Canadian Arctic Archipelago, Mainland and Hudson Bay.

Arctic char have the most northerly distribution of any freshwater fish. This species has a body shape typical of most salmonids and exhibits great variability in form and coloration. In the Cambridge Bay area, for example, spawning fish have an orange back, sides and belly, and the intensity of color is most pronounced in males. Arctic char may be anadromous, moving downstream to the sea in spring and returning in the fall or may remain permanently in freshwater.

9.2. Fisheries

Arctic char fisheries are important for the Inuit and in the subsistence economy of many circumpolar people. These fisheries are concentrated near communities and are predominately conducted using gillnets. In 2004, it was estimated that the subsistence harvest in the Cambridge Bay area was about 50% the size of the commercial harvest. The very first commercial fishing effort of Arctic char began in Cambridge Bay.

Fisheries are prosecuted in tow fashions using weirs (mainly at Cambridge Bay) and gillnets. Weirs may be traditional stone weirs or made using conduit pipe to form a fence guiding the char into a trap. Gillnets are typically set close to shore in the marine and estuarine areas and within lakes during the winter.

A number of commercial fisheries take place in various river systems throughout the Canadian Arctic, with the majority occurring in Nunavut such as Cumberland Sound and Cambridge Bay areas. There are also exploratory fisheries to examine potential for future commercial char fishing areas.

Commercial landings were 57 tonnes in 2012, 52 tonnes in 2011, 29.4 tonnes in 2010 and 31.8 in 2009.Value:Landed value was $186,000 in 2012, $175,550 in 2011, $118,000 in 2010 and $133,367 in 2009.

Abundance status and trends data are limited given the geographic distribution and nature of the fisheries. However, there are indications that the commercial stocks are stable.

Conservation measures: Arctic char fisheries in the Canadian Arctic are managed in co-operation with respective co-management partners. Conservation measures for commercial fisheries include minimum gillnet mesh size and total harvest levels.

The Arctic char fisheries in the Nunavut settlement area are co-managed by the DFO, the Nunavut Wildlife Management Board, Regional Wildlife Organizations, and Hunter and Trapper Organizations, in accordance with the Nunavut Land Claims Agreement, the Fisheries Act and its Regulations, and in some communities by local Hunter and Trapper Organization bylaws. This ensures that the best available information guides Arctic char fishery management decisions. Integrated Fishery Management Plans are also in development for the main Arctic char commercial fisheries.

9.3. Fisheries Assessment

DFO scientists, external experts and fish harvesters regularly review Arctic char stock assessments, and the results are published on the Canadian Science Advisory Secretariat Web site. Although these are data-poor stocks, the biological data collected from the fishery indicate a wide range of size and ages are present, with no loss of older age classes. This suggests that current levels of exploitation are likely sustainable.

Information about the condition and status of the oceans is also collected to better understand the effect of environmental conditions on char populations. For example, research activities have assessed: char biodiversity and trophic (feeding) variation in the Canadian North and its role in ecosystem structuring and function; the thermal ecology (temperature histories) of chars and how climate change might affect these; the link between climate change and the bioaccumulation of mercury; and changes in char populations as directly observed through community-based monitoring.

A number of approaches for the stock assessment of Arctic char have been used including basic catch-based models such as depletion-corrected average catch, status quo models), Bayesian and life history invariant-based surplus production models, age-structured models and zero-inflated generalized linear models [58, 59].

Commercial Arctic char fisheries in Nunavut are subject to a range of management measures designed to promote the sustainability and conservation of the char resource. Conservation measures include, but are not limited to minimum gillnet mesh size, total harvest levels and community-based monitoring. The management of these fisheries is complicated by the lack of harvesting data, the widespread distribution and biological complexity of Arctic char. New approaches using life history parameters as well as harvest and habitat information are being developed.

10. CANADIAN ARCTIC MAINLAND

10.1. FISHERIES

Subsistence, commercial and sport fishing are important activities on Canadian Arctic Mainland, both culturally and economically. Most of commercial fisheries are confined to the lakes under some specific quota system, while subsistence fisheries are confined to rivers and stream in the vicinity of the communities. There are two distinct fishing seasons, summer and winter. In Mackenzie River Basin, fish production rate is low because of late maturity and slow growth rate [62]; however, standing stocks of large size fish are high. The remote locations of commercial fisheries in Mackenzie River Basin are a limiting factor, and most of the commercial fisheries are not viable. Because of low species diversity and low productivity, Mackenzie River Basin is less resilient to anthropogenic impacts. Commercial fisheries in Mackenzie River Basin are dominated by whitefish, and other low-value fish, e.g., burbot, suckers, are usually discarded. Lake whitefish has the ability to respond to the exploitation and is the basis of its success (Healey 1980). Reliable data on the actual fish yield from Mackenzie River Basin is lacking. Fish stocks in the Mackenzie Great Bear sub-basin have been assessed but not on a regular basis. Harvesting of different fish species by recreational and commercial fishers is regularly monitored by DFO. Great Bear Lake is world renowned for its trophy lake trout. There was a decrease in the harvest of lake trout during the 1970s up to 1980s. However, conservation measures have been adopted including catch release practice and the use of barbless hooks. The current harvest of lake trout in Great Bear Lake is below the maximum sustainable yield [63, 64].

Subsistence fisheries occur in the Inuvialuit, Gwich'in, Sahtu and Deh Cho areas of lower Mackenzie River Basin [65]. Broad whitefish and lake whitefish are important species in subsistence catches along with inconnu and lake cisco. The subsistence fishing has declined in recent years, due to less dependency on dogs. Because of its high productivity, the lower Mackenzie River broad whitefish is able to withstand commercial and subsistence exploitation. Broad whitefish is one of the most important species to the aboriginal subsistence fisheries in the lower Mackenzie River and Delta [66, 67]. The Fisheries Joint Management Committee (FJMC) works jointly with DFO to co-manage all fish, fish habitat and marine mammals within the Inuvialuit Settlement Region and directly advise the Fisheries and Oceans on fisheries issues. The committee is responsible for collecting harvest information and making recommendations on subsistence quotas for fish and harvestable quotas for marine mammals. People from the communities of Tuktoyaktuk, Aklavik, Inuvik, Ft.

McPherson, Tsiigehtchic and Ft. Good Hope harvest broad whitefish for subsistence and commercial purposes. The Tetlit Gwich'in First Nation of Fort McPherson catches broad whitefish from the Peel River for domestic use [68]. Annual catches are around 10–12 thousand fish per year. Changing lifestyles and a less use of dogs for transportation have decreased the overall fish harvest [68]. The Rat River is associated with a key subsistence fishery for the Gwich'in of the Northwest Territories, the Rat River Dolly Varden char (*Salvelinus malma*). Rat River Dolly Varden char are a largely anadromous population, migrating from the Arctic Ocean up through the Mackenzie Delta to over-winter in the spring-fed pools of water known as Fish Hole on the headwater tributary Łuk Njik, Fish Creek [69]. Fisheries in Great Slave Lake Basin include subsistence, sport and commercial fishing. However, reliable up-to-date data are available for the water areas under commercial fisheries, for example Great Slave Lake. There are six fish management areas in Great Slave Lake with each area having its own management plan. Its east arm and certain inshore areas are closed to commercial fishing; however, east Arm is managed for a trophy lake trout fishery. Commercial fish catch in Great Slave Lake is the biggest fishery in the Mackenzie River Basin and catch about 1100 t of which 80% is lake whitefish and 10% is pike [70]. Exploitation of fish stocks in Great Slave Lake has affected fish populations in some cases, e.g., the lake trout population declined in west basin 40 years ago [71, 72]. There is a reduction in commercial fisheries since 1990 because of reduction in effort results from lifestyle changes. Information on fish stocks and harvest is not complete, and management usually adopt a precautionary approach towards management. There are also reports of reduction in inconnu populations in some of the major tributaries to Great Slave Lake (VanGerwen-Toyne et al. 2013). However, whitefish stocks and the commercial harvest of this species seem to be sustainable [72]. At present, commercial harvest is around 500 tonnes against a quota of 1800 tonnes. Eigty percentage of the catch is composed of lake whitefish with about 10% cull. Great Slave Lake Advisory Committee (GSLAC) advises DFO on fishery management issues and ensures long-term conservation of fish and fish habitat. All communities living in Great Slave Lake Basin have subsistence fisheries which are poorly monitored. Traditional knowledge surveys and recent evidence suggest that the subsistence harvest was less than 5% of the total for the commercial harvest [73].

Whitefish make up 68% of the overall subsistence catch. Kakisa and Tathlina lakes support important walleye fisheries which account for over 70% of the walleye harvested commercially each year in the N.W.T. [74]. All three important fisheries including commercial, subsistence and sport fishers are using the fish stocks in the Liard sub-basin. Watson, Frances and Simpson lakes are popular for lake trout and northern pike. Lake trout stocks are deteriorated in Watson Lake. White sucker, northern pike, burbot, walleye, lake whitefish, inconnu and Arctic grayling are important part of the subsistence fishery in the Liard sub-basin [75]. The main fishery management issues standing out in the Mackenzie River Basin include the potential for over-harvesting of migratory stocks which may also be fished in other areas, the potential for damage to harvested fish stocks from winterkill or summer warming, and the potential for adverse impacts from the development of pipeline, transportation, mining and hydroelectric developments.

10.2. Fisheries Assessment

A variety of assessment models and methods have been used for stock assessment. Survey and commercial catch rate indices have been used in Great Slave Lake and other fisheries. Trends in age structure and catch curve analysis is the most frequent approach, but surplus production models have been applied.

11. ALASKA

11.1. Fisheries

Arctic Alaska consists of the Chukchi and Beaufort Sea coasts (**Figure 9**). On the Chukchi Coast, the Noatak River produces a large run of chum salmon that maintain a Kotzebue-based commercial fishery. Many thousands of anadromous Dolly Varden over-winter the lower 300 km of the river and spawn in some of the river's tributary streams. This system is known for the large size of its Dolly Varden, and the current state record 8.9 kg (19.75 lbs.) was taken in 1991 from the Noatak River. Whitefish, Arctic grayling, burbot and northern pike are resident in the Noatak River. Inconnu use the lower reaches of the river for feeding during the spring of the year, but are not known to spawn there. Both the Selawik and Kobuk rivers support spawning populations of Inconnu in their upper reaches. (http://www.adfg.alaska.gov/index.cfm?adfg=ByAreaInteriorNorthwest.moreoverview).

Rivers on the Beaufort Sea coast have stocks of Arctic char that are harvested for subsistence and sport. The largest river is the Colville River where subsistence harvesting is done using gillnets. Arctic cisco (*Coregonus autumnalis*) and least cisco (*C. sardinella*) are harvested in the Colville River Delta near Nuiqsut, Alaska, after ice forms in the fall [76]. Arctic cisco targeted by the fall fishery derive from spawning stocks in the Mackenzie River of Canada. Young-of-the-year fish are recruited into the Colville

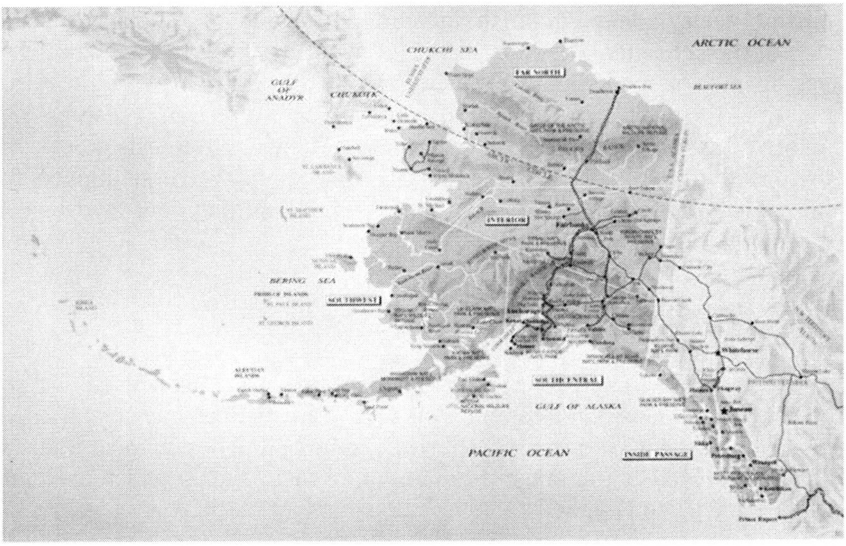

FIGURE 9. Continental Alaska and coastal seas.

region during August or September, aided by westerly coastal currents generated by predominantly easterly winds. In contrast, anadromous least cisco, harvested as the primary bycatch in the fishery, spawn and winter entirely in the Colville delta and lower river. Moulton et al. [76] reported on fishery monitoring for the 20-year period 1985–2004. During this period, effort in the subsistence fishery showed an increasing trend. Arctic cisco, the target species, averaged over 65% of the annual observed catch, and least cisco averaged 22%. From 1985 to 2002, total harvest of arctic cisco for the combined subsistence and commercial fisheries averaged 38,600 fish (15,958 kg) per year, ranging from a low of 5859 fish (2799 kg) in 2001 to 78,254 fish (31,340 kg) in 1993. During the same period, catches of least cisco averaged 18,600 fish (5819 kg), ranging from a low of 6606 fish (2014 kg) in 2001 to 33,410 fish (11,319 kg) in 1985. The subsistence fishery caught 56% of the total arctic cisco harvest and 42% of the least cisco harvest (in numbers of fish). In the six years for which estimates of both harvest and population level were available, total estimated annual harvest of arctic cisco within the Colville River Delta averaged 8.9% of the available fish, with yearly estimates ranging from 5.4 to 12.9%. For least cisco, the average annual removal rate was 6.8% (range 2.9–13.8%).

11.2. Fisheries Assessment

Assessments have used age-structured and hydro-acoustic models and environmental models in the case of Arctic cisco [77].

12. BEAUFORT SEA

12.1. Fisheries

The Beaufort Sea is shared by Alaska and Canada (**Figure 10**). On both sides, there is no commercial fishing on marine fish. Historically, there was a commercial fishery on the USA side for Arctic Cisco. The fisheries now are for subsistence with moderate harvests by Inuvialuit and Alaskan aboriginal peoples of marine mammals, Dolly Varden char, *Salvelinus malma*, and Arctic Cisco, *Coregonus autumnalis.*

History of commercial fisheries in Beaufort Sea is dated back prior to 1960s; however, an economically viable commercial fishery has not been implemented in the region. Most fisheries in the region have focused on harvesting large anadromous broad whitefish, *C. nasus*. These are caught by gillnets along the coast in summer or when they enter freshwater rivers to spend the winter. Broad whitefish grow to a large size and have higher-quality meat. However, because of limited local markets and high cost of transportation, a commercial fishery was not developed.

Dolly Varden (*Salvelinus malma*) char west of the Mackenzie River were once thought to represent a distinct form of Arctic char (*Salvelinus alpinus*). They inhabit the Rat River and its tributaries. They are harvested mostly for subsistence purpose. However, in two areas Big Fish River and Rat Rivers, they were harvested both by subsistence and commercial fisheries. Arctic char typically occur in river systems to the east of the Mackenzie River drainage (e.g., Hornaday River, Kuujjua River).A Fisheries Joint Management Committee (FJMC) works jointly with DFO to co-manage all fish, fish habitat and marine mammals. The FJMC has the power to directly advise the Minister of Fisheries and Oceans on fisheries issues. The committee is responsible for collecting harvest information and making recommendations on subsistence quotas for fish and marine mammals. The FJMC also monitor sports fishing in the area. Community consultations are essential part of fisheries management in the area. The FJMC has a research budget and provides funding to complement traditional knowledge and science to solve resource management issues. Canadian federal government and the Inuvialuit people of the western Arctic have signed an agreement in 2011 to maintain fishing limits in the Beaufort Sea as a step towards a comprehensive ocean management plan for the Beaufort Sea. At present, commercial fishing does not exist in the Beaufort Sea, but there were many experimental applications in fishing license since 2000. The melting of sea ice has opened many waterways of the Arctic that led to boats, commercial vessels and fishing boats coming into the region. Integrated management framework for fisheries in the Beaufort Sea requires future commercial fisheries to have the support of the Inuvialuit living in the area.

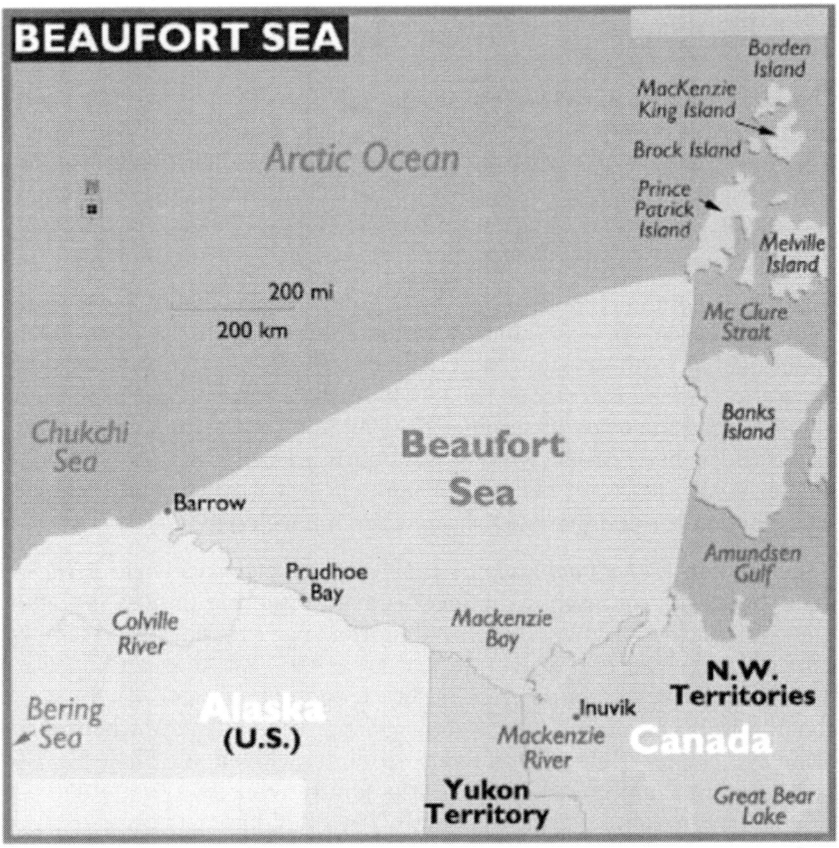

FIGURE 10. The Beaufort Sea.

Mathias [78] notes that main subsistence food for the Canadian Inuvialuit comes from beluga whale, broad whitefish, Arctic char, ringed seal, inconnu, humpback whitefish, and Arctic cisco. Fisheries productivity, or even biomass, on the Canadian Beaufort shelves, is unknown. There have been no systematic surveys of demersal and pelagic fish (other than larval fish) in the Canadian Beaufort Sea, although some locations have been fished with gillnets, beach seines, fyke nets and small mid-water trawls. In fact, Byers [[79], p. 40] stated: "The distribution and ecology of deep-water, off-shelf fish communities of the Canadian Beaufort Sea remain mostly unstudied. This is likely due, in part, to inaccessibility due to ice-cover and (until recently) a historical lack of industrial interest in abyssal regions." Mathias [78] noted that evidence suggests the Canadian shelves may be less productive than the Alaskan Shelf because they are less influenced by inflow from the productive Bering Sea and there may be other productivity differences brought about by the massive depositional environment caused by the Mackenzie River.

The Alaskan Shelf survey of the Beaufort Sea took place from the edge of Barrow Trough to ~109 km east and fished from 40 to 500 m, crossing the shelf break. A standardized bottom trawl survey quantified the distribution and density of demersal fish. Biomass estimates were produced for two depth strata (40–100 m and 100–500 m), and the two estimates were summed to provide a total biomass estimate of 18 kg/ha for a 6280-ha area of the shelf. Arctic cod made up 81% of the catch, while Bering flounder and walleye pollock together made up another 4%. The depths sampled were considered appropriate for a potential commercial fishery on the continental shelf and upper continental slope, but unlikely to occur in very shallow, near-shore areas.

On the same 6280-ha sampling site of the Alaskan Shelf, pelagic mid-water trawls and hydro-acoustic gear measured biomass estimates for Arctic cod of 19.2 kg ha–1 in the 40- to 100-m-depth range, and 53.7 kg ha^{-1} in the 100- to 500-m-depth range [80]. These densities may be compared with the estimate of unfished biomass for demersal Arctic cod from otter trawls, 18 kg ha^{-1} (=1.8 mt km^{-2}). The results of this fisheries survey on the Alaskan Shelf suggest that regardless of what species a commercial fishery targets, Arctic cod may constitute a significant bycatch [81]. Demersal fish were only 6% of the catch by mass; the overwhelming biomass was brittle stars. Arctic cod dominated fish catches, comprising 81% of fish biomass. Eelpouts made up 11% of the fish catch, while Greenland halibut were 0.7%, and walleye pollock and Pacific cod together (neither of which is reported from the Canadian Beaufort Sea) made up 2%.

12.2. Fisheries Assessment

Assessments of key marine mammal species, such as Bowhead Whale are done by aerial survey and a Potential Biological Removal (PBR) model. Assessment of Dolly Varden has been done using surplus production and age-structured models (X. Zhu, DFO personal communication).

13. CHUKCHI SEA

13.1. Fisheries

For most fish stocks within the Chukchi Sea, stock size was insufficient to support commercial activity (**Figure 11**). Three stocks: snow crab (*Chionoecetes opilio*); Arctic cod (*Boreogadus saida)*; and saffron cod (*Eleginus gracilis*) are of sufficient size to support commercial activity [82, 83].

FIGURE 11. The Chukchi Sea.

13.2. Fisheries Assessment

Little formal stock assessment has been done, but density-based estimates have been used for assessment. Stock size has been inferred from catch effort data from surveys.

14. SIBERIA

14.1. Fisheries

The nature of Siberian fisheries is relatively unknown because assessments, if any, are published in Russian and not disseminated broadly (**Figure 12**). It is likely that the fisheries are similar to the Canadian coastal and inland fisheries which are prosecuted using gillnets and are usually interception fisheries capturing various species during their migrations for spawning or over-wintering. Siberia does contain a number of large rivers and Chum Salmon, *Orcorhynchus keta*, exists as far along the coast as the Lena River. The fisheries are likely on Pacific salmon, broad whitefish, inconnu, Arctic Cisco and Arctic char.

FIGURE 12. Continental Russia featuring Siberia.

14.2. Fisheries Assessment

As noted above, published assessments in English are not known, and therefore, stock assessment models are unknown.

15. GENERAL ASPECTS OF ASSESSMENT

The Arctic region shows wide variation in the nature of fisheries and the practice of stock assessment. The differences are as great within the marine areas compared to the freshwater fisheries. For example, the Barents Sea has large-scale industrial fisheries with total harvest near to 2 million tonnes, mainly from bottom trawl harvest, whereas an area comparable in size and latitude, the Beaufort Sea, has a tiny fraction of that harvest from subsistence activities with no commercial fisheries. The key species are much different with the Barents Sea being dominated by groundfish, especially Gadiods, while the fisheries in the Beaufort Sea are mainly on marine mammals and anadromous fish such as Dolly Varden char. Stock assessment of the Barents Sea fisheries is highly sophisticated, deploying a variety of analytical population models such as extended survivor analysis. In contrast, with the exception of recent assessments of Dolly Varden char, the Beaufort Sea stocks are not modeled and only indicator values are used in assessment.

There is likely less variation in the nature and assessment of freshwater fisheries. A large proportion of freshwater fisheries are prosecuted as interception fisheries using gillnets. Assessment is mainly regarding demographic trends in size at age although recently, surplus production and age-structured models have been applied in analysis. Data-limited assessment tools hold much promise for freshwater assessments, and the incorporation of fisher and aboriginal traditional knowledge through methods such as the traffic light approach is likely to be important in the future.

16. CONCLUSION

16.1. Future Trends In Arctic Fisheries

16.1.1. Potential and realized effects of climate change

Climate change will have the most profound effect on the aquatic ecosystems of the Arctic [82, 83]. It has already been noted that the range of demersal species in the Barents Sea has shifted northward in recent years as there is less and less multiyear sea ice. Shift in Barents Sea stocks northwards. Ice-based ecologies of animals such as seals and Arctic cod in the Canadian Archipelago will shift even further northwards, and lower overall population sizes will occur. Access to fishing areas in the central offshore will be increased with unknown results in terms of harvest and management. Some species will increase growth and colonization, and increases in abundance of species more typical of temperate areas will occur.

16.1.2. Use of data-limited and traffic light approaches

Traditional assessment methodologies with a full range of fishery-dependent and fishery-independent metrics may not be possible for many of the small widely spread artisanal fisheries of the Arctic. However, there has been much new research in the area of data-limited fisheries assessment [58, 83]. These methodologies may be well suited to areas where sampling for detailed time series of demographic traits and other metrics is difficult. As well, for community-based fisheries where traditional ecological knowledge may exceed the knowledge based on scientific data collection approaches such as the traffic light approach that uses the sum of series of relatively "soft" indicators may be more effective for stock assessment than traditional methods from industrial fisheries.

16.1.3. Development and application of advanced population models

For areas with large-scale industrial fisheries such as the Barents Sea, ongoing development of advanced population models will surely happen in the future. The population models will likely include both fishery and stock statistics but environmental factors as well. Bayesian approaches to stock assessment are likely to grow in importance to be able to take advantage of prior knowledge and provide multiple options to managers.

16.1.4. Development and application of ecosystem models

Ecosystem models such as ECOPATH-ECOSYM may gradually mature so that they can be used for the development of annual quotas—taking into account the environmental conditions as well as stock and fishery. The incorporation of ecosystem-based models will be important to predict the effects of climate variation on yield and should be part of the fisheries manager's toolkit.

Ecosystems, fisheries and stock assessment are highly variable around the Arctic region. Understanding of Arctic fisheries is as complex as any other part of the world, if not more so. There will be great advances tailored to the Arctic fisheries assessment as understanding becomes clearer with more integration of scientific and traditional ecological knowledge.

REFERENCES

1. Hønneland, G. (2005) Barents Breaking. Norwegian Foreign Policy in the North after the Cold War. HøyskoleForlaget. ISBN: 82-7634-649-9
2. Bogstad, B., Kovalev, Y., Lindemann, D., Aanes, S., Aglen, A., Alonso, C., Alpoim, R., Aranda, M., Bernruether, M., Bertelsen, M., Borisov, V., Bulatov, O., Bulgakova, T., Chetyrkin, A., Drevetnyak, K., Filin, A., Fotland, A., Janusz, J., Jakobsen, T., Gjosaeter, H., Howell, D., Nedreaas, K., Planque, B., Prozorkevich, D., Russkikh, A., Santurtin, M., Smirnov, O., Stainsen, J-E., Sunnana, K., Tallman, R., Titov, O., and N. Yaragina. (2014) Report of the Arctic Fisheries Working Group (AFWG), 21–29 April 2014, Lisbon, Portugal. 572 pp.
3. Sirenko, B.I. ed. (2001) List of species of free-living invertebrates of Eurasian Arctic seas and adjacent deep waters. Russian Academy of Science, Zoological Institute, Moscow.
4. Antipova, T.V. (1975) Distribution of benthos biomass in the Barents Sea. Trudy PINRO, 35, pp. 121–124.
5. Wiborg , K.F. (1962) Some observations on the Iceland scallop. *Chlamys islandica*.
6. Sundet, J.H. and Vahl, O. (1981) Seasonal changes in dry weight and biochemical composition of the tissues of sexually mature and immature Iceland scallops, *Chlamys islandica*. Journal of the Marine Biological Association of the United Kingdom, 61(04), pp. 1001–1010.
7. Golikov, A.V., Sabirov, R.M., Lubin, P.A. and Jørgensen, L.L. (2013) Changes in distribution and range structure of Arctic cephalopods due to climatic changes of the last decades. Biodiversity, 14(1), pp.28–35.
8. Borges, T.C. (1990) A contribution to the biology of the ommastrephid squid, Todarodes sagittatus, with emphasis on the possible application of morphometric data to population studies. Norwegian college of Fisheries Science at the University of Tromsø. Tromsø, University of Tromsø, Doctor scientiarum.
9. Bjørke, H. (1995) Norwegian investigations on *Gonatus fabricii* (Lichtenstein). ICES.
10. Vikebø, F., Jørgensen, C., Kristiansen, T. and Fiksen, Ø. (2007) Drift, growth, and survival of larval Northeast Arctic cod with simple rules of behavior.
11. Opdal, A.F., Vikebø, F. and Fiksen, Ø. (2011) Parental migration, climate and thermal exposure of larvae: spawning in southern regions gives Northeast Arctic cod a warm start.
12. Jorgensen, T. (1992) Long-term changes in growth of North-east Arctic cod (*Gadus morhua*) and some environmental influences. ICES Journal of Marine Science., 49:263–277.
13. Ponomarenko, I. (1984) Survival of bottom-dwelling young cod in the Barents Sea and its determining factors. Reproduction and Recruitment of Arctic cod, p. 210.
14. Von Quillfeldt, C., and Dommasnes, A. (2005). Report of the working group on fish stocks and fisheries (including production and energy flows). The Scientific Basis for Environmental Quality Objectives (EcoQOs) for the Barents Sea Ecosystem. Norway, 0–2.
15. Gjøsæter, H., Bogstad, B. and Tjelmeland, S. (2009) Ecosystem effects of the three capelin stock collapses in the Barents Sea. Marine Biology Research, 5(1), pp.40–53.
16. Mehl, S. (1991) The Northeast Arctic cod stock's place in the Barents Sea ecosystem in the 1980s: an overview. Polar research, 10(2), pp. 525–534.

17. Dolgov, A., Johannesen, E., Olsen, E. and Heino, M. (2010) Trophic ecology of blue whiting in the Barents Sea. IIASA. Interim Report . IIASA, Laxenburg, Austria, IR-10-035
18. Folkow, Lars; Haug, Tore; Nilssen, Kjell Tormod; Nordøy, Erling Sverre. Estimated food consumption of minke whales Balaenoptera acutorostrata in northeast Atlantic waters in 1992-1995. (fulltekst) NAMMCO scientific publications (2000); Volum 2. ISSN 1560-2206.s 65 - 80.s doi: 10.7557/3.2972.
19. Nilssen, K.T., Pedersen, O.-P., Folkow, L., and Haug, T. (2000). Food consumption estimates of Barents Sea harp seals. NAMMCO Scientific Publication Series, 2: 9–28.
20. Lindstrøm, U., Nilssen, K.T., Pettersen, L.M.S. and Haug, T. (2013). Harp seal foraging behaviour during summer around Svalbard in the northern Barents Sea: diet composition and the selection of prey. Polar biology 36: 305–320.
21. Haug,T., Nilssen, K.T., Lindblom, L. et al. (2007). Diets of hooded seals (Cystophora cristata) in coastal waters and drift ice waters along the east coast of Greenland. Marine Biology 3: 123–133.
22. Dingsør , G.E. (2015) Bottom-trawl indices from ecosystem survey for cod and haddock, using a design-based approach. WD17, WKARCT.
23. Shepherd, J.G. (1991) Extended survivors analysis: an improved method for the analysis of catch-at-age data and abundance indices. ICES Journal of Marine Science, 56(5), pp. 584–591.
24. Darby, C.D. and Flatman, S. (1994) Virtual population analysis: version 3.1 (Windows/Dos) user guide .Great Britain, Ministry of Agriculture, Fisheries and Food, Directorate of Fisheries Research. p. 85.
25. Pope, J.G. and Shepherd, J.G. (1982) A simple method for the consistent interpretation of catch-at-age data. Journal du Conseil, 40(2), pp.176–184.
26. Loeng, H., and Drinkwater, K. (2007). An overview of the ecosystems of the Barents and Norwegian Seas and their response to climate variability. Deep Sea Research Part II: Topical Studies in Oceanography, 54(23), 2478–2500.
27. Lindstrøm, U., Nilssen, K.T., Pettersen, L.M.S. and Haug, T. (2010). Harp seal foraging behaviour in the northern Barents Sea: diet composition and the selection of prey. Polar biology 33: 305–320.
28. ICES. (2015) Report of the Working Group on Widely Distributed Stocks (WGWIDE), 25 August-31 August 2015, Pasaia, Spain. ICES CM 2015/ACOM:15. 588pp
29. Nielsen, A. and C.W. Berg. (2014). Estimation of time-varying selectivity in stock assessments using state-space models. Fisheries Research, 158:96–101.
30. Torrissen, O., Olsen, R.E., Toresen, R., Hemre, G.I., Tacon, A.G.J., Asche, F., Hardy, R.W., Lall, S. (2011) Atlantic Salmon (Salmo salar): the "Super-Chicken" of the Sea? Reviews in Fisheries Science. 19, 257–278.
31. Buch, E.M. (2000) A monograph on the physical oceanography of the Greenland waters. Danish Metrological Institute. Scientific Report 00–12, 405 pp.
32. Walczowski , W. and Piechura, J. (2007) Pathways of the Greenland Sea warming. Geophysical Research Letters, 34(10).
33. Cheung, W.W., Dunne, J., Sarmiento, J.L. and Pauly, D. (2011). Integrating ecophysiology and plankton dynamics into projected maximum fisheries catch potential under climate change in the Northeast Atlantic. ICES Journal of Marine Science: Journal du Conseil, p.fsr012.
34. Pálsson, Ó.K., Gislason, A., Guðfinnsson, H.G., Gunnarsson, B., Ólafsdóttir, S.R., Petursdottir, H., Sveinbjörnsson, S., Thorisson, K. and Valdimarsson, H. (2012) Ecosystem structure in the Iceland Sea and recent changes to the capelin (*Mallotus villosus*) population. ICES Journal of Marine Science: Journal du Conseil, 69(7), pp.1242–1254.
35. Møller, P. R., Nielsen, J. G., Knudsen, S. W., Poulsen, J. Y., Sünksen, K., & Jørgensen, O. A. (2010). A checklist of the fish fauna of Greenland waters. Zootaxa, (2378), 1–84.
36. ICES. (2015) Report of the North-Western Working Group (NWWG), 28 April 5 May, ICES HQ, Copenhagen Denmark. ICES CM 2015/ACOM:07. 717 pp.

37. Folkow, L.P. and Blix, A.S. (1999) Diving behaviour of hooded seals (*Cystophora cristata*) in the Greenland and Norwegian Seas. Polar Biology, 22(1), pp. 61–74.
38. ICES (2013) Report of the ICES Working Group on Harp and Hooded Seals (WGHARP), PINRO, Murmansk, Russia, 26–30 August 2013. ICES CM 2013/ACOM: 20, 55 pp.
39. Christiansen, J.S., Mecklenburg, C.W. and Karamushko, O.V. (2014) Arctic marine fishes and their fisheries in light of global change. Global change biology, 20(2), pp.352–359.
40. ICES (2004) Report of the ICES Advisory Committee on Fishery Management and Advisory Committee on Ecosystems, 2004. ICES Advice, vol. 1, no. 2. 1544 pp.
41. McKinney, M.A., Iverson, S.J., Fisk, A. et al. (2013) Global change effects on the long-term feeding ecology and contaminant exposures of East Greenland polar bears. Global Change Biology, 19, 2360–2372.
42. Boertmann, D. and Mosbech, A. (eds.) (2011) The western Greenland Sea, a strategic environmental impact assessment of hydrocarbon activities. Aarhus University, DCE—Danish Centre for Environment and Energy, 268 pp. Scientific Report from DCE—Danish Centre for Environment and Energy no. 22.
43. Hamilton, J.M., and Wu, Y. (2013) Synopsis and trends in the physical environment of Baffin Bay and Davis Strait. Canadian Technical Report of Hydrography and Ocean Sciences. 282: vi + 39 p.
44. Coad, B.W. and Reist, J.D. (2004) Annotated list of the Arctic Marine Fishes of Canada. Canadian Manuscript Report of Fisheries and Aquatic Sciences. 2674: iv + 112 p.
45. Jørgensen, O.A., Hvingel, C., Møller, P.R. and Treble, M.A. (2005) Identification and mapping of bottom fish assemblages in Davis Strait and southern Baffin Bay. Canadian Journal of Fisheries and Aquatic Sciences, 62(8), pp.1833–1852
46. Jørgensen, O.A., Hvingel, C., Treble, M.A. (2011) Identification and mapping of bottom fish assemblages in Northern Baffin Bay. Journal of Northwest Atlantic Fishery Science 43:65–78
47. DFO (2014) Greenland Halibut (*Reinhardtius hippoglossoides*) Northwest Atlantic Fisheries Organization Subarea 0—Effective 2014. http://www.dfo-mpo.gc.ca/fm-gp/peches-fisheries/ifmp-gmp/groundfish-poisson-fond/halibut-fletan-eng.htm
48. Jørgensen, O.A. and Treble, M.A. (2015) Assessment of the Greenland Halibut Stock Component in NAFO Subarea 0 + Division 1A Offshore + Divisions 1B-1F. Serial No. N6457 NAFO SCR Doc.15/032.
49. Nygaard, R. (2015) Fisheries and catches of Greenland Halibut Stock Component in NAFO Division 1A Inshore in 2014. Serial No. N6466, NAFO SCR Doc. 15/039.
50. Aschan, M., Powles, H., Angel, J. (2012) MSC Assessment Report for the Canadian Offshore northern shrimp (Pandalus borealis) trawl fishery – Shrimp Fishing Area 1. Client: Canadian Association of Prawn Producers (CAPP) and the Northern Coalition (NC). Version: Final Report Draft. 176 pp.
51. Arboe, N.H. and Kingsley, M.C.S. (2013) The Fishery for Northern Shrimp (Pandalus borealis) off West Greenland, 1970–2013. Serial No. N6219 NAFO SCR Doc. 013/058.
52. DFO (2007) Integrated Fisheries Management Plan Northern Shrimp Northeast Newfoundland, Labrador Coast and Davis Strait. http://www.dfo-mpo.gc.ca/fm-gp/peches-fisheries/ifmp-gmp/shrimp-crevette/shrimp-crevette-2007-eng.htm.
53. Harris, L.N. and Tallman, R.F. (2010) Information to support the assessment of Arctic Char, Salvelinus alpinus, from the Isuituq River system, Nunavut. DFO Canadian Science Advisory Secretariat Research Document. 2010/063. vi + 37 p.
54. Roux, M.J., Tallman, R.F. and Lewis, C.W. (2011) Small-scale Arctic charr Salvelinus alpinus fisheries in Canada's Nunavut: management challenges and options. Journal of Fish Biology, 79(6), pp.1625–1647
55. Moore, J.-S., Harris, L.N., and Tallman, R.F. (2014) A review of anadromous Arctic char (*Salvelinus alpinus*) migratory behavior: implications for genetic population

structure and fisheries management. Canadian Manuscript Report of Fisheries and Aquatic Sciences. 2014/3051: vi + 27 p.

56. GINR 2010.

57. Tallman., R.F, VanGerwen-Toyne, M., Harris, L., Gallagher, C., Carmichael, T., Howland, K. and Zhu, X. (2012) Stock Assessment of anadromous charrs [genus Salvelinus] in the Central and Arctic Region. Canadian Manuscript Report of Fisheries and Aquatic Sciences. 3007: v + 26 p.

58. Tallman, R.F., Xinhua Zhu, Janjua, Y., Toyne, M., Roux, M.J., Harris, L., Howland, K. L., and Gallagher, C. (2013) Data limited assessment of selected North American anadromous charr stocks. Journal of Ichthyology, 53(10):867–874.

59. Zhu, X., Day, A.C., Carmichael, T.J., Tallman, R.F. (2014) Hierarchical Bayesian modeling for Cambridge Bay Arctic Char, *Salvelinus alpinus* (L.), incorporated with precautionary reference points. DFO Canadian Science Advisory Secretariat Research Document. 2014/096. v + 35 p.

60. Stewart, D.B. and Lockhart, W.L. (2004) Summary of the Hudson Bay Marine Ecosystem. Overview. Canada Department of Fisheries and Oceans, Central and Arctic Region, Winnipeg, Manitoba.

61. Tallman, R.F. and Kristofferson A.V. (1995) A virtual population analysis using ADAPT of the population dynamics of the Diana River Arctic Charr, Salvelinus alpinus. IN: Cosens, S.E. (ed) Report of the Arctic Fisheries Stock Assessment Committee for 1995. Department of Fisheries and Oceans.

62. Bodaly, R. A., Reist, J.D., Rosenberg, D.M., McCart, P.J. and Hecky, R.E. (1989) Fish and fisheries of the Mackenzie and Churchill River basins, northern Canada. In Proceedings of the International Large River Symposium. Canadian Special Publication of Fisheries and Aquatic Sciences, vol. 106, pp. 128–144. 1989.

63. Muir, A.M., Leonard, D.M. and Krueger, C.C. (2013) Past, present and future of fishery management on one of the world's last remaining pristine great lakes: Great Bear Lake, Northwest Territories, Canada. Reviews in Fish Biology and Fisheries, 23(3), pp. 293–315.

64. Janjua, M.Y., Tallman R.F. and Howland K.L. (2015) Preliminary analysis of trophic relationships in Great Bear Lake using Ecopath model. Canadian Technical Report of Fisheries and Aquatic Sciences. 3137: vi + 24 p.

65. Stewart, D.B., and G. Low. (2000). A review of information on fish stocks and harvests in the Deh Cho area, Northwest Territories. Can. Manuscr. Rep. Fish. Aquat. Sci. 2549: iv + 73 p.

66. Tallman, R.F. and Reist, J.D. (1997) The proceedings of the Broad Whitefish workshop: the biology, traditional knowledge and scientific management of Broad Whitefish in the lower Mackenzie River. Canadian Technical Report of Fisheries and Aquatic Sciences. 2193:63–74.

67. Martin, Z. (2010) Adaptation and habitat selection during the migration of an Arctic anadromous fish, Broad Whitefish (*Coregonus nasus* (Pallas 1776). M.Sc. Thesis. University of Manitoba, Winnipeg.

68. VanGerwen-Toyne M. and Tallman R. (2000) The Peel River Fish Study, 1998–1999 with emphasis on broad whitefish (Coregonus nasus). DFO, Winnipeg.

69. Gwich'in Renewable Resources Board (2010) Gwich'in Traditional Knowledge: Rat River Dolly Varden Char. Gwich'in Renewable Resources Board, Inuvik.

70. Read C.J. and Taptuna W.E.F. (2003) Data from the commercial fishery for lake whitefish, Coregonus clupeaformis (Mitchill), on Great Slave Lake, Northwest Territories, 1999/00 to 2001/02. Canadian Data Report of Fisheries and Aquatic Science, 1111: 54.

71. Low, G., Stewart D.B., Day A.C. and Taptuna W.E.F. (1999) Comparison of Fish Harvests From the East Arm of Great Slave Lake, N.W.T., By Itinerant Sport Anglers in 1986 and 1994. Canadian Technical Report Fisheries and Aquatic Sciences 2263.Department of Fisheries and Oceans, Winnipeg.

72. Tallman R. F. and Friesen M. K. (2007) A review of trends in length and age of lake whitefish (*Coregonus clupeaformis*) harvested from Great Slave Lake between 1972

and 1995. Canadian Manuscript Report of Fisheries and Aquatic Sciences. 2819: 27 p.

73. Janjua, M.Y. and Tallman R.F. (2015) A mass-balanced Ecopath model of Great Slave Lake to support an ecosystem approach to fisheries management: preliminary results. Canadian Technical Report of Fisheries and Aquatic Sciences. 3138: vi + 32 p.
74. Roberge, M.M., Low, G. and Read, C.J. (1988) An assessment of the commercial fishery and population structure of walleye in Tathlina Lake, Northwest Territories. Canadian Technical Report of Fisheries and Aquatic Science 1594: 54 p.
75. Louie, W.H., Hardisty, E., and MacDonald, D.D. (1995) Acquisition of traditional environmental knowledge in the lower Liard River Basin. Environment and Renewable Resources Directorate, Indian and Northern Affairs Canada, Ottawa.
76. Moulton, L.L., SeAvey, B. and Pausanna, J. (2010) History of an under-ice subsistence fishery for arctic cisco and least cisco in the Colville River, Alaska. Arctic, pp. 381–390.
77. Fechhelm, R.G., Streever, B. and Gallaway, B.J. (2007) The Arctic cisco (*Coregonus autumnalis*) subsistence and commercial fisheries, Colville River, Alaska: a conceptual model. Arctic, pp. 421–429.
78. Mathias, J. (2013) Canada/Inuvialuit Fisheries Joint Management Committee Technical Report 2013-01: viii + 111 p.
79. Byers, M., (2013), International Law and the Arctic, Cambridge Series on International and Comparative Law, 314 pages.
80. Logerwell, E., K. Rand, S. Parker-Stetter, J. Horne, T. Weingartner, and B. Bluhm. (2010): Beaufort Sea Marine Fish Monitoring 2008: Pilot Survey and Test of Hypotheses. Final Report. Minerals Management Service, U.S. Department of the Interior, BOEMRE 2010-048, 263 pp.
81. NPFMC 2009.
82. Hollowed, A.B. and Sigler, M.F. (2012) Fish and Fisheries in the Chukchi and Beaufort Seas: Projected Impacts of Climate Change. Alaska Fisheries Science Center, National Marine Fisheries Service, National Oceanic and Atmospheric Administration, 7600 Sand Point Way NE, Seattle, WA 98115 2 Alaska Fisheries Science Center, National Marine Fisheries Service, National Oceanic and Atmospheric Administration, 17109 Point Lena Loop Road, Juneau, AK 99801.
83. Hollowed, A.B. and Sundby, S. (2014) Insights.
84. Øigård , T.A., Haug, T. and Nilssen, K.T. (2014) From pup production to quotas: current status of harp seals in the Greenland Sea. ICES Journal of Marine Science: Journal du Conseil, 71(3), pp. 537–54.
85. Skjoldal, H. R., Dalpadado, P., and Dommasnes, A. (2004) Food webs and trophic interactions. In The Norwegian Sea Ecosystem, 1st edn, pp. 263–288. Eds. H. R. Skjoldal R. Sætre, A. Fernö, O.A. Misund and I. Røttingen. Tapir Academic Press, Trondheim, Norway. 55.
86. Tallman, R., Cadigan, N, Cass, A., Duplisea, D., Healey, B., Trzcinski, K. and Wade, E. (2011) Techniques for the Provision of Advice in Information-Poor Situations. DFO Canadian Science Advisory Secretariat Research Document. 2011/106. vi + 42 p.
87. Gordon, J.D.M., Bergstad, O.A., Figueiredo, I. and G. Menezes. (2003) Deep-water Fisheries of the Northeast Atlantic: I. Description and current Trends. J. Northw. Atl. Fish. Sci. Vol: 31; 37–150.
88. Bowering, W.R. and Brodie, W.B. (1995) Greenland halibut (Reinhardtius hippoglossoides). A review of the dynamics of its distribution and fisheries off eastern Canada and Greenland. Pp. 113–160. In: Hoppr, A.G. (ed.). Deep-water fisheries of the North Atlantic Oceanid slope. – NATO ASI Series.
89. ICES. (2006) Report of the Working Group on the Biology and Assessment of Deep-Sea Fisheries Resources. ICES CM 2006/ACFM:07.
90. VanGerwen-Toyne, M., Walker-Larsen, J., and Tallman, R. F. (2008) Monitoring spawning populations of migratory inconnu and coregonids in the Peel River, NWT:

the Peel River fish study 1998–2002. Canadian Manuscript Report of Fisheries and Aquatic Sciences 2851, Winnipeg.

91. VanGerwen-Toyne, M., Day, A.C., Taptuna, F., Leonard, D., Frame, S., and Tallman, R. (2013) Information in support of Assessment of Buffalo River Inconnu, (*Stenodus leucichthys*), Great Slave Lake, Northwest Territories, 1945–2009. DFO Canadian Science Advisory Secretariat Research Document. 2012/069. vii + 81 p.

CHAPTER 3

The Brown Seaweeds Fishery in Chile

Julio A. Vásquez*

[1]Department of Marine Biology, Faculty of Marine Science, Center for Advanced Studies in Arid Zones (CEAZA), Universidad Católica del Norte, Coquimbo, Chile

ABSTRACT

Chilean fishery of brown algae includes species belonging to the genus *Lessonia*, *Durvillaea*, and *Macrocystis*, which can be found along the coast, ranging latitudes from 18° to 55°S. The exploitation of these seaweeds is done mainly in the Northern coast because the environmental conditions of this region decrease initial production costs. Brown algae are exploited from natural populations and exported to international markets as row material, source of alginates, widely utilized in diverse manufacturing processes and industries. International demand for Chilean kelps has produced sustained increase in harvest during the last decade, reaching more than 390,000 dry tons/year. This chapter approaches the most relevant aspects of the brown seaweed fishery in Chile which covers a wide range of the Southeast Pacific coast, considering the number of commercial species, its abundance and distribution, knowledge achieved on their ecology and biology regarding management, and conservation of these resources, and finally, provides tools for stakeholders and policy makers directed to sustainable management of natural kelp beds occurring in the cold temperate seas.

Keywords: Brown algae, kelp, fishery, coastal environment, management

1. INTRODUCTION

Chile, a narrow and long country with over 4500 km of continental coastline, has an ancient tradition in the use of sea resources. Numerous algae, shellfish, and fish species have been incorporated in the diet and every day habits of his inhabitants, since prehistoric times. The astonishing evidence found at the archaeological site Monte Verde, dated 12,500 years BP and located near Puerto Montt (41°S), provides evidence of Pre-Clovis human settlement in South America, and exhibits ancient use of macroalgae, probably as food and for medicinal purposes [1, 2]. The diet of coastal human communities incorporated brown and red algae as significant components along the last 500 years, especially in those coastal populations situated South beyond 30°S [3].

As sources for alginates production, brown algae in Chile are exploited from natural populations and exported to world markets as row and dried commodity for alginates extraction [4, 5]. The local national gel industry, as well as invertebrate aquaculture production, utilizes only a minor fraction of the annual harvest [6]. During the last decade, a sustained increase of harvesting has been taking place, because of the international demand for Chilean kelp; production has reached more than 390,000 dry tons associated with an economic return of more than US$ 90 million [5, 7]. Chilean brown algae of economic importance belong to genus *Lessonia*, *Durvillaea*, and *Macrocystis*, and they occur throughout the coast from 18° to 55°S [8]. Even this wide distribution range, the exploitation of these resources is done mainly in the Northern region of the country between 18° and 32°S, because environmental conditions, such as remarkable air high temperature and dryness of the coastal desert, which decrease production costs of drying process close to zero and, consequently, the total processing cost before their commercialization [6].

In Chile, the harvest of brown algae is also matter of social relevancy since more than 15,000 people depend more or less directly on the exploitation and collection of this marine resource [9]. As established by local law, only authorized and registered artisanal fisherman are allowed to harvest such kelp [6]; however, enforcement measures and control are difficult to put into effect because of the topography of coastal territory where these natural populations of kelp occur, but also due to their extension and accessibility [10]. From the point of view of their ecological role, kelps have been defined as engineer species in the coastal marine ecosystems; they are key species which participate maintaining and preserving foci of high biological and genetic diversity [11, 12]. Also, these species are sensitive to disturbances from both natural and/or anthropogenic origin [13, 12].

2. SPECIES IN THE FISHERY

Geographic distribution and occurrence of commercial brown seaweed are associated with high-energy environments in the Southeast Pacific (**Figure 1**). *Lessonia* species can be intertidal and subtidal as well; they form belts along exposed rocky coasts (**Figure 1A, B, D**); *Macrocystis* forms shallow kelp beds ranging from intertidal zone to *ca.* 15 m depth in Northern latitudes (**Figure 1C, G**); it is gradually replaced toward Southern areas by *Durvillaea* (**Figure 1F**), which dominates the intertidal zone in wave-exposed areas [8]. In South direction and beyond 42°S, *Macrocystis* is the most abundant and dominant kelp species [15]; *Lessonia* species have almost continuous distribution along the whole Chilean continental coast; instead, the distribution of *Macrocystis* is fragmented into populations that form patches in Northern Chile [11, 16, 13], contrasting with its continuous coastal belts distribution from 42°S toward Cape Horn (55° 58°S); *Macrocystis* distribution includes the Chilean Southern fjord zone, surrounding Cape Horn, and ascending in North direction by the Argentinian exposed coast until Chubut area in the Atlantic (41°S). Several elements are combined to determine the observed kelp distribution patterns; they result from multifactorial interaction of complex life-history strategies of the involved species with environmental factors, such as spatial and temporal variations in water movement, nutrient availability, light, and temperature [17, 16, 12].

Chilean kelp species commercially exploited are as follows: *Lessonia trabeculata*, *Lessonia berteroana, Lessonia spicata* (members of *Lessonia nigrescens* complex, for details see [18–20]); these are all species having two morphotypes during their natural life cycle, which represent very different ecological roles and requirements, in both environmental and physiological aspects. Large and conspicuous (so, harvestable) sporophytes alternate with microscopic few cell and benthic organisms which are the gametophytes.

3. BIOLOGICAL AND ECOLOGICAL ASPECTS

Studies on the distribution and abundance of Chilean commercial brown seaweed were scarce and locally restricted until the end of 2000 [21–29, 30, 31]. Other than this, the use of non-comparable methodologies in the few studies carried out, which approached biomass stocks information, did

FIGURE 1. Intertidal beds of *Lessonia nigrescens* (A, B), subtidal beds of *Macrocystis pyrifera* (C), subtidal *Lessonia trabeculata* (D), Harvest on the beach of *Lessonia nigrescens* (E), *Durvillaea antarctica*, intertidal beds of *Macrocystis pyrifera* (G).

not allow extrapolation and inter-annual comparisons of the available total biomass. Similar situation occurred in relation to distribution studies of the involved species, both in temporal and spatial gradients [4]. Harmonization of methods or collection of information, as well as systematization of them, is essential aspects of the population ecology of commercial species, in order to propose adequate regulations and policies that guarantee a sustainable management of these resources.

One of the few extensive and intensive evaluations of biomass distribution of *Lessonia trabeculata, Lessonia nigrescens,* and *Macrocystis pyrifera* was done during 2004–2005, for the geographic extension ranging between 26° and 32°S; its main goal was to determine the distribution and abundance of the mentioned species in more than 700 km of coast. During the study, 140 sampling stations were established separated by 4.5 linear km, excluding *a priori* sandy beaches and soft bottom subtidal areas because they represent inadequate substrates for kelp's spores settlement [6, 5]. *L. trabeculata* was found in rocky subtidal habitats between 0 and 30 m depth in the whole study area. Depending on the extension of the rocky platform into the subtidal zone, local biomass up to 50,000 wet tons was recorded. Estimation of abundance indicated a standing stock of approximately 800,000 wet tons of *L. trabeculata* in the study area. *L. nigrescens* was distributed in a continuous pattern along the rocky intertidal zone of the whole study area, with local biomass (sampling station) registered between 50 and 4000 wet tons. Using GIS abundance polygons were constructed which revealed a standing stock of more than 100,000 wet tons of *L. nigrescens* in the study area [6]. In contrast, *Macrocytsis* evidenced fragmented distribution within the study area. Local populations were small and estimated biomass fluctuated between 2 and 12 wet tons. The sum of local biomass allowed estimation of a standing stock that did not exceed 200 wet tons for the entire study area at the time [6, 5].

In contrast to the lack of ecological information used to determine fishery regulations, several studies arise since the middle 1980s approaching different aspects of kelp knowledge, such as biology and population ecology [12, 13, 17, 32, 33], genetics and taxonomy [18–20, 34], enhancement and cultivation [35–38], new and novel uses for their natural by products, for medical nanotechnology, for example (see [39–41]).

Because of *Lessonia nigrescens* represents more than 70% of total landings of brown Chilean seaweed fishery, several studies have been focused on this species. Studies on genetics, molecular biology, and population ecology show that *L. nigrescens* is a species complex composed by two cryptic species: *Lessonia berteroana* and *Lessonia spicata* [18, 42, 43]; *L. berteroana* is distributed from the South of Perú (*ca.* 15°S) to approximately 30°S, and *L. spicata* occurs form 30°S toward South [20]. In the last decade, the kelp harvest in Chile has been sustained principally by the *Lessonia nigrescens* complex, especially in the area known as latitudinal break for biodiversity distribution which location is considered approximately at 30°S [19].

Exploited and unexploited species belonging to *Lessonia nigrescens* complex form Northern Chile were compared using morphological and demographic parameters such as density, biomass, recruitment, and population structure [5]. These are traits which allow estimation of the effect of harvesting in other natural populations of brown algae [44–47], and the assessment of the impact of natural disturbances [48, 16, 13] and other anthropogenic activities on such species [37]. In this context, the morphological and demographic parameters used for evaluate *L. nigrescens* could be useful as ecological indicators for: (1) to evaluate the consequences of good harvesting practices agreed upon by fisherman, (2) to compare the effect of harvesting in areas with different administration policies, (3) to monitor the sustainability of exploited kelps, and (4) to propose to competent authorities the precautionary and/or recovery adequate measures for sustainable managing of dynamics population of commercially important brown algae.

4. COLLECTION, HARVEST, AND LANDINGS

Largely and until year 2000, the whole brown algae fishery in Chile was based on the collection of natural mortality kelp from coastal populations. A sudden and significant increase of its demand, other than as raw material for alginate source, also as food source for cultivated abalones introduced in Chile in the last 16 years, triggered the harvest of kelp species. Since then, the Chilean brown seaweed fishery becomes an extractive fishery in which converge four main factors: (1) still the international market request for alginic acid source; (2) use for feeding the emergent local farming of several kelp-consuming organisms under controlled conditions; (3) the switch of fishers toward the harvest of commercial brown seaweeds as consequence of the collapse of other benthic fisheries, and (4) the strong impact on local economy produced by the international fluctuation of the price of copper. This metal constitutes the Chilean main resource, representing more than 60% of the internal gross product (PIB, as its acronym in Spanish); it provides direct and/or indirect jobs for thousands of people in the country, being Chile the first copper producer around the world. Local economy is extremely sensitive to fluctuations of international cooper price, and the fall (currently crush) of it provokes significant unemployment, particularly of non-specialized workforce; as one of its primary consequences, unemployed people are forced to migrate to coastal areas where they can develop subsistence economy based on precarious jobs represented by the collection and harvesting of brown seaweeds [9, 6].

During the last 35 years in Chile brown algae, landings have fluctuated between 40,000 and 390,000 tons/year, showing sustained increase since 2000 (**Figure 2**); *L. nigrescens* and *L. trabeculata* comprise more than 90% of the total production of them, whereas *Macrocystis* and local consumption of *Durvillaea antarctica*, contribute only marginally to total landings [61].

The significant rise of kelp demand, as commodity source of alginic acid around the world during the last years, explains the increase of kelp extraction. A smaller fraction of this increment is consequence of the yield reduction of kelp used for milling, because of higher humidity contents of lately processed plants if compared with those from previous years [7, 50]. From a different perspective, during 1997–1998 when a severe ENSO event occurred, Chilean exports of brown algae showed a considerable peak, probably related to significant mortalities generated by this large scale oceanographic event. The warming of the ocean surface, simultaneously to decreasing of nutrients concentrations, both associated with "El Niño," has strong impact over kelp populations and thousands of dead plants are cast ashore by waves which end collected by fishermen [16, 23, 60].

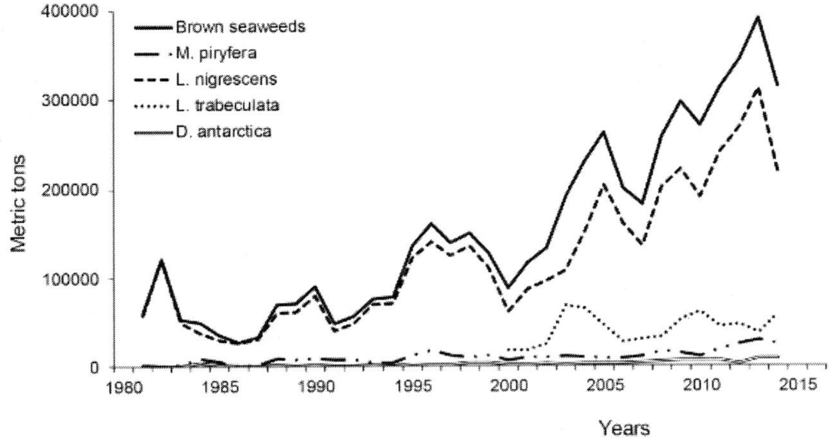

FIGURE 2. Temporal landing (1980–2014) of brown seaweeds of economic importance in Chile. Source: National Fishery Service (Chile) [61].

Since 2005, the abalone cultivation industry exhibits a remarkable and sustained growth, especially in Northern Chile; this commercial activity consumes close to 4800 tons of fresh alga/year, mainly *Macrocystis;* neither the utilization nor the economic yield of this activity is comparable, until date, to demand of Chilean kelp as source for alginate extraction.

5. MANAGEMENT

Chilean authorities have implemented a management and conservation strategy program for economically important brown algae, considering its economic, social, and ecological importance, and also the significant increase of kelp harvest. The expectative of this program is focused on surveillance of available and harvestable biomass, evaluation of strength of harvesting (Capture per Union Effort-CPUE), and characterization of the productive chain based on these primary producers. As a result of this strategy, carried on since 2010, plus several years of kelp knowledge achieved, recommendations have been established for the management of kelp sustainability. The premise is *"how to harvest is more important than how much you harvest"* [5]. This program has been implemented in three Chilean conservation tools performed at the country level and which represent three different conservation strategies: (1) marine protected areas (marine park, marine reserve), (2) open-access areas for artisanal fishermen (OAA), where they collect and harvest marine resources, and (3) management areas for exploitation of benthic resources (MAEBR), where organized fisherman have some territorially rights assigned over a sector of the coast. The conservation strategy involved in MAERB is based on co-management (see [50–55]).

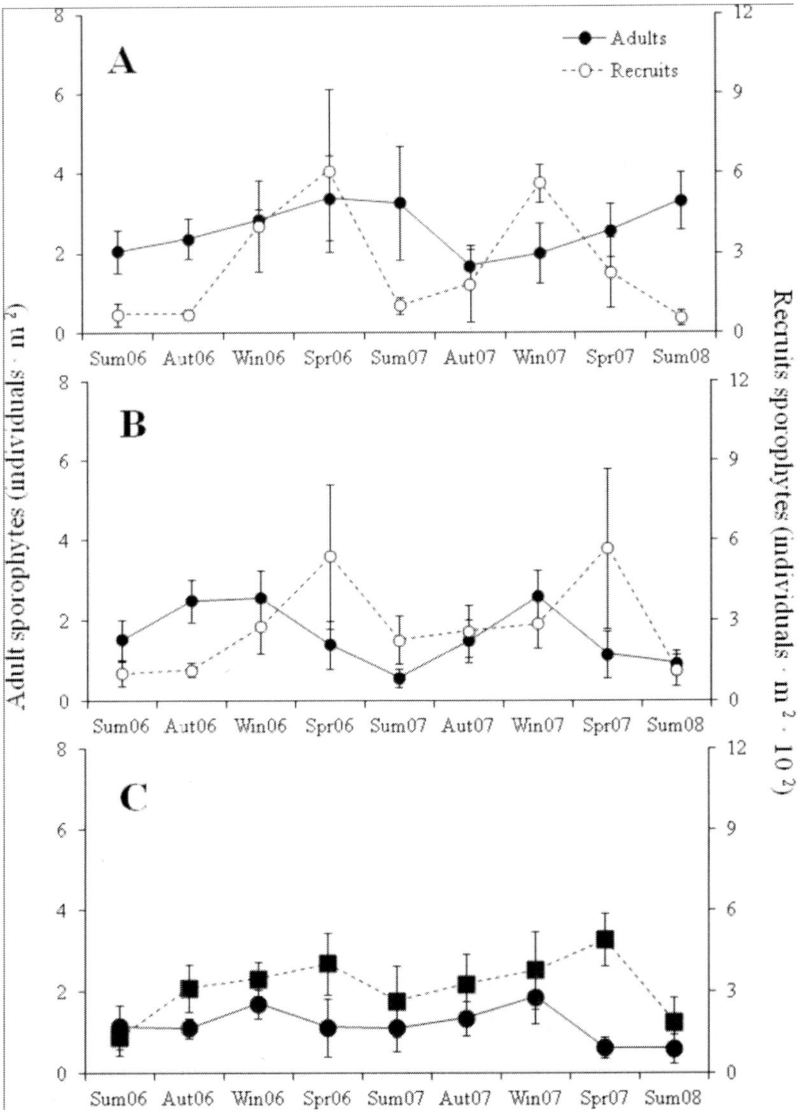

FIGURE 3. Temporal variation in the density of *Lessonia nigrescens* complex adult plants and recruits in kelp beds located in marine protected areas MPC (A), management areas for exploitation of benthic resources (MAEBR), and open-access area (OAA). Mean + 2SE [49].

The main practical recommendations of the program for the sustainability of brown seaweeds are focused on selective harvesting of adult sporophytes and maintenance of a permanent stock of individuals able to reproduce, recruitment facilitation, decrease of grazing by benthic invertebrates, and permitting the sustainability of kelps and the conservation of its associated biodiversity [6, 5, 56, 62]. Considering all aspects mentioned, the following bio-ecological recommendations must be applied to kelp beds subjected to significant, frequent, and intense harvesting: (1) to harvest the whole plant, including the holdfast. (2) To harvest plants with a basal diameter larger than 20 cm. (3) To harvest one out of every three plants, with preference for the biggest specimens, thereby thinning the population. (4) For the particular case of *Macrocystis*, to cut the canopy to one meter below the surface.

According to administration regime (Conservation Strategy) assigned by competent authority to natural populations, the density of both adult plants and juvenile recruitment of *Lessonia nigrescens* is subjected to temporal variation (**Figure 3**). In marine protected areas like marine parks or marine reserves (MPA), the annual renewal of kelp populations exhibits a seasonal cycle wherein the natural mortality of adults is compensated by intense juvenile recruitment (**Figure 3A**). In MAEBR, the density of adult plants decreases during the maximum harvest period, which is preferably carried on during spring and summer; in these conditions, the annual cycle of kelp renewal is maintained by recruitment of juvenile plants post-harvest (**Figure 3B**). In OAA, where the kelp harvest occurs all along the year, adult plants density decreased significantly (**Figure 3C**). Thus, constant releasing of substrate by permanent harvesting facilitates sustained juvenile recruitment, which significantly increases the density of recruits throughout the year; this last population renewal process takes places independently of seasonal variation in opposition to what was observed in *Lessonia* populations belonging to Conservation Strategies of MAEBR and MPA (**Figure 3**).

The density of adult plants is greatest in populations inside MPA in contrast to those inside OAA (**Figure 3**). As previously exposed in MAEBR, the seasonal harvest of *Lessonia* decreases the density of adults; however, this impact of harvesting is significantly lower than which is observed in OAA (**Figure 3**). In these open-access areas, the density of adult plants is less because of high harvesting pressure produces contraction of the stock of reproductive plants, and thus negatively affects the kelp renewal. There exists an exception to this pattern observed in OAA constituted by those populations where the difficulty of access to the coastline generates a barrier to permanent harvest. In this case, topography constraints to access generate effects that mimic a sort of natural co-management; in this way, the pattern of size distribution of *Lessonia* plants in OAA is quite similar to the pattern observed in MAERB [49].

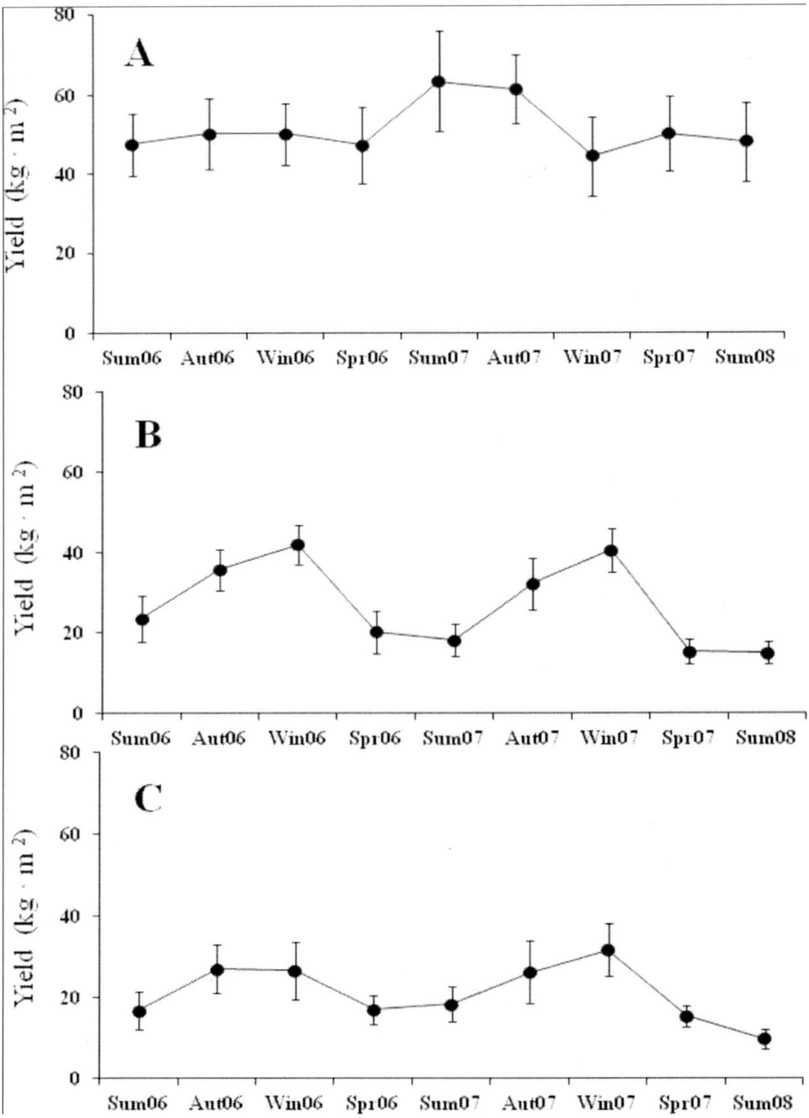

FIGURE 4. Temporal variation of *Lessonia nigrescens* yield (biomass kg m^2) in kelp beds located in marine protected areas MPC (A), management areas for exploitation of benthic resources (MAEBR), and open-access area (OAA). Mean + 2SE [49].

In MPA, *Lessonia* yield (kg of biomass/m^2) is constant throughout an annual cycle and close to 50 kg/m^2 (**Figure 4A**). By contrast, the available biomass in MAEBR evidences marked seasonality with an annual cycle of renewal of the kelp post-harvest during spring and summer (**Figure 4B**). A similar tendency is observed in OAA, even if with significantly less available biomass which does not surpass 25 kg/ m^2 (**Figure 4C**). Available biomass is biggest in MPA populations and lowest in OAA populations. In MAEBR and OAA, the available biomass is 50 and 65% lower, respectively, than stocks in MPA [50]. The available biomass in MAEBR represents the permissible limit for sustainable exploitation of the kelp forests and is evidence of the adequate application of the management plan. On the other hand, the level of available biomass in OAA is an indicator of highly exploited populations, and its management program would help to establish sustainability parameters dealing with a strong harvest pressure [49].

The size structure of *Lessonia nigrescens* populations, according with the morphological variables suggested [6], varies according to the management and conservation measures applied (**Figure 5**). In MPA, 20% of the populations are recruits and 35% are juveniles, and the rest of the population is adults with large-sized plants (**Figure 4A**). In MAEBR and OAA, recruits represent 35% of the whole population, while juvenile plants represent 45 and 55%, respectively (**Figure 5B, C**). In MAEBR, the fraction of adult plants is renewed by the growth of juvenile plants during the annual cycle, while in OAA, strong harvesting pressure facilitates recruitment and colonization of free primary substrate. In MPA, the portion of adult plants available for harvesting is in general, 45% of the population. In MAEBR, the harvestable portion corresponds to 25% of the total available biomass. By contrast, in OAA, the available biomass for commercial harvesting did not exceed 10% of the total plants in the whole *Lessonia* population studied in Northern Chile [5, 49].

After 25 years of observation and assessment of *Lessonia* populations, after thousands of hours of field monitoring and abundant literature produced, and based on bio-ecological knowledge accomplished, the main concept associated with recommendation for sustainable management of *Lessonia nigrescens*complex is "*how to harvest is more important than how much you harvest*" [6]. This resource management practice has been spread among fisherman during the last 15 years and adopted voluntarily as an alternative to the traditional precautionary method, in which the fishing authority imposes a capture quota, arises from the total available biomass [5, 49]. However, fulfilling the premise previously mentioned requires good

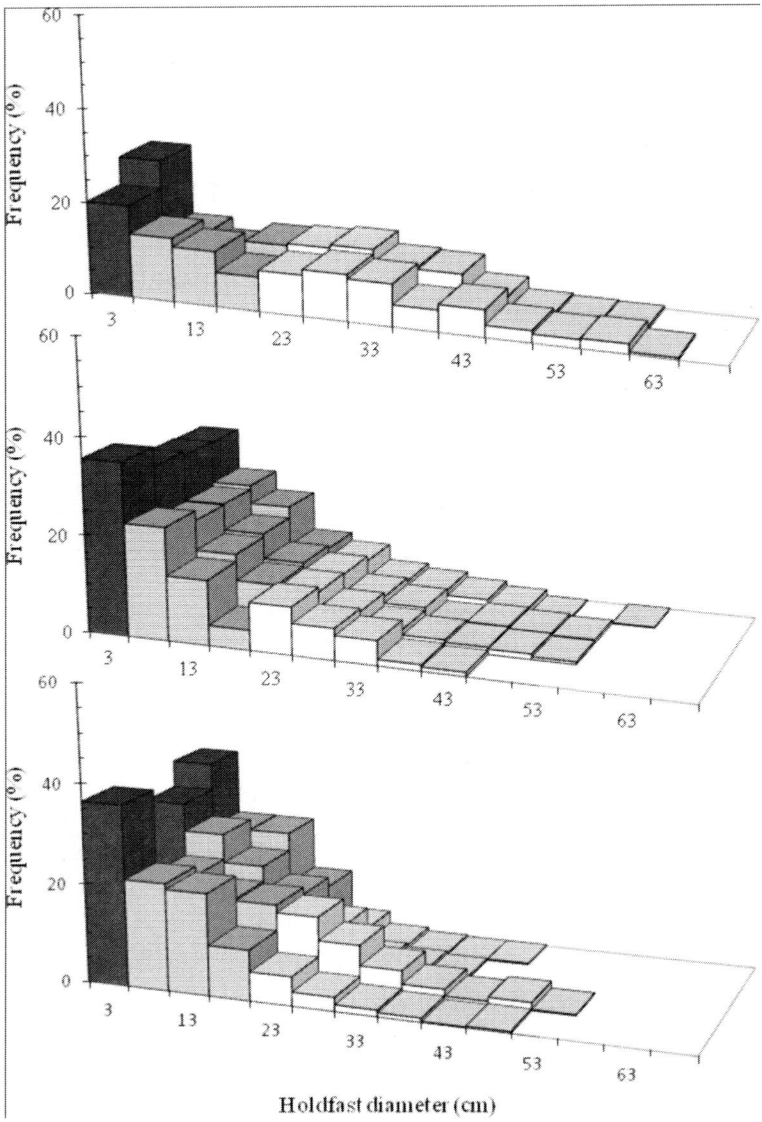

FIGURE 5. Population structure of *Lessonia nigrescens* in kelp beds located in marine protected areas MPC (A), management areas for exploitation of benthic resources (MAEBR), and open-access area (OAA). The black bars indicate recruits (<5 cm long), the gray bars indicate juvenile plants (without reproductive structures, <20 cm holdfast diameter), and white bars indicate adult plants (with reproductive structures, >20 cm in holdfast diameter) [49].

practices on the part of artisanal fishermen, who are the only authorized users to harvest brown algae in Chile. The correct application of the management recommendations of MAEBR seems to be strongly related to the social capital that co-management generates [54], a concept that does not make sense in OAA where harvesting activity is individual and difficult to enforce (see [52–54]). The wide latitudinal extension of the littoral zone and several restrictions to it access increase the cost of enforcement and decrease the efficacy of control of the fishing authority [10]. Therefore, as in many other cases if not all of them, the construction of participative awareness is a key factor in the conservation of natural populations of Chilean kelps and the sustainability of this resource [5, 49].

As last recommendations, it will be necessary to make significant progress in areas such as: (a) perfection of the capacities of commercial management by using social capital, (b) optimization of control mechanisms and enforcement considering the idiosyncrasy of Chilean artisanal fishermen, (c) improvement of information flow between and among the different actors in the productive chain and the authorities, and (d) establishment of controlled extraction of brown algae by using management plans from territorial perspective.

A participative, adaptive, and multidisciplinary management plan requires ecological indicators that permanently monitor administrative measures agreed upon by the direct users of *Lessonia* resource. Considering that these ecological indicators need to be validated [55], they can be selected from administrative measures regarding brown algae such as volume harvested, capture per amount of effort, and minimum legal size of capture [50]. Landing of harvested volume is an easy indicator to track and verify but requires an efficient, participative recording system, in real time, allowing the reaching and use of information at the right moment (**Table 1**). Capture per amount of effort and minimum legal size are indicators that are comparatively more complex to monitor and enforce, because they depend on the participation of scientific observers and also on the interest of fishermen to generate such records (**Table 1**). These indicators are useful tools to assign harvest quotas, establish rotation areas, or to establish extractive or biological bans [49].

TABLE 1. Resource variables proposed to monitor the harvest of Lessonia nigrescens complex.

Resource variable	**Description**	**Time regime**	**Decision policy**	**Verification source**	**Additional requirements**	**Investment items**
Landing (kg)	Fishing / harvesting area	Permanent (daily)	Once fishing quota is reached, stop the harvest	Artisanal fishing	Implementation of an electronic registration system for harvest/ landing	Implementation, maintenance of electronic equipment. Training
Capture per unit effort (CPUE)	Capture per unit effort (kg/h/ fisherman) per bed or area. Fishing gear: "barreta"	Permanent (Monthly)	Once CPUE >150 kg/fisherman/h (Fishing ban, extraction area rotation, change fishing gear)	Scientific survey, Landing register, Fishermen statistics	Implementation of CPUE registration system by area	Monitors for recording landing information. Training
Minimum legal size of capture (MLS)	Morphological variable: holdfast diameter. MLS 20 cm	Permanent (seasonal)	Once MLS of holdfast diameter ≤ 20 cm. (Change harvesting area, fishing ban)	Scientific survey, landing register, fishermen statistics	Implementation of a registration system of MLS by area	Monitors for recording landing information. Training

The effect of harvesting in OAA has been explained by the absence of precautionary management measures in a scenario of high demand for biomass [9, 14, 44, 46]. Thus, management based on the ecosystem approach requires ecological indicators sensitive to harvesting pressure, which allow establishment of decision criteria that are easy to observe, communicate, and measure by both scientific observers and artisanal fishermen [49]. Demographic attributes, such as density of adult plants, biomass per unit of area, recruitment, and size structure all constitute indicators that satisfy these characteristics, are easy to obtain and can be evaluated along spatial and temporal gradients (**Table 2**) [49].

TABLE 2. Demographic variables proposed to monitor the harvest of Lessonia nigrescens complex.

Demographic variable	**Description**	**Time period**	**Decision policy (criteria)**		**Verification source**	**Additional requirements**	**Investment items**
			Harvest	**No-harvest**			
Density of adult plants >20 cm holdfast diameter	Number of plants per m^2	Permanent (seasonal)	Once adult plant density ≥2.0 plants m^2	Once adult plant density ≥1.5 plants m^2	Landing from fishermen, Scientific survey	Implementation of a registration system by harvesting area, zone or Region	Training for registration system. Scientific survey
Biomass	kg/m^2		Once biomass	Once biomass			

Demographic variable	Description	Time period	Decision policy (criteria)		Verification source	Additional requirements	Investment items	
			Harvest	**No-harvest**				
			≥25 kg m_2	<20 kg/m^2				
Recruitment	Number of recruits m^2 ≤1 cm holdfast diameter		Once number of recruits ≤5 plants/m^2	Once number of recruits > 40 plants/m^2				
Size structure of populations in natural beds	Population size structure using holdfast diameter as morphological indicators		Once standing crop ≥30 % standing stock	Once standing crop <20 % standing stock				

Based on demographic indicators, the rule establishes that the harvest in OAA should begin when the abundance and biomass of a population per unit of area is close to biomass or demographic levels detected in an un-intervened population (e.g., MPA): There is a minimal density of recruits, the portion of adult plants must be above 40% of the total population, and the percentage of remaining adult plants in the area should be enough to generate post-harvest recruitment (**Table 2**). Afterward, once the population reaches levels of abundance and biomass per unit of area similar to those found in the population under intense harvesting pressure (e.g., OAA), its sustainability will depend on following elements: (1) stability of recruitment frequency, (2) maintenance of a stock of reproductive individuals, and (3) stability of harvesting frequency. Once these indicators exceed the harvesting period should end and should be followed by a recess period (ban or quotas), until adequate pre-harvest values would be reached (**Table 2**). Thus, the installation of a permanent monitoring program of the populations of *Lessonia nigrescens* complex in OAA and in MAEBR, using demographic indicators, will allow as follows: (a) validation of the application of management plans, (b) detection of the deleterious effects on population dynamics produced by exogenous disturbances in the harvest, (c) respect the necessary period to renew the forest to optimal harvesting levels, and if necessary, (d) determination of extraction quotas by sector, and (e) establishment of extraction bans in a justifiable, participative, and localized way.

6. CONCLUDING REMARKS

The landings of brown seaweeds in Chile [61] reach 390,000 wet tons/year being the world's largest landings from natural populations. This fishery is managed under the concept of "good practices," based on biological and ecological knowledge of the species [6, 12, 14, 23, 25, 32, 48, 57–59]. Most of the brown macroalgae are known as foundational species of marine ecosystems [12]; they constitute the basis of coastal food webs [14, 26, 60, 62], contribute significantly to the total biomass of the ecosystem [23, 32], and are highly connected with all trophic levels [61]; they provide shelter, food, nursery, and breeding areas [23, 32, 6]. Indiscriminate harvest of a foundational species as *L. nigrescens* can generate a significant negative impact on the ecosystem with unknown effects. In this context, ecological indicators proposed [49] are tools for stakeholders and policy makers, enabling greater sustainability of exposed rocky shores in cold temperate seas of the world where these kelp of economic importance are key dominant organisms in cover, biomass, and ecological role.

7. ACKNOWLEDGEMENTS

This synthesis is the result of numerous research projects financed by public and private institutions and with the participation of many co-workers, students, and technicians. I deeply appreciate the support of CONICYT-Chile, FONDECYT, FONDEF-Huam AQ12I0001, and COPRAM-Chile.

REFERENCES

1. Dillehay T (1989) Monte Verde. Late Pleistocene Settlement in Chile. Volume 1. Paleoenvironment and site context. Smithsonian Institution Press, Washington & London.
2. Dillehay T (1997) Monte Verde. Late Pleistocene Settlement in Chile. Volume 2. The archaeological context and interpretation. Smithsonian Institution Press, Washington & London.
3. Vásquez JA & Vásquez LA (2016) Algas Marinas Chilenas: aspectos etnográficos y socioculturales. Anales del VI Congreso Internacional de Etnobotánica, Córdoba (Spain), November 17–21, 2014.
4. Vásquez JA & Fonck E (1994) Algas productoras de ácido algínico en Sudamérica: diagnóstico y proyecciones. In: Documento de Campo N° 13 Situación actual de la industria de macroalgas productoras de ficocoloides en América Latina y el Caribe. FAO-Italia. Programa Cooperativo Gubernamental: 17–26.
5. Vásquez JA, Piaget N & Vega JMA (2012) Chilean *Lessonia nigrescens* fishery in northern Chile: how do you harvest is more important than how much do you harvest. Journal of Applied Phycology 24:417–426.
6. Vásquez JA (2008) Production, use and fate of Chilean brown seaweeds: re-sources for a sustainable fishery. Journal of Applied Phycology 20 (5):457–467.
7. Vásquez JA, Zuñiga S, Tala F, Piaget N, Rodriguez DC & Vega JMA (2014) Economic evaluation of kelp forest in northern Chile: values of good and service of the ecosystem. Journal of Applied Phycology 26:1081–1088.
8. Hoffmann AJ & Santelices B (1997) Flora Marina de Chile Central. Ediciones Universidad Católica de Chile, Santiago. Chile 356 pp.
9. Vásquez JA & Westermeier R (1993) Limiting factors in optimizing seaweed yield in Chile. Hydrobiologia 260/261:313–320.
10. Frangoudes K (2011) Seaweeds Fisheries Management in France, Japan, Chile and Norway. Cahiers Biologie Marine 52:517–525.
11. Vásquez JA (1992) *Lessonia trabeculata*, a subtidal bottom kelp in northern Chile: a case of study for a structural and geographical comparison. In: U. Seeliger (Ed.), Coastal Plant Communities of Latin America. Academic Press Inc., San Diego: 77–89.
12. Graham MH, Vásquez JA & Buschmann AH (2007) Global ecology of the giant kelp Macrocystis: from ecotypes to ecosystems. Oceanography and Marine Biology: An Annual Review 45:39–88.
13. Vásquez JA, Vega JMA & Buschmann AH (2006) Long term variability in the structure of kelp communities in northern Chile and the 1997–98 ENSO. Journal of Applied Phycology 18:505–519.
14. Thiel M, Macaya EC, Acuña E, Arntz WE, Bastias H, Brokordt K, Camus PA, Castilla JC, Castro LR, Cortés M, Dumont CP, Escribano R, Fernandez M, Gajardo JA, Gaymer CF, Gomez I, González AE, González HE, Haye PA, Illanes JE, Iriarte JL, Lancellotti DA, Luna-jorquera G, Luxoro C, Manriquez PA, Marín V, Muñoz P, Navarrete SA, Perez E,

Poulin E, Sellanes J, Sepúlveda HH, Stotz W, Tala F, Thomas A, Vargas CA, Vásquez JA & Vega A (2007) The Humboldt Current System of northern-central Chile: oceanographic processes, ecological interactions and socioeconomic feedback. Oceanography and Marine Biology: An Annual Review 45:195–344.

15. Buschmann AH, Moreno C, Vásquez JA & Hernandez-Carmona MC (2006) Reproduction strategies of *Macrocystis pyrifera* (Phaeophyta) in southern Chile: the importance of population dynamics. Journal of Applied Phycology 18:575–582.
16. Vega JA, Vásquez JA & Buschmann AH (2005) Population biology of the subtidal kelps *Macrocystis integrifolia* and *Lessonia trabeculata*(Laminariales, Phaeophyceae) in an upwelling ecosystem of northern Chile: interannual variability and El Niño 1997–1998. Revista Chile Historia Natural 78:33–50.
17. Muñoz V, Hernández-González MC, Buschmann AH, Graham MH & Vásquez JA (2004) Variability in per capita oogonia and sporophyte production from giant kelp gametophyte (*Macrocystis pyrifera*, Phaeophyceae). Revista Chile Historia Natural 77:639–647.
18. Tellier F, Tapia J, Faugeron S, Destombe C & Valero M (2011) The *Lessonia nigrescens* species complex (Laminariales, Phaeophyceae) Shows strict parapatry and complete reproductive isolation in a secondary contact zone. Journal of Phycology 47:894–903.
19. Tellier F, Vega JMA, Broitman B, Vásquez JA, Valero M & Faugeron S (2011) The importance of having two species instead of one in kelp management: the *Lessonia nigrescens* species complex. Cahiers Biologie Marine 52:455–465.
20. González A, Beltrán J, Hiriart-Bertrand L, Flores V, de Reviers B, Correa JA & Santelices, B (2012) Identification of cryptic species in the *Lessonia nigrescens* complex (Phaeophyceae, Laminariales). Journal of Phycology 48:1153–1165.
21. Alveal K, Romo H & Avila M (1982) Estudios del ciclo de vida de Macrocystis pyrifera de Isla Navarino, Chile. Gayana Botanica 39:1–12.
22. Borazo de Zaixo AL, Piriz ML & Romanello EE (1983) Posibilidades de desarrollo de la industria alguera de la Provincia de Santa Cruz (República Argentina). Informe preparado para el proyecto "Cinco Provincias Argentinas OEA" 187 pp.
23. Santelices B, Castilla JC, Cancino J & Schmiede P (1980) Comparative ecology of *Lessonia nigrescens* and *Durvillaea antartica* (Phaeophyta) in central Chile. Marine Biology 59:119–32.
24. Santelices B & Lopehandía J (1981) Chilean seaweeds resources: a quantitative review of potential and present utilization. In: 10th Proc Int Seaweed Symp: 725–730.
25. Santelices B (1982) Bases Biológicas para el manejo de *Lessonia nigrescens* (Phaeophyta, Laminariales) en Chile Central. Monografías Biológicas 2:135–150.
26. Santelices B (1989) Algas marinas de Chile: Distribución, Ecología, Utilización, Diversidad. Ediciones Universidad Católica de Chile, Santiago de Chile, 399 pp.
27. Werlinger C & Alveal K (1988) Evaluación de algas en ambientes restringidos del Golfo de Arauco (Chile): Punta Fuerte Viejo a Río Tubul. Gayana Botánica 45:461–474.
28. Vásquez JA (1991) Variables morfométricas y relaciones morfológicas de *Lessonia trabeculata* Villouta & Santelices, 1986 en poblaciones submareales del Norte de Chile. Revista Chile Historia Natural 64:271–279.
29. Vásquez JA (1993) Natural mortality of giant kelp *Macrocystis pyrifera* affecting the fauna associated with its holdfasts. Pacific Science 47:180–184.
30. Vásquez JA (2015) Evaluación de poblaciones de *Lessonia nigrescens* ("huiro negro") en cuatro localidades del norte de Chile (II y III Región). Informe de Circulación restringida para Exportaciones M2. 76 pp.
31. Westermeier R, Murúa P, Patiño DJ, Muñoz L, Atero C & Müller DG (2014) Repopulation techniques for *Macrocystis integrifolia* (Phaeophyceae: Laminariales) in Atacama, Chile. Journal of Applied Phycology. 26:511–518.

32. Santelices B & Ojeda FP (1984) Effects of canopy removal on the understory algal community structure of coastal forests of *Macrocystis pyrifera*from southern South America. Marine Ecology Progress Series 14:165–173.
33. Vásquez JA & Buschmann AH (1997) Herbivory-kelp interactions in subtidal Chilean communities: a review. Revista Chile Historia Natural 70:41–52.
34. Macaya EC & Zuccarello GC (2010) Genetic structure of the giant kelp *Macrocystis pyrifera* along the southeastern Pacific. Marine Ecology Progress Series 420:103–102.
35. Vásquez, JA & Tala F (1995) Experimental repopulation of *Lessonia nigrescens* (Phaeophyta, Laminariales) in intertidal areas of northern Chile. Journal of Applied Phycology 7:347–349.
36. Gutiérrez A, Correa T, Muñoz V, Santibáñez A, Marcos R, Cáceres C & Buschmann AH (2006) Farming of the giant kelp *Macrocystis pyrifera* in southern Chile for development of novel food products. Journal of Applied Phycology 18:259–267.
37. Correa JA, Lagos N, Medina M, Castilla JC, Cerda M, Ramírez M, Martínez E, Faugeron S, Andrade S, Pinto R & Contreras L (2006) Experimental transplants of the large kelp *Lessonia nigrescens* (Phaeophyceae) in high-energy wave exposed rocky intertidal habitats of northern Chile: experimental, restoration and management applications. Journal of Experimental Marine Biology and Ecology 335:13–18.
38. Westermeier R, Patiño DJ & Müller DG (2007) Sexual compatibility and hybrid formation between the giant kelp species *Macrocystis pyrifera* and *Macrocystis integrifolia* (Phaeophyceae: Laminariales) in Chile. Journal of Applied Phycology 19:215–221.
39. Wolff R, Zimmermann D, Weber M, Feilen P, Ehrhart F, Salinas-jungjohann RM, Katsen A, Behringer M, Geßner P, Plieb L, Steinbach A, Spitz J, Vásquez JA, Schneider S, Bamberg E, Weber MM, Zimmermann U & Zimmermann H (2005) Real-time 3-D dark-field microscopy for the validation of the cross-linking process of alginate microcapsules. Biomaterials 26:6386–6393.
40. Zimmermann H, Zimmermann D, Reuss R, Feilen PJ, Manz B, Katsen A, Weber M, Ihmig FR, Ehrhart F, Geßner P, BehringerM, SteinbachA, Wegner LH, Sukhorukov VL, Vásquez JA, SchneiderS, Weber MM, Volke F, Wolff R & Zimmermann U (2005) Towards a medically approved technology for alginate-based microcapsules allowing long-term immunoisolated transplantation. Journal of Material Science: Materials in Medicine 16:491–501.
41. Zimmermann H, Wahlisch F, Baier C, Westhoff M, Reuss R, Zimmermann D, Behringer M, Ehrhart F, Katsen-Globa A, Giese C, Marx U, Sukhorukov VL, Vásquez JA, Jakob P, Shirley SG & U Zimmermann (2007) Physical and biological properties of barium cross-linked alginate membranes. Biomaterials 28:1327–1345.
42. Oppliger LV, Correa JA, Faugeron S, Beltrán J, Tellier F, Valero M & Destombe C (2011) Sex ration variation in the *Lessonia nigrescens* complex (Laminariales, Phaeophyceae): effect of latitude, temperature and marginality. Journal of Phycology 47:5–12.
43. Oppliger LV, Correa JA, Engelen AH, Tellier F, Vieira V, Faugeron S & Destombe C (2012) Temperature effects on gametophyte life-history traits and geographic distribution of two cryptic kelp species. Plos one 7 (6):e39289.
44. Thompson SA, Knoll H, Blanchette CA & Nielsen KJ (2010) Population consequences of biomass loss due to commercial collection of the wild seaweed *Postelsia palmaeformis*. Marine Ecology Progress Series 413:17–31.
45. Omoregie E, Tjipute M & Murangi J (2010) Effects of harvesting of the Namibian kelp (*Laminaria pallida*) on the re-growth rate and recruitment. African Journal of Food, Agriculture, Nutrition and Development 10:2542–2555.
46. Ugarte R (2011) An evaluation of the mortality of the brown seaweed *Ascophyllum nodosum* (L.) Le Jol. produced by cutter rake harvests in southern New Brunswick, Canada. Journal of Applied Phycology 23:401–407.

47. Ugarte R & Sharp G (2012) Management and production of the brown algae *Ascophyllum nodosum* in the Canadian maritimes. Journal of Applied Phycology 24:409–416.
48. Martínez EA, Cárdenas L & Pinto R (2003) Recovery and genetic diversity of the intertidal kelp *Lessonia nigrescens* (Phaeophyceae) 20 years after El Niño 1982/83. Journal of Phycology 39:504–508.
49. Vega JMA, Broitman B & Vásquez JA (2014) Monitoring the sustainability of *Lessonia nigrescens* complex (Laminariales, Phaeophyta) in northern Chile under strong harvest pressure. Journal of Applied Phycology 26:791–801.
50. Gelcich S, Edwards-Jones G, Kaiser MJ & Watson E (2005a) Using discourses for policy evaluation: the case of marine common property rights in Chile. Society and Natural Resources 18:377–391.
51. Gelcich S, Edwards-Jones G & Kaiser MJ (2005b) Importance of attitudinal differences among artisanal fishers towards co-management and conservation of marine resources. Conservation Biology 19:865–875.
52. Gelcich S, Godoy N & Castilla JC (2009). Artisanal fishers' perceptions regarding coastal co-management policies in Chile and their potentials to scale-up marine biodiversity conservation. Ocean Coastal Management 52:424–432.
53. Marín A, Gelcich S, Castilla JC & Berkes F (2012) Exploring social capital in Chile's coastal benthic co-management system using a network approach. Ecology and Society 17:13.
54. Garcia SM & Cochrane KL (2005) Ecosystem approach to fisheries: a review of implementation guidelines. ICES Journal of Marine Science 62:311–318.
55. Vásquez JA (2013) Implementación de un sistema piloto de auto-atención electrónica, para la acreditación de origen de recursos pesqueros. Informe Final Proyecto 11BPC 10060. INNOVA-CORFO Chile. 125 pp.
56. Vásquez JA & Santelices B (1990) Ecological effects of harvesting *Lessonia* (Laminarials, Phaeophyta) in central Chile. Hydrobiologia 204/205:41–47.
57. Camus P, Vásquez E, González E & Galaz L (1994) Fenología espacial de la diversidad intermareal en el norte de Chile: patrones comunitarios de variación geográfica e impacto de los procesos de extinción-recolonización post El Niño 82–83. Medio Ambiente 12 (1):57–68.
58. Vásquez JA (1995) Ecological effects of brown seaweed harvesting. Botanica Marina 38:251–257.
59. Seeley RH & Schlesinger WH (2012) Sustainable seaweed cutting? The rockweed (*Ascophyllum nodosum*) industry of Maine and the Martitime Provinces. Annals New York Academy of Science 1249:84–103.
60. Vásquez JA & Vega JMA (2004) El Niño 1997–1998 en el norte de Chile: efectos en la estructura y en la organización de comunidades submareales dominadas por algas pardas. In: Avaria S, Carrasco J, Rutland J and Yañez E (Eds.). El Niño-La Niña 1997–2000 sus efectos en Chile. CONA. Valparaíso. Chile. 119–135.
61. Anuario Estadística de Pesca. Subsecretaría de Pesca y Acuicultura, Ministerio de Economía Fomento y Reconstrucción. Gobierno de Chile. www.sernapesca.cl.
62. Vásquez JA & Santelices B (1984) Comunidades de macroinvertebrados en discos de adhesión de *Lessonia nigrescens* en Chile central. Revista Chile Historia Natural 57:131–154.

CHAPTER 4

Trawl Selectivity in the Barents Sea Demersal Fishery

Eduardo Grimaldo[1*], Manu Sistiaga[1], Bent Herrmann[2] and Roger B. Larsen[3]

[1]SINTEF Fisheries and Aquaculture, Trondheim, Norway

[2]SINTEF Fisheries and Aquaculture, North Sea Science Park, Hirtshals, Denmark

[3]Norwegian College of Fisheries and Aquatic Sciences, University of Tromsø, Tromsø, Norway

ABSTRACT

This chapter provides a general overview of the Barents Sea demersal trawl fishery. First, it reviews historical catch levels and current biomass status of four commercially important demersal species (cod, haddock, Greenland halibut, and redfish) and includes an overview of their management plan that has been carried out by the Joint Norwegian–Russian commission. Then, it presents the evolution of the technical regulations for improving size selectivity in this fishery and describes current challenges in gear selectivity. Later, this chapter describes the concept of size selectivity, introduces the selective parameters that define a selection curve, and progressively introduces different parametric models that describe the selection process. The most common experimental methods and gear used to collect selectivity data are described, and their advantages and disadvantages are discussed. Finally, this chapter describes an alternative, or a complementary method, to the conventional estimation of trawl selectivity—the FISHSELECT method. This method is based on morphology measurements and fish penetration models to estimate the selective properties of different mesh shapes and sizes at different mesh openings, which are later used to provide simulation-based prediction of size selectivity. FISHSELECT has already been applied to four important species of the Barents Sea Demersal Fishery, and the results have in all cases showed to be coherent with the results obtained from sea trial results.

Keywords: selectivity, trawl, bottom trawl, Barents Sea, demersal fishery, FISHSELECT

1. INTRODUCTION

The main target species for the Norwegian bottom trawling fleet north of 62° N are the Northeast Arctic cod (*Gadus morhua*), haddock (*Melanugrammus aeglefinus*), and saithe (*Pollachius virens*); all of these species belong to the gadoid family. During the past 20 years, the catches of cod have averaged 625 649 tonnes per year, with the largest landing being in 2014 (986 449 tonnes) [1]. During the same period, the catches of haddock and saithe averaged, respectively, 165 222 and 162 578 tonnes per year, with the largest landing for haddock registered in 2012 (315 627 tonnes) and for saithe in 2006 (212 822 tonnes) [2, 3]. Of these catches, approximately 83% of cod, 90% of haddock, and 97% of saithe were harvested by Norwegian and Russian vessels; the rest was caught by vessels from Iceland, Greenland, and the EU [1–3]. In 2014, nearly 70% of the total catch of cod, 76% of haddock, and 49% of saithe were fished by bottom trawlers. The rest of the quota for these species was fished by gillnetters, longliners, jiggers, and demersal seiners [1–3]. Other commercially important species, mostly caught as bycatch in this fishery, are redfish (*Sebastes marinus*) and Greenland halibut (*Reinhardtius hippoglossoides*).

A key strategy of the Norwegian–Russian fishing legislation is to minimize the capture of undersized fish. When more than 15% of the catch (by numbers) is undersized, the fishery is closed. Moreover, the discard of dead or dying fish from regulated species is forbidden in this (Norway–Russian) fishery. The aim with size selection is to reduce the capture of undersized fish while also reducing the loss of fish above the minimum legal size (MLS). To achieve this goal, the use of sorting grids with a minimum bar spacing of 55 mm is currently mandatory in all trawls fishing for gadoids in this fishery.

Assessments of the Joint Norwegian–Russian Fisheries Commission show that the Northeast Arctic cod stock is in a fairly good state. The spawning biomass in 2013 was estimated at 2 million tonnes, while the total stock biomass was estimated to be around 3.5 million tonnes [1]. The International Council for the Exploration of the Sea (ICES), which is a global organization that develops science and advice to support the sustainable use of the oceans, advises the Joint Russian–Norwegian Fisheries

Commission to set a total allowable catch (TAC) of 805 000 tonnes in 2016 [1].

The Northeast Arctic haddock and saithe stocks are considered to be in good condition. ICES advises that when the Joint Russian–Norwegian Fisheries Commission management plan is applied, landings in 2016 should not exceed 223 000 tonnes for haddock and 140 000 tonnes for saithe [2, 3].

2. TECHNICAL REGULATIONS FOR IMPROVING SIZE SELECTIVITY

In 1979, the Northeast Atlantic Fisheries Commission (NEAFC) agreed to increase the mesh size of codends made of cotton, polyamide, and polyester from 120 to 125 mm. As of 1 January 1980, the minimum legal size of cod and haddock that could be kept was set at 39 and 35 cm, respectively; in addition, the catches of undersized fish could not exceed 15% (by weight) of the catch [4]. In 1982, the NEAFC increased the commercial sizes of cod and haddock to 42 and 39 cm, respectively, and established that the bycatch of undersized fish should not exceed 15% (by numbers) of the catch. Norway also unilaterally increased the minimum mesh size of trawl codends to 135 mm regardless of the trawl material [4]. In 1983, a discard ban was introduced into the fishery [5]; it stated that vessels were obliged to land also the catches of all undersized fish. A surveillance program established during 1983–1984 showed that substantial areas need to be closed to the commercial fishery due to the high proportions of undersized cod and haddock in the catch. Typical catches in these areas could contain 30–50% of juvenile fish (by numbers). Therefore, the temporary closure of certain fishing areas led to a less efficient fishery because the fleet had to search for other areas to fish on [6].

Between 1985 and 1991, researchers in Norway conducted selectivity experiments that aimed at improving size selectivity of bottom trawl codends. These experiments considered different mesh configurations, round straps, different top and bottom panels, codend twine materials, and sorting grid systems [7, 8]. The methods found to be most practical were the shortened lastridge ropes (12–15% reduction) and the escape grids (Sort-X system). The shortened lastridge ropes gave similar size selection estimates as square mesh codends. However, square mesh codends were associated with meshing of fish, and especially redfish, and maneuverability challenges on deck [9].

Even though the Sort-X sorting grid represented a revolutionary method for improving selectivity in trawl codends, in 1990 the Norwegian Directorate of Fisheries, in 1990, increased the minimum size of cod to 47 cm and of haddock to 44 cm. According to this institution, an alternative mean of resource management had to be enforced in the fishery because the development of technical codend modifications had still not given the expected results [10].

Between 1990 and 1996, experiments with sorting grids that aimed at reducing the catch of young fish were conducted in Norway and Russia [11, 12]. In 1995, the Joint Russian–Norwegian selectivity experiments on cod in the Barents Sea proved that the selectivity of the Russian single-steel grid system "Sort-V" and that of the double steel grid system "Sort-X" were similar [12]. Based on the results of these experiments, as of 1 January 1997 both types of grids with 55-mm interbar space began to be used in cod and haddock fisheries in some limited areas of the Barents and Norwegian Seas (**Figure 1**). Later, the Institute of Marine Research of Norway satisfactorily adapted the Sort-V system to Norwegian trawlers [13, 14]. The new system, which is known as the Single grid and only includes some slight changes from the Sort-V system, was legalized by the Norwegian Directorate of Fisheries in 2000.

Since their introduction, the Sort-X, Sort-V, and Single grid have been associated with crew safety problems, especially when handling the grid sections in bad weather. During the late 1990s, experiments with grids made of other materials, such as plastics, fibreglass, nylon, and rubber, were promoted. The intention behind these experiments was to offer the industry grids that could be more easily handled, were more user-friendly, and were cheaper to purchase [15]. The systems studied were the plastic Sort-X, the Eurogrid, and the Flexigrid. The plastic Sort-X was made of high-density polyethylene (HDPE) materials and weighed considerably less than the steel version. The Eurogrid was made of massive nylon and was developed primarily for the trawl fishery in the North Sea [16]. Although these grids were designed to be more user-friendly alternatives to other grids on the market because of their low weight, flexibility, and the possibility of storing them in net drums, none of them was successfully introduced into the fishery. During three years of experimental work, the Flexigrid (made of Polyamide (PA) bars and rubber frames) (**Figure 1**) proved to have a selection capacity similar to that of the Sort-X [17–20]. This new grid system was lighter, smaller, more flexible, and therefore easier to handle on deck. Thus, the Flexigrid system was legalized in the beginning of 2002.

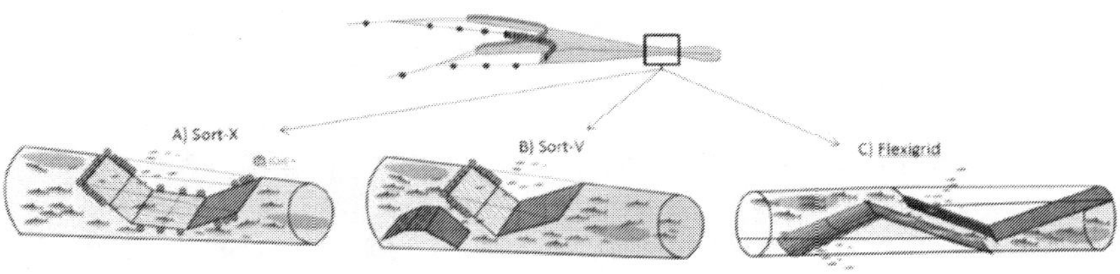

FIGURE 1. Mandatory sorting grids in the Barents Sea Demersal Fishery: (A) Sort-X, (B) Sort-V, and (C) Flexigrid.

Between 2004 and 2008, further experiments with escape panels were performed in the Barents Sea [21]. Despite the results showing similar selective performance to mandatory sorting grids [22, 23], escape panels were not considered as an alternative selection device for the Barents Sea Demersal Fishery.

Today, all four sorting grids described above are legal with a minimum grid bar spacing of 55 mm: the Sort-X, Sort-V, Single grid, and Flexigrid. The Sort-V and the Single grid system are still used by some vessels, but the system employed by the majority of vessels today is the Flexigrid. The two main reasons for fishermen to prefer the flexigrid system are that (a) it is a safer grid to use on deck, especially in bad weather, and (b) the retention of fish above the minimum size for cod and haddock is higher for the flexigrid than for the Sort-V/Single grid [24]. This second argument became especially important when in 2011 the minimum legal size was reduced from 47 to 44 cm for cod and from 44 to 40 cm for haddock. This reduction in minimum size resulted from the Russian–Norwegian joint meetings and the aim of standardizing the regulations for the Barents Sea. This reduction in minimum sizes was also accompanied by a reduction in the minimum codend mesh size. Thus, in addition to the grid, trawlers can today use codends with a minimum diamond mesh size of 130 mm.

The rapid increase in the biomass of Atlantic cod in the period 2008–2014 has brought some unusual challenges for the Barents Sea trawlers, which have often met really high concentrations of cod in the fishing grounds. From the beginning, these high concentrations at sea were quickly reflected in reduced sorting capacity of sorting grids. The high catch rates encountered by the grid sections have led not only to reduced selection performance of the grids but also to more serious problems like breakage of grids and grid sections. Underwater recordings have shown that fish accumulates both in front and behind the grid [25] and does not fall back in the codend. Because the catch sensors are placed in the codend, the skipper does not receive any information on the ongoing catch accumulation in the gear. The consequence of this process is unwanted big catches that exceed the vessel's processing capacity and lead to reduced quality of the catch [25].

The main reason for the accumulation of fish in front and behind the grid is related to the reduction in water flow created by the presence of the grid section and especially the sorting grid(s) [26]. In the past years, there have been efforts to increase the water flow in the grid sections to try to mitigate the fish accumulation problem. If the water flow is too high, grids lose their selective properties and fish can flow through the section without having a chance to contact the grid. Sorting grid sections have traditionally been constructed as two-panel constructions, but in an attempt to improve the flow conditions in the section new four-panel constructions have been tested in the past years. Gjøsund et al. [26] showed that a four-panel Single grid section allows higher water flow through the section than an equivalent two-panel grid section. Because the tests carried out at sea also indicate that at high catch rates four-panel sorting grid sections with larger cross-section areas perform better than two-panel grid sections (Sort-X, Sort-V, and Flexigrid) (own data, unpublished), the use of these types of sections was legalized in 2014 and 2015.

3. DESCRIPTION OF SIZE SELECTIVITY

Selective fishing refers to the ability of a fishing method to target and capture fish by size and/or species. Size selective fishing in the Barents Sea demersal trawl fishery aims at avoiding catching fish below the minimum landing sizes that enters the fishing gear while simultaneously having high probability of retaining fish above the minimum size. In the Barents Sea demersal trawl fishery, the selective devices in the fishing gear consist of a sorting grid followed by a size selective codend. The grid enables releasing fish up to a certain size through the grid, provided that they are able to find their way out between the bars. Fish that do not manage to escape through the grid fall back into the codend where they have an additional chance to escape through the meshes, provided that they are not too big to be able to pass through them. It is obvious that based on this system only fish below a certain size will have a chance to escape through the grid or codend meshes. The smaller the fish the easier they will fit through between the grid bars and through the codend meshes. Therefore, it is to be expected that the retention probability for a fish entering the gear will increase with the size of the fish. For the description of the size selection in a fishing gear, it is convenient to use a parametric mathematical model that is able to describe the retention probability r of the fish as a function of its length l. Traditionally, as for many other fisheries and trawl constructions the combined size selection of the grid and codend for the species in the

Barents Sea demersal trawl fishery has been described by the *logit* size selection model:

$$r(l, L50, SR) = \frac{\exp\left(\frac{\ln(9)}{SR} \times (l - L50)\right)}{1.0 + \exp\left(\frac{\ln(9)}{SR} \times (l - L50)\right)} \quad (1)$$

In a *logit* size selection model, the retention probability will increase monotonously with the length of the fish. The size selection for this model is fully defined based on the value of the two parameters *L50*and *SR*. *L50* can be interpreted as the length at which a fish will have 50% probability of being retained by the fishing gear conditioned that it enters the gear. *SR* can be interpreted as the difference length between fish with, respectively, 75% and 25% of being retained in the fishing gear conditioned they enter it. **Figure 2** illustrates the *logit* size selection model and the meaning of the selection parameters L50 and SR.

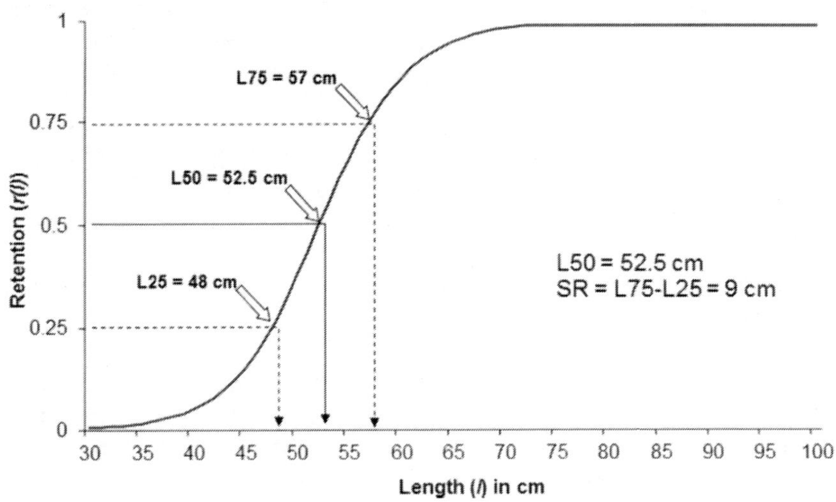

FIGURE 2. A selection curve that describes the *logit* size selection model.

Based on the values of *L50* and *SR* for a *logit* size selection model, the length *Li* of a fish with *i* % of being retained can be calculated by:

$$Li = L50 + \frac{SR}{\ln(9)} \times \ln\left(\frac{0.01 \times i}{1.0 - 0.01 \times i}\right) \quad (2)$$

Several studies have aimed at quantifying size selection for cod and haddock in the selective systems deployed in the Barents Sea demersal trawl fishery in terms of the values for L50 and SR using the *logit*size selection model [11, 13, 14, 16, 21, 22].

However, the *logit* size selection model presents some limitations which are important when considering the size selection systems used in the Barents Sea demersal trawl fishery. First, it does not explicitly reflect the dual nature of the size selection of a system consisting of a size selective grid preceded by a size selective codend. This can lead to a poor description of the size-dependent retention probability in the fishing gear in cases where the size selective properties of the grid and the codend differ considerably. Second, the *logit* model does not enable quantifying the individual contributions of the grid and the codend to the combined size selection in the gear. This is an important limitation when trying to improve size selection based on implementing gear modifications. Third, the *logit* model is not able to account for that not necessary all fish are able to make contact with the grid during their drift towards the codend. Besides potentially leading to a poor overall description of the size selection in the gear, this limitation also disables the quantification effect on the sorting efficiency of the grid by, for example, changing grid area size or making modifications to the lifting panel ahead of a Sort-V/Single grid section.

Based on the above-mentioned limitations, recent studies on the size selection in the Barents Sea demersal trawl fishery [30] have applied a more complex dual size selection model r_{dual} that accounts for all of these limitations:

$$r_{dual}\left(l, C_{grid}, L50_{grid}, SR_{grid}, L50_{codend}, SR_{codend}\right) = \left(1.0 - C_{grid} + C_{grid} \times logit\left(l, L50_{grid}, SR_{grid}\right)\right) \times logit\left(l, L50_{codend}, SR_{codend}\right) \quad (3)$$

The dual size selection model contains five parameters, C_{grid}, $L50_{grid}$, SR_{grid}, $L50_{codend}$, and SR_{codend}. The first three describe the size selection in the grid while the last two describe the size selection in the codend. The parameter C_{grid} quantifies the proportion of fish entering the grid section that makes contact with the grid and is size selected by it. For the fish that makes contact with the grid, their size selection is described by a *logit* model with the parameters $L50_{grid}$ and SR_{grid}. For the codend, the size selection is described by an additional *logit* model with the parameters $L50_{codend}$ and SR_{codend}.

The grid contribution to the combined size selection is described by Eq. (3) (the term in the first brackets) and is named as the *clogit* size selection model [31]:

$$r_{clogit}\left(l, C_{grid}, L50_{grid}, SR_{grid}\right) \equiv 1.0 - C_{grid} \times \left(1.0 - r_{logit}\left(l, L50_{grid}, SR_{grid}\right)\right) \quad (4)$$

Model (4) quantifies the length-dependent retention probability for that (conditioned it entered that part of the trawl) a fish is still inside the trawl after it has passed through the section of the trawl where the grid is installed. Based on Eq. (4), Eq. (3) can be simplified to:

$$r_{dual}\left(l, C_{grid}, L50_{grid}, SR_{grid}, L50_{codend}, SR_{codend}\right) =$$
$$r_{clogit}\left(l, C_{grid}, L50_{grid}, SR_{grid}\right) \times r_{logit}\left(l, L50_{codend}, SR_{codend}\right) \quad (5)$$

4. EXPERIMENTAL METHODS FOR COLLECTING SELECTIVITY DATA

The selectivity of fishing gear should be measured under conditions that mimic commercial fishing. Ideally, the sea trials should be undertaken on commercial fishing grounds, at commercial fishing depths, at commercial catch rates and sizes, and during the commercial fishing season. It is well known that variations in the environmental conditions or fish composition in the fishing grounds can influence the selective performance of fishing gear. This is not only because environmental factors affect fish behaviour towards fishing gears but also because selective fishing gears can have limitations with regard to fish densities entering it for example.

Success in selectivity trials at sea rely first on a good experimental design. It is important to determine when and where the experiments will be carried out so that the trials represent ordinary fishing conditions. Also, the choice of a sampling method to be used is crucial because it will determine what kind of additional gear one will have to use during the trials. Finally, the choice of a sampling method should be done considering the analytical and practical advantages/disadvantages of each of the available methods.

Different methods for measuring selectivity in towed fishing gears are described in reference [27]. These authors divided the methods into two categories: the Paired-gear method and Covered-gear method. In the Paired-gear method, two gears of equal overall dimensions are towed alternatively or alongside each other. In one of the gears (the test gear), the selectivity device to be tested is installed, whereas the other gear (the control gear) is built in small unselective meshes (**Figure 3**). Thus, the selective properties of the tested device are calculated, assuming that the small mesh size gear captures a size and species composition that is equal to the one that has entered the test gear.

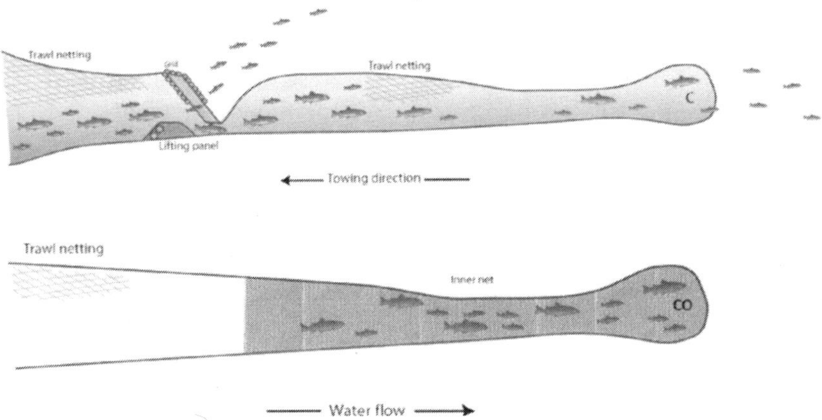

FIGURE 3. Illustration of the Paired-gear method showing the test (above) and the control (below) gear.

The trouser trawl, twin trawl, parallel haul, and alternate haul methods are examples of the Paired-gear method [27]:

- Trouser trawl method: In a single trawl, the belly is divided into two extension pieces and codends in a way that one of them acts as the control gear and the other one acts as the test gear.
- Twin trawl method: In this case, the gear is composed of two trawls. In one of the trawls, the gear to be tested is installed while the other one is used as control.
- Parallel haul method: In this case, two vessels tow one single trawl each alongside each other. In one of the trawls, the gear to be tested is installed while the other one is used as control.
- Alternate haul method: Here one vessel alternates the use of the test and the control gear.

The covers used in the Covered-gear method can vary a lot in size and shape, depending on the gear they need to cover [24]. The most important properties to look for when choosing a cover are that it needs to cover the selective device completely, it should keep the desired geometry and not be an obstacle for fish escaping through the tested device (avoid masking effect), and it should not (or minimally) reduce water flow. Square mesh covers are therefore most often used to avoid geometry problems and masking of the selective device. Still, measurements of the water flow show that the flow inside codends can be substantially reduced even when using these types of square mesh covers [28].

In the Barents Sea, the compulsory selectivity devices fishermen are allowed to use are composed of a grid section followed by a 130-mm codend. Thus, the selectivity process in the gear is a dual selection process [30] where a selection process in the grid section is followed by a selection process in the codend. To measure the selectivity of such dual selection devices, one can either install a single cover over both devices (**Figure 4**) [11] or install an independent cover over each of the selective devices (**Figure 5**) [29]. The challenge of using independent covers with respect to using a single cover is that there is an additional compartment to be considered and that the practical operations on board with multiple covers can be more challenging.

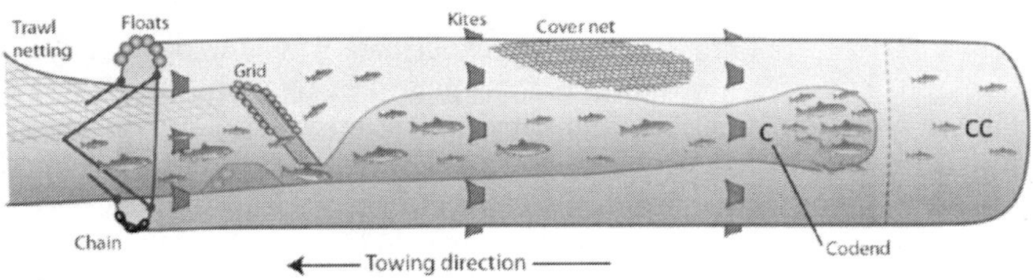

FIGURE 4. Single covered method that covers the grid section and codend.

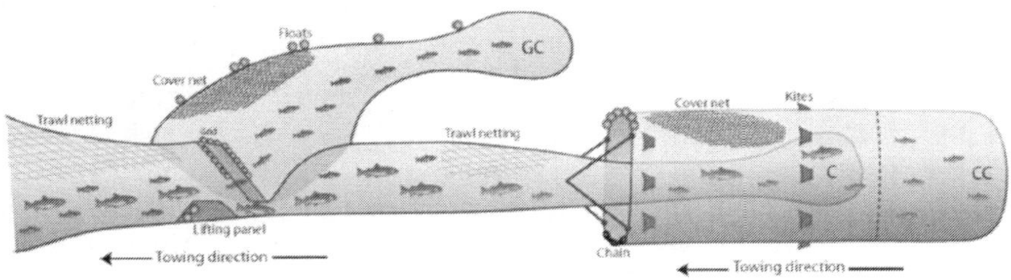

FIGURE 5. Dual cover method.

If the objective of the experiment is to collect selectivity data in order to assess only the size selection in the grid section, then the covered-grid method can be applied (**Figure 6**) [25].

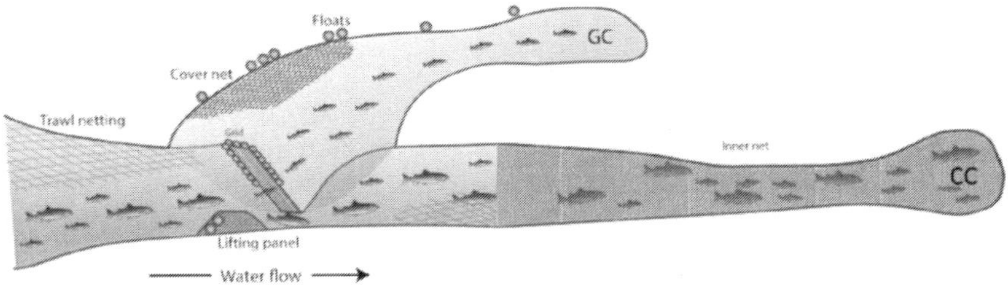

FIGURE 6. The Covered-grid method.

In some experiments, the use of covers can be challenging, for example, they cannot be installed in a way that can guarantee that all escapees are collected, there is limited space on the vessel, etc. In such circumstances, the Paired-gear method is applied.

When carrying out trawl selectivity experiments, it is important that the trials are carried out within a limited period of time and in an area that holds a fairly similar species and size distribution of fish. Large variations in the availability of fish can create loss of precision in the results, whereas large variations in the size distribution of fish can lead to large between-haul variation in the selectivity parameters L50 and SR. Methods like the Covered-gear method, which provides direct information about the fish escaping from the selective device being tested, are more robust regarding variability in both abundance and size distribution of fish. Other methods, like, for example, the alternate haul method, are more sensible to changes in fishing area as one needs to assume that the size and species distribution entering the gear is the same for the test and control hauls carried out.

Measuring the size selective properties of towed fishing gears requires length measuring the fish collected in the different compartments of the gear (test codend and cover(s) or control codend). The number of fish that needs to be caught and length measured to achieve a certain precision in the selectivity results have for years been an issue among fisheries scientists. Especially in the cases where the catches exceed the number of fish that can be measured, it is important to know how many fish one should measure and how an eventual subsample should be taken. An ideal subsample is a random sample that represents well the size distribution of the fish in the catch. Normally this subsample is taken from the catch in random batches to avoid potential accumulations of specific sizes of fish that can occur, for example, in the collection bins of the vessels. For each species, the fish in the fraction of the catch that is not length measured needs to be counted to calculate the sampling ratio in the catch.

Diverse sampling strategies in experimental trawl selectivity gear were studied by Millar [34], who concluded that sampling the same number of fish from the different compartments was the most efficient sampling method. Also, how the precision in selectivity results varied depending on the amount of fish caught and length measured with both the Paired-gear and the Covered-gear methods was investigated in reference [35]. The results of the investigation showed that the uncertainty in the selection parameters L50 and SR decreased with increasing number of fish measured, and that this relationship could be described by a power model. The results also demonstrated that the sampling effort needed to achieve a specific uncertainty level for the selection parameters was always lower for the Covered-gear method compared to the Paired-gear method (in many cases the number of fish that would need to be measured to maintain a specific uncertainty level was around 10 times higher for the Paired-gear method than for the Covered-gear method). The results of these studies illustrate again the importance of carrying out proper experimental design before starting the sea trials. Both the potential limitations of the operations that need to be carried out on board and the advantages and disadvantages of the different sampling methods need to be always considered.

For each haul conducted with one of the experimental collection methods described above (**Figures 3–6**), the catch in each compartment (codend, cover(s), control) is length measured species by species. These length measurements are sorted into so-called length classes that in the Barents Sea are typically 1 cm wide. In a dataset, each haul consists of count data that show how many fish of those collected in each of the compartments belong to the same length class. If subsampling is applied only to a proportion of the fish in each compartment, the sampling factor for each compartment is also provided together with the count data. **Table 1** shows an example of a data file resulting from one haul. In this case, the data was collected using the single covered method (**Figure 4**), and 54% of the fish in codend and 62% of the fish in the cover were length measured. The rest of the fish in each compartment was counted. Analysis of such data (**Table 1**) forms the basis for estimating the selectivity of trawl gear by the methods described in the next section.

TABLE 1. Illustration of a dataset (one single haul) collected using the covered codend method.

Length class (cm)	Number of fish in the codend	Number of fish in the cover
25.5	0	0
26.5	0	0
27.5	0	0
28.5	0	0
29.5	0	0
30.5	2	43
31.5	2	23
32.5	3	64
33.5	1	118
34.5	7	116
35.5	5	279
36.5	7	137
37.5	5	179
38.5	10	180
39.5	13	106
40.5	12	266
41.5	2	61
42.5	7	50
43.5	9	52
44.5	12	26
45.5	16	21
46.5	12	61
47.5	14	9
48.5	21	18
49.5	16	8
50.5	26	3

51.5	13	11
52.5	21	0
53.5	10	5
54.5	22	0
55.5	16	1
56.5	11	2
57.5	10	0
58.5	20	1
59.5	7	0
60.5	9	0
61.5	8	0
62.5	7	0
63.5	5	0
64.5	0	0
65.5	2	0
SAMPLING	0.54	0.62

5. METHODS TO ESTIMATE SIZE SELECTION PARAMETERS

The accepted scientific method for estimation of trawl size selectivity is based on a Maximum Likelihood Estimation (MLE) [27]. The maximization problem this represents is by convenience converted into a minimization problem simply by adding a minus sign in front of the function that else would have been a maximization. Further, to simplify the formulation, the natural logarithm (*ln*) is taken to the function prior to minimization as this leads to a simpler formulation that is easier to minimize. This step is in valid since the *ln* of a function, and the function itself, has the same minimum for the same model parameter values. The function to minimize depends both on the selection system for which the data has been collected and the level of detail intended, and as described in the previous section on which of the experimental data collection has been applied. Conditioned that the size selection model applied in the estimation is able to describe the size selection processes occurring sufficiently well, the MLE estimation provides the model parameters (see Section 3) that make the collected experimental size selection data most likely. If the intention is to assess the combined size selectivity in a single haul j based on the covered gear method (**Figure 4**), based on the *logit* model Eq. (1), the following equation is minimized with respect to $L50_j$ and SR_j:

$$-\sum_l \left\{ \begin{array}{l} nc_{jl} \times \ln\left(\dfrac{qc_j \times r_{logit}(l, L50_j, SR_j)}{qc_j \times r_{logit}(l, L50_j, SR_j) + qcc_j \times (1.0 - r_{logit}(l, L50_j, SR_j))} \right) + ncc_{jl} \\ \times \ln\left(\dfrac{qcc_j \times (1.0 - r_{logit}(l, L50_j, SR_j))}{qc_j \times r_{logit}(l, L50_j, SR_j) + qcc_j \times (1.0 - r_{logit}(l, L50_j, SR_j))} \right) \end{array} \right\} \quad (6)$$

Where nc_{jl} is the number of fish in length class l that is length measured in the haul j in the codend; and ncc_{jl} is the number of fish in length class l that is length measured in the haul j in the cover. qc_j and qcc_j are the corresponding sampling rates. The summation in Eq. (6) is over length classes l.

In case the Paired-gear method (**Figure 3**) is applied for the data collection, the function to minimize would be:

$$-\sum_l \left\{ \begin{array}{l} nc_{jl} \times \ln\left(\dfrac{qc_j \times SP_j \times r_{logit}(l, L50_j, SR_j)}{qc_j \times SP_j \times r_{logit}(l, L50_j, SR_j) + qco_j \times (1.0 - SP_j)} \right) + nco_{jl} \times \\ \ln\left(\dfrac{qco_j \times (1.0 - SP_j)}{qc_j \times SP_j \times r_{logit}(l, L50_j, SR_j) + qco_j \times (1.0 - SP_j)} \right) \end{array} \right\} \quad (7)$$

Compared to Eq. (6), Eq. (7) includes an additional parameter SP_j that needs to be estimated together with $L50_j$ and SR_j. SP_j is the so-called split parameter that quantifies the proportion of fish entering the test side of the gear. nco_{jl} is the amount of fish in length class l that is length measured in the control codend and qco_j is the corresponding sampling rate.

Several of the size selection studies carried out in the Barents Sea demersal trawl fishery have applied Eq. (6) or (7) to estimate the combined size selection for individual hauls. In most cases, a mean size selection for the group hauls is subsequently estimated in a second estimation step using the size selection results obtained in the individual hauls and following an estimation procedure described in references [24, 30, 32]. This procedure accounts for both the uncertainty in the estimated size selectivity in the individual hauls (often named within-haul variation) and the between-haul variation in the size selection.

However, in most of the recent studies the dual nature of the size selection process in the gears has been explicitly accounted for by basing the analysis on the dual selection model described by Eq. (5). Further, in most of these studies the final aim has been to estimate the size selection averaged over a group of hauls. Therefore, the process involves summing data over hauls in the estimation process.

If the single covered data collection method (**Figure 4**) is applied, this would lead to minimizing the following equation:

$$-\sum_{j=1}^{m}\sum_{l}\left\{\begin{array}{l}\frac{nc_{jl}}{qc_j}\times\ln\left(r_{dual}(l,C_{grid},L50_{grid},SR_{grid},L50_{codend},SR_{codend})\right)+\frac{ncc_{jl}}{qcc_j}\times\\ \ln\left(1-r_{dual}(l,C_{grid},L50_{grid},SR_{grid},L50_{codend},SR_{codend})\right)\end{array}\right\} \quad (8)$$

In the case of Eq. (8), the minimization is carried out in five dimensions to estimate C_{grid}, $L50_{grid}$, SR_{grid}, $L50_{codend}$, and SR_{codend}. The outer summation is over the *m* hauls conducted.

If the Paired-gear data collection method (**Figure 3**) is applied, then the estimation is conducted in six dimensions because the average split *SP* also needs to be estimated. In this case, the function to minimize becomes:

$$-\sum_{j=1}^{m}\sum_{l}\left\{\begin{array}{l}\frac{nc_{jl}}{qc_j}\times\ln\left(\frac{SP\times r_{dual}(l,C_{grid},L50_{grid},SR_{grid},L50_{codend},SR_{codend})}{SP\times r_{dual}\left(l,C_{grid},L50_{grid},SR_{grid},L50_{codend},SR_{codend}\right)+1-SP}\right)+\frac{nco_{jl}}{qco_j}\\ \times\ln\left(\frac{1-SP}{SP\times r_{dual}\left(l,C_{grid},L50_{grid},SR_{grid},L50_{codend},SR_{codend}\right)+1-SP}\right)\end{array}\right\} \quad (9)$$

If the dual covered data collection method (**Figure 5**) is applied, the precision in estimating the size selection of the gear can be improved. In this case, the equation to be minimized becomes:

$$\sum_{j=1}^{m}\sum_{l}\left\{\begin{array}{l}\frac{ng_{jl}}{qg_j}\times\ln\left(1-r_{clogit}\left(l,C_{grid},L50_{grid},SR_{grid}\right)\right)+\frac{ncc_{jl}}{qcc_j}\times\\ \ln\left(r_{clogit}\left(l,C_{grid},L50_{grid},SR_{grid}\right)\times\left(1-r_{logit}\left(l,L50_{codend},SR_{codend}\right)\right)\right)+\frac{nc_{jl}}{qc_j}\times\\ \ln\left(r_{clogit}\left(l,C_{grid},L50_{grid},SR_{grid}\right)\times r_{logit}\left(l,L50_{codend},SR_{codend}\right)\right)\end{array}\right\} \quad (10)$$

If the only objective is to assess the average size selection in the grid section, then the Covered-grid data collection method (**Figure 6**) can be applied. In this case, the function to minimize will be:

$$\sum_{j=1}^{m}\sum_{l}\left\{\begin{array}{l}\frac{ng_{jl}}{qg_j}\times\ln\left(1-r_{clogit}\left(l,C_{grid},L50_{grid},SR_{grid}\right)\right)+\frac{ncc_{jl}}{qcc_j}\times \\ \ln\left(r_{clogit}\left(l,C_{grid},L50_{grid},SR_{grid}\right)\right)\end{array}\right\} \quad (11)$$

When estimating the average size selectivity, based on minimizing Eq. (8)–(10) or (11), the uncertainty on the parameters and size selection curve is often estimated using a double bootstrap method as described in references [30, 31, 33].

The ability of a size selection model to describe the experimental size selection data is frequently evaluated based on calculating a p-value for model deviance versus the degrees of freedom [31]. This p-value quantifies the probability to by coincidence obtain at least as big discrepancy between model and data as observed. Thus, this p-value should not be below 0.05 for the model to be able to describe the experimental size selection data well. However, a low p-value can also be a result of over-dispersion in the data, which in case of having a low p-value should be inspected before rejecting the size selection model.

6. THE FISHSELECT METHODOLOGY

FISHSELECT is a framework of methods, tools, and software developed to determine whether or not a fish is able to penetrate a certain mesh [36]. Through computer simulation, FISHSELECT enables the estimation of the selectivity parameters L50 and SR for a certain species and selection device by comparing the morphological characteristics of the former and the shape and size of the latter. This methodology has been successfully used to estimate mesh selectivity of the most relevant demersal fish species in the Barents Sea: cod (*G morhua*), haddock (*M. aeglefinus*), Greenland halibut (*R. hippoglossoides*), and redfish (*Sebastes ssp.*) [31, 37–39]. The results obtained with the application of FISHSELECT have proved the reliability of the method as they are coherent with the results registered from earlier sea trials [37–39]. Thus, if we consider the flexibility the method offers compared to the traditional selectivity measuring methods, the value of the method becomes obvious.

By means of the FISHSELECT tools and software, one can predict the selectivity of a specific fishing gear and species. This can also be achieved with the more traditional selectivity study methods; however, FISHSELECT is unique that once the morphology analyses are carried out for a certain species, one can predict the selectivity of this species for endless mesh shapes (including grids), sizes, and opening angles (OAs). Some of the features or advantages of FISHSELECT with respect to the traditional sea trial selectivity studies are that:

I. New estimations do not need additional fieldwork.

II. FISHSELECT provides fast answer on the selectivity parameters that can be expected from an eventual change in the gear. In a similar way, if one wishes to change the selection properties of a gear to achieve certain selectivity for a species or multiple species, FISHSELECT can fast predict the changes necessary in the gear to achieve those selectivity objectives.

III. The method gives an overall estimation of as many species as one wishes. It is interesting to note that most often the gear will be designed as a compromise on the selective properties of several species [37] that different cleaner-fish species with various body sizes and shapes are harvested at the same time.

IV. The results obtained from FISHSELECT can also be used to aid in the interpretation of sea trial selectivity results.

Because FISHSELECT is based on the relationship between fish length and weight, and the shape of the fish's cross-section at different points of its body, factors like the condition of the fish in the different seasons need to be taken into consideration. When collecting the morphological data for a certain species, it is important to cover the whole length spectra for the fish as through its growth not all the body parts grow proportionally. Also, if the fish included in the measurements is captured through different seasons, it would help covering a wider spectra of the different shapes the species can acquire. The wider the spectra of fish covered and the higher the amount of fish included in the measured fish pool, the lower the uncertainty in the predictions.

The application of FISHSELECT is divided into four steps:

Step 1: morphological data

In the first step, the fish length (in mm), weight (in g), and its morphology at different cross sections (CSs) need to be measured. The number and position of the CSs taken are different for each fish species, and are decided based on earlier experiences considering the positions likely to determine if a fish will be able to escape through meshes or grids of different sizes and shapes. For fish, in general, three cross sections have been measured per fish [38]. For cod and haddock, for example, these three cross sections were located at the end of the opercula (maximum girth of the head), at the foremost point of the first dorsal fin, and maximum girth of the fish [37]. The cross sections are measured by means of a mechanical sensing tool named morphometer. The morphometer consists of an aluminum frame and measuring aluminum sticks (2.5 mm wide) that can be shifted horizontally and fixed at a desirable position (**Figure 7**). The shape formed in the morphometer is later converted into a digital image using a flatbed scanner. The image resulting from the scanner is finally digitized using the image analysis tools implemented in the FISHSELECT software tool [36].

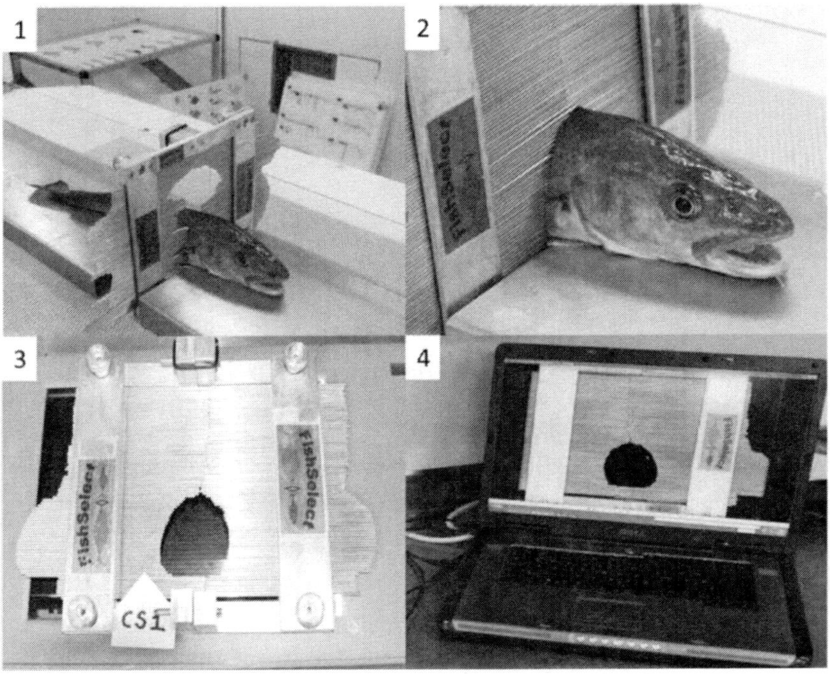

FIGURE 7. Illustration of the use of a morphometer on cod (*Gadus morhua*).

Step 2: Fall-through experiments

Fall-through experiments are carried out to decide if a fish can in principle physically pass through a certain rigid shape (subjected to the force of gravity only) (**Figure 8**). These shapes are perforated in 5-mm-thick solid nylon plates [39]. The shapes tested normally include diamonds, hexagons, and rectangles; however, there are no restrictions to the shapes one can test and use in FISHSELECT.

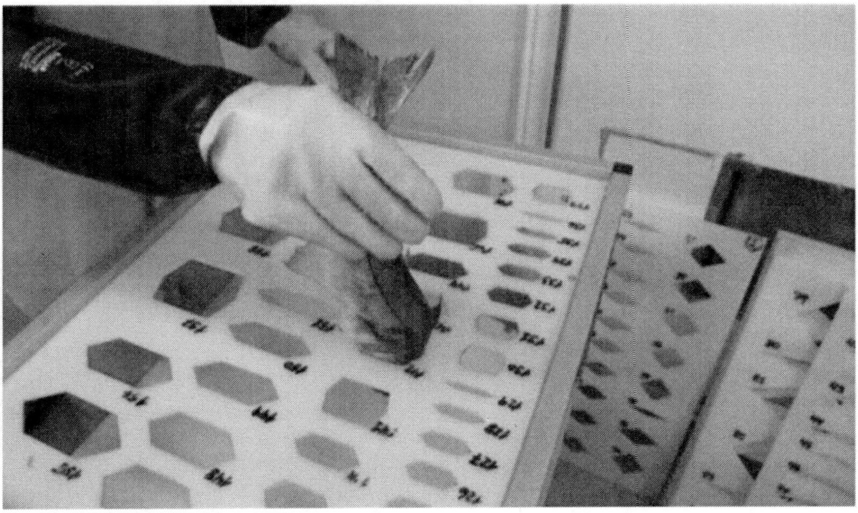

FIGURE 8. Illustration of the fall-through procedure on haddock (*Melanugrammus aeglefinus*). Each interchangeable plate contains a number of different mesh shapes where each fish is tested. All fish included in the study need to be tested in all meshes.

Step 3: Cross-section modeling

The CS shapes registered with the morphometer need to be modeled so that they can be further analyzed in FISHSELECT. The software has more than 100 different models available to model each of the different cross sections of the fish (**Figure 9**). One needs to first determine which of the available models seems to represent each of the CSs well enough and later fit all these relevant models to each digitized shape. Each of the models is tested on each of the CSs registered for each fish, and the model with the lowest AIC [40] is chosen for further analysis in FISHSELECT. Once the parameters in the model defining each of the cross sections are the length of the fish, one can create virtual populations with defined CSs.

Step 4: search for penetration model

Fish can be compressed both dorsoventrally and laterally. Thus, different compression models need to be tested for each CS as a first step to establish an optimal penetration model for the species tested (**Figure 10**). The optimal penetration model is established by comparing the penetration results of each compression model tested with the fall-through results. The degree of agreement (DA) between the simulated and experimental fall-through results is then used to choose an optimal penetration model (see

[36, 37] for the mathematical expression and further information about DA).

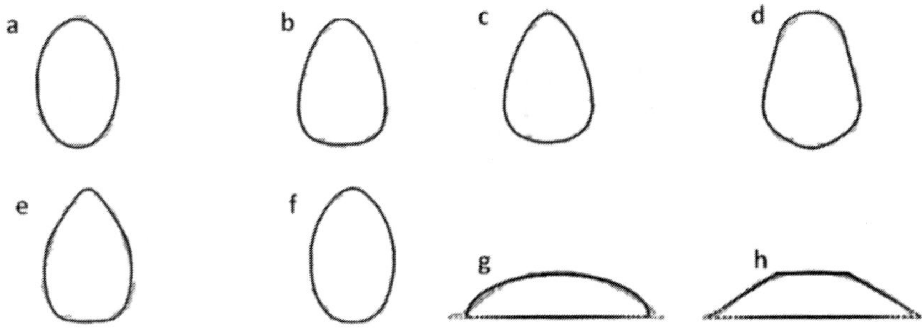

FIGURE 9. Illustration of some of the parametric shapes available in FISHSELECT: Shapes like (a)–(f) are typically used for roundfish while (g)–(h) are normally used on flatfish species.

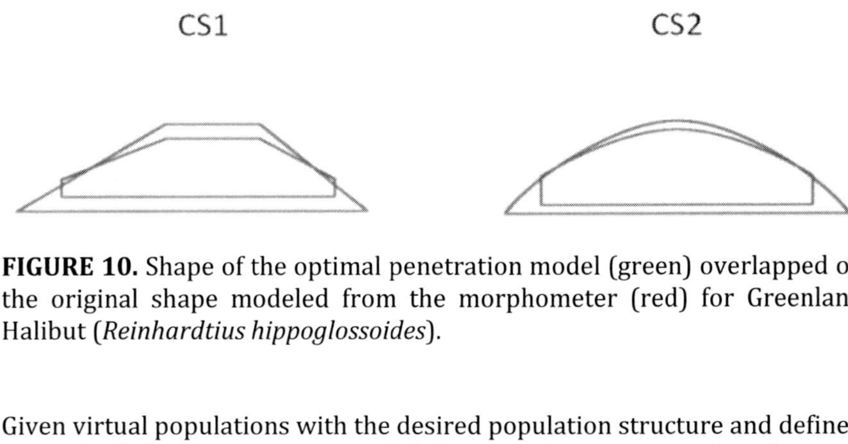

FIGURE 10. Shape of the optimal penetration model (green) overlapped on the original shape modeled from the morphometer (red) for Greenland Halibut (*Reinhardtius hippoglossoides*).

Given virtual populations with the desired population structure and defined CSs and a defined penetration model, the size selective properties of a range of mesh shapes and sizes can be predicted in FISHSELECT by simulation. The outcome of the method consists of L50 and SR estimations for all of the included mesh sizes, shapes, and OAs.

7. FISHSELECT results for Northeast Arctic demersal species

Cod, haddock, redfish, and Greenland halibut are the four most important commercial species in the Barents Sea demersal trawl fisheries. FISHSELECT has already been applied to all four species, and the results have in all cases showed to be coherent with the results obtained from earlier sea trials [31, 37–39], . Because the compulsory gear in the Barents Sea demersal trawl fishery is composed of a sorting grid followed by a size selective diamond mesh codend, the FISHSELECT studies present results for both different bar spacing grids and different diamond mesh size codends. **Figure 11** shows selectivity results obtained with the different bar spacing grids for both cod and haddock [37]. **Figure 12** shows the predicted and observed L50 versus mesh size for *Sebastes* spp. **Figure 13** shows historical selectivity results for Greenland halibut compared to FISHSELECT predictions. Finally, **Figure 14** shows the variation in L50 for redfish with varying mesh size and OA for diamond meshes.

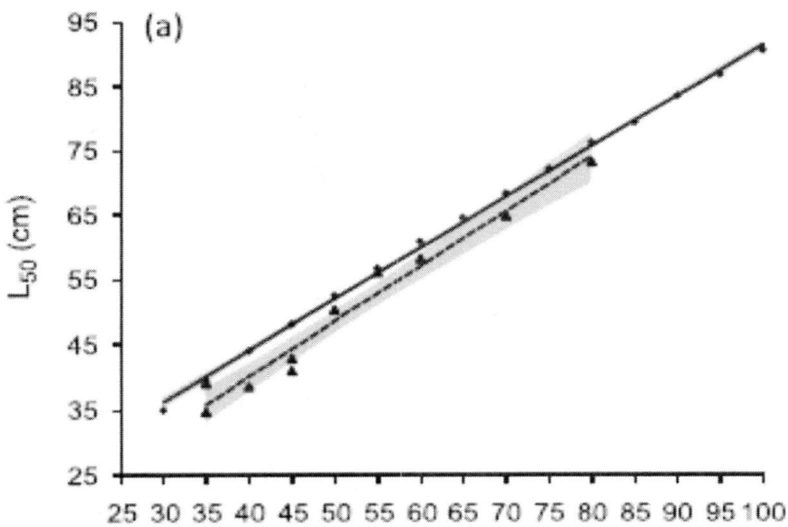

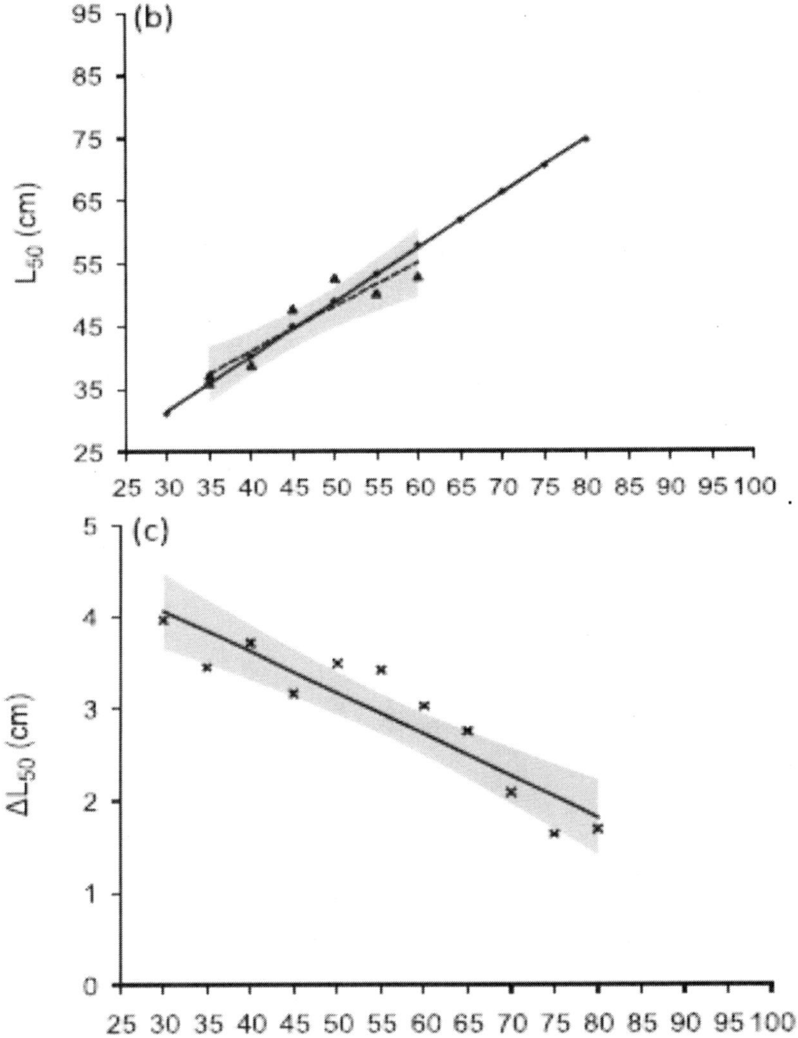

FIGURE 11. Selectivity results obtained with the different bar spacing grids for both cod (*G. morhua*) (a) and haddock (*M. aeglefinus*) (b). The solid line is a trend line fitted to the FISHSELECT results (diamonds), which are presented for up to the maximum fish size included in the data collection process. The broken line is a trend line added to the sea trial selectivity results (triangles) obtained for the different bar spacing grids in [41]. Panel (c) shows the L50 differences and a trend line for these differences between cod and haddock. The gray areas in the three panels represent the 95% confidence limits for the added trend lines. Source: [37].

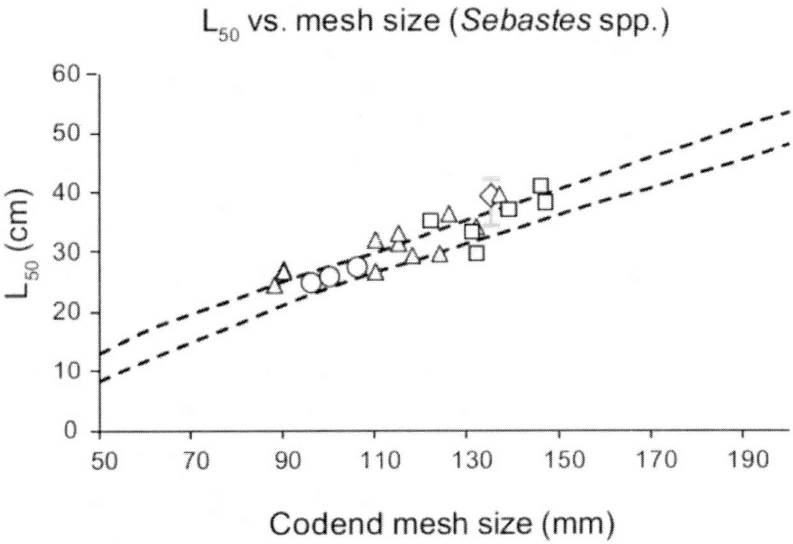

FIGURE 12. Predicted and observed L50 versus mesh size relationships for redfish (*Sebastes* spp.): Predicted band for codend L50 for different mesh sizes based on the FISHSELECT analysis of the data collected for *Sebastes marinus* (stippled curves); new results from sea trials for *S. marinus* (diamonds); previous results for *S. marinus* (squares); previous results for *Sebastes mentella* (triangles) and *Sebastes mentella/f aciatus* (circles). Source: [38].

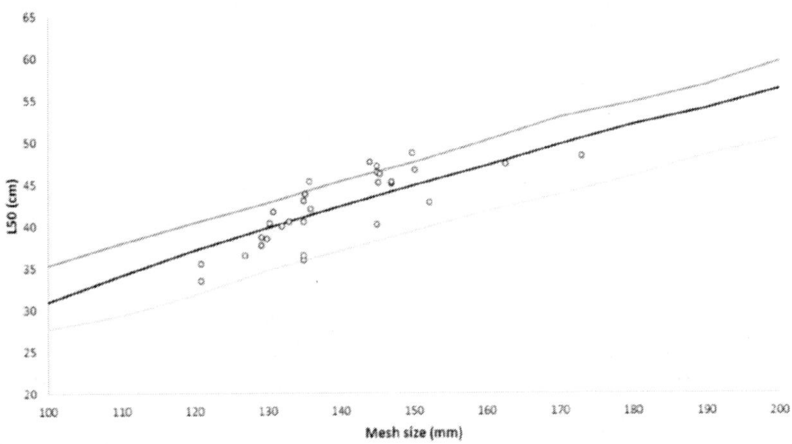

FIGURE 13. Historical data for Greenland halibut (*R. hippoglossoides*) codend selectivity (circular marks) plotted together with FISHSELECT estimations (lines). Source: [39]

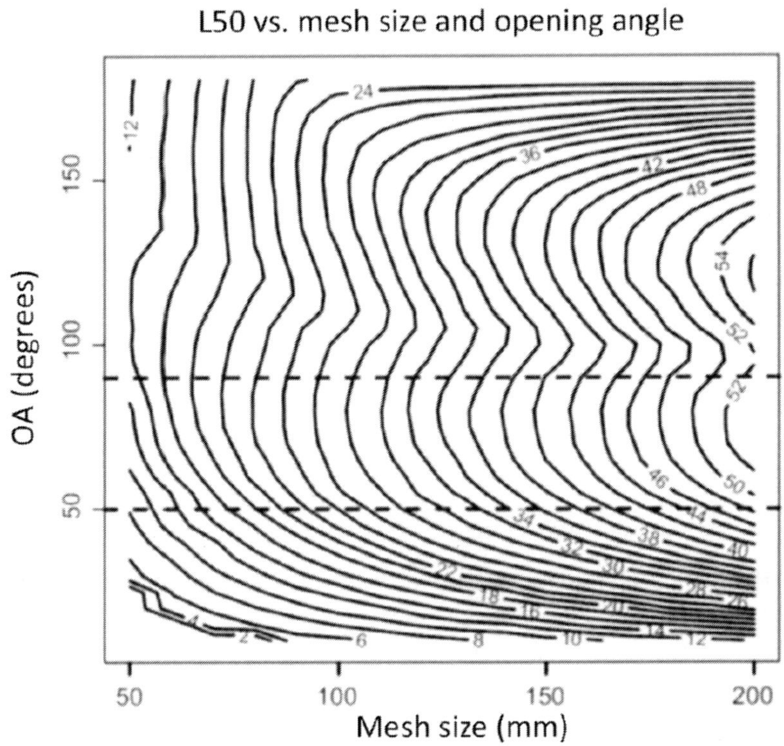

FIGURE 14. Design guide showing the variation in L50 for redfish (*S. marinus*) with varying mesh size and OA for diamond meshes. Source: [38].

8. ACKNOWLEDGEMENTS

Most of the information presented in this work has been obtained through diverse research projects that have been carried out in the past years. We gratefully acknowledge the financial support from the Norwegian Research Council, the Norwegian industrial research fund (FHF), and the Norwegian Directorate of Fisheries. The initial development of FISHSELECT was financially supported by a project under the development program for sustainable fisheries financed by the European Union and Danish Ministry of Food, Agriculture, and Fisheries.

REFERENCES

1. ICES [2015]ICES Advice: Cod (*Gadus morhua*) in Subareas I and II (Northeast Arctic) [http://www.ices.dk/sites/pub/Publication%20Reports/Advice/2015/2015/cod-arct.pdf]. 12 June 2015.[Accessed: 26 January 2016]
2. ICES [2015] ICES Advice: Haddock (*Melanogrammus aeglefinus*) in Subareas I and II (Northeast Arctic) [http://www.ices.dk/sites/pub/Publication%20 Reports /Advice/2015/2015/had-arct.pdf]. 12 June 2015.[Accessed: 26 January 2016]
3. ICES [2015] ICES Advice: Saithe (*Pollachius virens*) in Subareas I and II (Northeast Arctic) [http://www.ices.dk/sites/pub/Publication%20Reports/Advice/2015/2015/sai-arct.pdf]. 12 June 2015.[Accessed: 26 January 2016]
4. Lisovsky, S.F. [2001] Evolution of technical measures regulating fishery on cod and haddock in the Barents Sea and Norwegian Sea. In: Shevelev, M., Lisovsky, S.F. (Eds.) Technical regulations and by-catch criteria in the Barents Sea fisheries. Proceedings of the 9th PINRO-IMR Symposium, Murmansk 14–15 August, 2001.
5. Kvamme, C., Frøysa, K.G. [2008] Assessing the effects on stocks of selectivity changes in a fishery. Fish. Res.; 93: 283-292.
6. Huse, I., Ånondsen, S., Ellingsen, H., Engås, A., Furevik, N., Graham, N., Isaksen, B., Jørgensen, T., Løkkeborg, S., Nøttestad, L. & Soldal, A.V. [2002] A desk-study of diverse methods of fishing when considered in perspective of responsible fishing, and the effect on ecosystem caused by fishing activity. Bergen, July 2002. TemaNord 501.
7. Isaksen B., Valdemarsen, J.W. [1986] Selectivity experiments with square mesh codends in bottom trawls. ICES CM1986; B:28.
8. Larsen, R.B. [1989] A review on the application and selectivity of square mesh netting in trawls and seines. In: Carr, H.A. (Ed.) Proceedings of the square mesh workshop. World symposium on fishing gear and fishing vessel design. St. John's Canada, November 1989; 13–36.
9. Isaksen, B. [1991] A brief review of codend selectivity investigations in Norway from 1985-1991. In: Workshop on codend selectivity. June 1991. Marine Laboratory, Aberdeen Scotland.
10. Løbach, T. [1991] Gear selectivity: Report of the Nordic conference on gear selectivity. Køge, 28-29 May 1991.TemaNord 513: 115–124.
11. Larsen, R.B., Isaksen, B. [1993] Size selectivity of rigid sorting grid in bottom trawls for Atlantic cod (*Gadus morhua*) and haddock (*Melanogrammus aeglefinus*). ICES Mar. Sci. Symp.; 196: 178–182.
12. Isaksen, B., Lisovsky, S., Larsen, R.B., Sakhnoe, V., Gamst, K. & Misund, R.[1996] Joint Russian - Norwegian selectivity experiments on cod (*Gadus morhua*) in the Barents Sea 1995. Fisken og Havet.; 1996; 16: 55pp.
13. Isaksen, B., Gamst, K., Kvalsvik, K. & Axelsen, B. [1998] Comparison of selectivity and handling characteristics between the Sort-X and the single grid (Sort-V). Experiments performed on board the commercial vessel "Anny Kræmer" in Bear Island, 15-28 August 1997. Norwegian Institute of Marine Research. 1998; 15pp. (In Norwegian).
14. Isaksen, B., Gamst, K., Kvalsvik, K. & Kristiansen, A.[1998] Single sorting grid in to panel trawls for cod. Comparison of selectivity and handling characteristics between Sort-X and the single grid (Sort-V). Norwegian Institute of Marine Research. 1998; 19pp. (In Norwegian).
15. Misund, R. [2001] Development of sorting grids in the Norwegian fishery – A review. In: Shevelev, M. and Lisovsky, S. (Eds.), Technical regulations and by-catch criteria in the Barents Sea fisheries. Proceedings of the 9th PINRO-IMR Symposium, Murmansk, 14-15 August, 2001.

16. Jørgensen, T., Galbraith, D., Graham, N., Hansen, K., Holst, R., Larsson, P.O., Mortreux, S., Sangster, G.I., Soldal, A.V. & Tschernij, V.[2002] Design, handling and selectivity properties of the Eurogrid. In: ICES-FAO Working Group on Fishing Technology and Fish Behaviour, ICES CM 2002/B:01. Sáte, France, 6-8 June 2002.
17. Angell, S., Andersen, M., Lilleng, D. & Torsvik, O.K. [2001] Comparison of the selectivity and handling characteristics between a 50 mm Sort-X and 50 mm Flexigrid in the bottom trawl fishery for saithe. Experiments on board"Bliki", 9–20 March 2001. SINTEF Fisheries and Aquaculture. STF80 A013032. 2001; 23pp. (In Norwegian).
18. Angell, S., Lilleng, D., Hassel, K. & Røttingen, Å.[2001] Selectivity experiments with a 55mm Flexigrid in the bottom trawl fishery for saithe. Experiments on board "Bliki", 3–12 April 2000. SINTEF Fisheries and Aquaculture. STF80 A00050.2001; 16pp. (In Norwegian)
19. Angell, S., Lilleng, D., Kolbeinshavn, A., Vollstad, J. & Larsen, R.B. [2001] Selectivity experiments with a 55 mm Flexigrid in the bottom trawl fishery for saithe. Experiments on board "Bliki", 21 September – 2 October 2000.SINTEF Fisheries and Aquaculture. STF80 A0113030. 2001; 18pp. (In Norwegian)
20. Angell, S., Lilleng, D., Vollstad, J. & Grimaldo, E.[2001] Selectivity experiments with 55 and 60 mm Flexigrid in the bottom trawl fishery for cod. Experiments on board "Bliki" 11– 23 June 2001. SINTEF Fisheries and Aquaculture. STF80 A013040. 2001; 19pp. (In Norwegian).
21. Grimaldo, E., Larsen, R.B. & Holst, R. [2007] Exit Windows as an alternative selective system for the Barents Sea Demersal Fishery for cod and haddock. Fish. Res.; 91: 295–305.
22. Grimaldo, E., Sistiaga, M. & Larsen, R.B. [2008] Evaluation of codends with sorting grids exit windows, and diamond meshes: size selection and fish behaviour. Fish. Res.; 91: 271–280.
23. Grimaldo, E., Larsen, R.B., Sistiaga, M., Madsen, N. & Breen, M. [2008] Selectivity and escape percentages during three phases of the towing process comparing codends fitted with different selection systems. Fish. Res.; 95: 198–205.
24. Sistiaga, M., Herrmann, B. & Larsen, R.B. [2009] Investigation of the paired-gear method in selectivity studies. Fish. Res.; 97: 196–205.
25. Grimaldo, E., Sistiaga, M., Herrmann, B., Gjøsund, S.H. & Jørgensen, T. [2015] Effect of lifting panel on selectivity of a compulsory grid section (Sort-V) used by the demersal trawler fleet in the Barents Sea cod fishery. Fish. Res.; 170: 158–165.
26. Gjøsund, S.H., Grimaldo, E., Sistiaga, M. & Hansen, K.[2013] Water flow measurements in two and four-panel grid sections. SINTEF Fisheries and Aquaculture. Report A24698. 2013 (In Norwegian).
27. Wileman, D., Ferro, R.S.T., Fonteyne, R. & Millar, R.B.[1996] Manual of methods of measuring the selectivity of towed gears. ICES Coop. Res. Rep. No. 215; 126 pp.
28. Breen, M., Sangster, G., O'Neill, B., Kynoch, R., Jones, E. & Soldal, A.[2002] Evidence of sampling induced biases in mortality estimates from experiments investigating mortality in fish escaping from towed fishing gears. ICES CM 2002/V: 25.
29. Kvamme, C., Isaksen, B. [2004] Total selectivity of a commercial cod trawl with and without a grid mounted: grid and codend selectivity of North-east Arctic cod. Fish. Res.; 68: 305–318.
30. Sistiaga, M., Herrmann, B., Grimaldo, E. & Larsen, R.B.[2010] Assessment of dual selection in grid based selectivity systems. Fish. Res.; 105: 187–199.
31. Herrmann, B., Sistiaga, M., Larsen, R.B. & Nielsen, K. [2013] Size selectivity of redfish (Sebastes spp.) in the Northeast Atlantic using grid-based selection systems for trawls. Aquat. Living Resourc.; 26: 109–120.
32. Fryer, R.J. [1991] A model of between-haul variation in selectivity. ICES J. Mar. Sci.; 48: 281–290.
33. Sistiaga, M., Herrmann, B., Grimaldo, E. & Larsen, R.B. [2015] Effect of lifting the sweeps on bottom trawling catch efficiency: a study based on the Northeast arctic cod (*Gadus morhua*) trawl fishery. Fish. Res.; 167: 164–173

34. Millar, R.B. [1994] Sampling from trawl gears used in size selectivity experiments. ICES J. Mar. Sci.; 51: 293–298.
35. Herrmann, B., Sistiaga, M., Santos, J. & Sala, A. [2015] How many fish need to be measured in trawl selectivity studies? In: ICES. 2015. Second Interim Report of ICES-FAO Working Group on Fishing Technology and Fish Behaviour (WGFTFB), 4-7 May 2015, Lisbon, Portugal. ICES CM 2015/SSGIEOM:22. 183pp.
36. Herrmann, B., Krag, L.A. & Frandsen, R.P. [2009] Prediction of selectivity from morphological conditions: methodology and a case study on cod (*Gadus morhua*). Fish. Res.; 97: 59–71.
37. Sistiaga, M., Herrmann, B., Nielsen, K.N. & Larsen, R.B. [2011] Understanding limits to cod and haddock separation using size selectivity in a multispecies trawl fishery: an application of FISHSELECT. Canadian J. Fish. Aquat. Sci.; 68: 927–940.
38. Herrmann, B., Sistiaga, M., Nielsen, K.N. & Larsen, R.B. [2012] Understanding the size selectivity of redfish (*Sebastes spp.*) in North Atlantic trawl codends. J. Northw. Atlan. Fish. Sci.; 44: 1–13.
39. Herrmann, B., Sistiaga, M., Larsen, R.B., Nielsen, K.N. & Grimaldo, E. [2013] Understanding sorting grid and codend size selectivity of Greenland halibut (*Reinhardtius hippoglossoides*). Fish. Res.; 146: 59–73.
40. Akaike, H. [1974] A new look at the statistical model identification. IEEE Trans. Autom. Control.; 19(6): 716–723.
41. Sistiaga, M., Grimaldo, E., & Larsen, R.B. [2008] Size selectivity patterns in the north-east Arctic cod and haddock fishery with sorting grids of 55, 60, 70 and 80 mm. Fish. Res.; 93: 195–203.

CHAPTER 5

Using Taxes to Manage a Multigear Fishery: An Application to a Spanish Fishery

M. Dolores Garza-Gil*, Manuel Varela-Lafuente and Juan C. Surís-Regueiro

[1] *Department of Applied Economics, University of Vigo, Vigo, Spain*

ABSTRACT

When fishing gears alter the composition of fish populations or modify the recruitment rate, it is advisable to include the degree of their fishing selectivity in the analysis. Fishing selectivity can cause two different management problems: interspecies selectivity or by-catch of fish stocks for which no quota has been set by the regulator. The case study is the Spanish fishery of hake (*Merlucius merlucius*), where the fleet operates using two main gears; most of the vessels are trawlers but a few boats use longlines and other fixed gears. Fishery management by means of effort taxes and how the degree of intraspecies selectivity may affect the resource and tax levels are analyzed. The results show that the tax level will depend on the social value of the marine stock, the marginal productivity of each fleet's effort, and the effect that the fishing activity of each one has on the growth of the hake biomass.

Keywords: European hake, fisheries management, multigear fishery, tax, Spanish fishery, fishing selectivity

1. INTRODUCTION

From an economic point of view, fishery resources are assets that provide flows of income over time but show certain characteristics. These are linked with the renewable character of fish stocks, the institutional structure under which the activity takes place, and the existence of externalities in the use of a resource. Bioecological rules are essential to determine the functions of production and meet the necessary biological restrictions in an objective function optimization. However, the institutional conditions in the fish stock exploitation establish who is entitled to capture that resource and under what circumstances, and this is essential to understand and predict the behavior of the economic agents involved in the economic activity (the fishermen) and properly drive any regulatory intervention.

Concern for the implications associated with the extraction of marine resources is relatively recent; scarcity problems were largely associated with nonrenewable natural resources until the mid-twentieth century. From then on, the fishing economy has developed quickly. This can be explained by the increasing concerns for the conservation of resources to the perception of degradation of nature and the environment. The effects of the decisions taken at the Third Conference of UN on Law of the Sea in the mid-1970s also have influenced this development, as it recognized the extension of fishery jurisdiction to 200 miles from coastal line and transforming the status of fishery resources from free access to the exclusive property of coastal states.

Marine resource exploitation is one of the typical examples of the tragedy of the commons in which the logic of individual maximization of benefits leads to a continual increase in pressure on the resources and their consequent overexploitation. As the population has expanded, the problem of a lack of resources has become more evident. Society has increasingly valued natural and environmental resources. Key institutional figures have become more necessary for establishing more efficient and sustainable management of natural resources to prevent a tragedy of the commons. Thus, the study of the commons is relevant when analyzing common ownership or open access systems, but its conceptual significance goes far beyond these concrete systems because it represents the starting point in the search to understand the rise and formation of institutions.

These characteristics pose specific management problems for those who need to build theoretical formalization different from those used for the rest of economic assets and those who must be focused on the determination of optimal trajectories for the exploitation of the renewable natural resources sustainably over time. The marine resources must be managed in a rational way, especially if the welfare of future generations is taken into account in the decision-making process.

In a fishery where two or more fleets are using several fishing technologies or gears, it is useful to assume that fishing activity influences the net natural dynamics of the marine resources through the catches, whereas the natural growth function depends on the fish biomass and environmental conditions, and these are taken as stable and constants over time in the specialized literature [1–4]. However, in some fisheries (as the Spanish hake fishery), several fishing technologies could alter the composition of fish populations or modify the recruitment rate [5]. In this case, it is advisable to include the degree of their fishing selectivity in the study. The selectivity could cause two different management problems: interspecies selectivity or by-catch of fish stocks for which no quota has been set by the regulator [6–9].

The case study is the Spanish fishery of European hake *(Merlucius merlucius*) in Ibero-Atlantic grounds. The Spanish fleet involved in this fishery operates uses two main gears; most of vessels are trawlers, but a few boats use longlines and other fixed gears (majority gillnets). Trawlers harvest mainly young individuals of hake of a lower size than that corresponding to sexual maturity (although it too catches mature fish). The other fishing technology (artisanal fleet) catches only mature fish. Based on this, we focus on the intraselectivity problem. We introduce in the analysis of the management of the fishery by means of effort taxes [10–16].

On the contrary, and given that the International Council for the Exploration of the Sea (ICES; this institution analyzes the stock situation and proposes management measures to the European regulator) and the European Commission (EC) recommend that one of the two technologies involved in the hake fishery (in particular, trawling fleet) improves the level of fishing selectivity and aim to individuals of a larger size, we pose several scenarios and study how the levels of hake stock and the tax applied to each group of vessels would be affected. The results obtained show that the optimum tax level depends not only on the social value of the marine resource and the marginal productivity of each fleet's effort but also on the effect that the fishing activity of each one has on the growth of the hake biomass. Furthermore, and as the fleet that is less conservationist with the stock (trawlers) improves the degree of selectivity of its technology, the equilibrium fishing effort level for this fleet increases and the optimum tax falls, to the detriment of the stationary values corresponding to the other fleet.

The particular issue with which this chapter is concerned is how the degree of intraspecies selectivity may affect the hake stock and tax levels. The chapter is structured as follows: the Spanish fishery is described in Section

2. A simple management model applied to the fishery is analyzed in Section 3. The primary results are summarized in Section 4. Lastly, the chapter concludes with the discussion presented in Section 5.

2. DESCRIPTION OF THE FISHERY

The *M. merlucius* species is listed within the group of demersal beings and therefore a fish stock of long life. Although it is distributed in the area located between the coast north of Morocco and the North Sea, the ICES valued it separately since 1979, distinguishing two biological units: Northern stock (corresponding to zones IV, VI, and VII and divisions VIIIa and VIIIb; see **Figure 1**) and Southern stock (divisions VIIIc and IXa). Thus, these two stocks are considered by European regulators as two different management units. This is due to the existence of two well-differentiated recruitment areas: one on the west coast of France (Northern stock) and the other on the coast northwest of the Iberian Peninsula (Southern stock).

The fishery we are studying is European hake in ICES divisions VIIIc and IXa, better known as the Southern stock of European hake. The juvenile individuals of European hake mainly feed on zooplankton and decapod prawns (*Nephrops norvegicus*). Larger hake feed predominantly on fish, with blue whiting (*Micromesistius poutassou*) being the most important prey in waters deeper than 100 m. Horse mackerel (*Trauchurus trauchurus*) and mackerel (*Scomber scombrus*) are the most important prey species in shallower waters. Hake are known to be cannibalistic species located at the top of the food chain. European hake recruitment processes lead to patches of juveniles found in the localized areas of the Iberian continental shelf. European hake concentrations could vary in density according to the strength of the year class; however, they remain generally stable in size and spatial location. The ICES estimates that the spatial patterns could be related to environmental conditions. On the eastern shelf of the Cantabrian Sea, years of large inflow of the shelf-edge current have produced low recruitment rates due to larvae and pre-recruits being transported away from spawning areas. The recent high recruitment has not yet been linked to an environmental process.

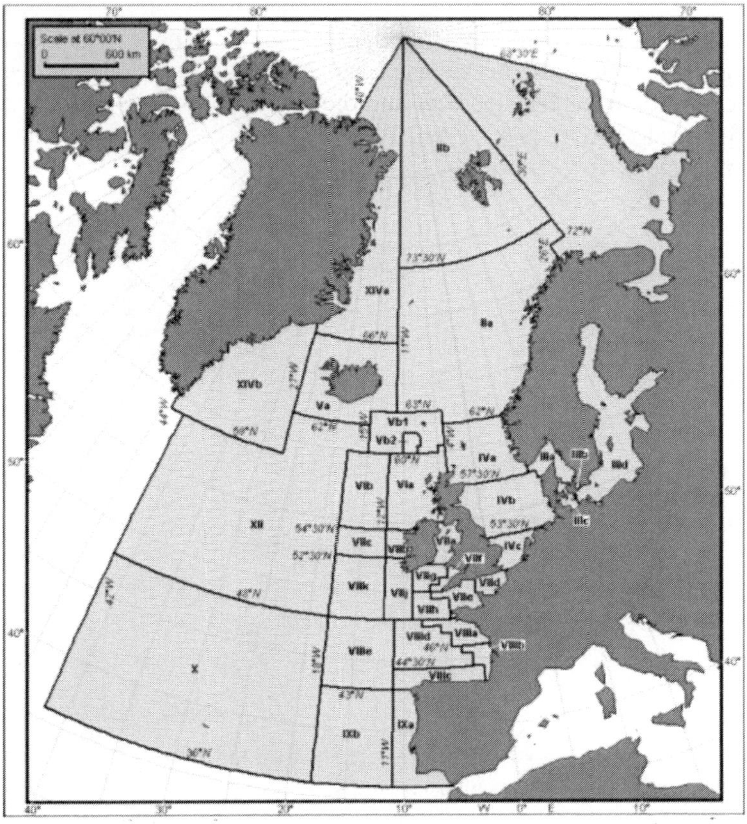

FIGURE 1. ICES zones. Source: Spanish Oceanographic Institute.

European hake in ICES divisions VIIIc and IXa is caught in a mixed fishery by trawlers and artisanal vessels. The trawling fleet is homogeneous and uses mainly two gears: pair trawl and bottom trawl. The artisanal fleet is quite heterogeneous and uses a wide variety of fixed gears, mainly large and small fixed gillnets and longlines. The amount of hake in the landings of Spanish trawlers is low in relative terms. However, trawling vessels provide by 55% of the total Spanish hake landings for last years. These fishing gears affect the hake biomass in different ways. Trawling, although it catches individuals of all ages, has a negative impact on young individuals preventing them from reaching adulthood. The more traditional method, however, affects mainly mature fish and is less damaging to the hake stock.

Trawl fleet is one of the most important fleets among those operating on the Spanish Atlantic continental shelf in terms of landings value. The standard vessel has approximately 145 GRT of fishing capacity and 330 kW of engine power, is close to 28 m long, has 9 crew members, and has an average age of 20 years. The main target species are hake, megrim, anglerfish, lobster, and horse mackerel. The longline and gillnet fleet is less important than the trawler fleet and the standard vessel has approximately 35 GRT and 150 kW, is close to 20 m long, has 5 crew members, and has an average age of 18 years.

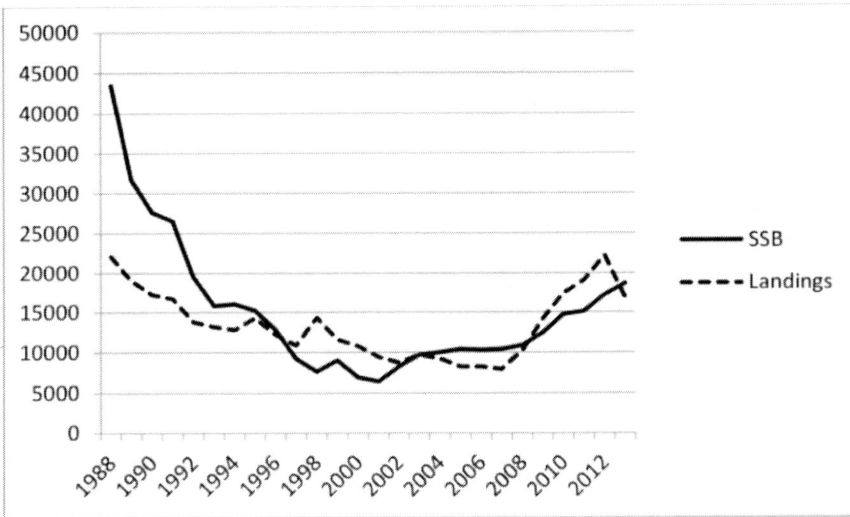

FIGURE 2. Spawning stock biomass (SSB) and landings. Data in tons. 1988–2013. Source: Own compilation from ICES.

The European Union (EU), within the framework of the Common Fisheries Policy (CFP), manages European hake fishery with total allowable catch (TAC), mainly set based on biological criteria. In addition to TACs, EU implements minimum sizes of catches for hake since 1987 and closed areas. The Spanish Government sets a closed list of vessels of each fishing fleet for the last decades. Furthermore, and in the face of the poor biological situation of the stock (see **Figure 2**), since 2006, a recovery plan has been implemented, aimed at recovering the spawning biomass above precautionary biomass and reducing fishing mortality to 0.27 [17]. To do so, the EC, while continuing with the establishment of downward TAC, proposes to reduce the effort exercised in the fishery and includes the improvement in the selectivity of some of the fishing methods.

Regarding the Southern stock of European hake, we have obtained information from the ICES on the spawning biomass for the period 1985 to 2014. **Figure 2** shows how the hake biomass has decreased to such an extent in the late 1990s, as it reached only 25% of that which existed in the early 1980s, falling well outside the biological safety limits in spite of the recovery experienced in the last 3 years [18]. This hake biomass evolution indicates that the resource is being exploited to excess. With respect to the total catches, we can see that it has shown a decreasing trend in the said period and in keeping with the deterioration of the fish biomass (see **Figure 2**).

The trends in both variables show that the measures adopted by the EU were not sufficient to avoid the overexploitation of hake stock and the resource is still being overfished in the last years. Therefore, it is necessary to introduce a regulatory mechanism to manage the hake fishery in a sustainable way to avoid the overexploitation of resource and depletion of the fish stock.

3. METHOD

If the regulator of fishery establishes a tax on effort (τ_i), both fleets will assume an increase in the unit cost of the effort and will be faced with the following problem:

$$\underset{e_i}{Max} \int_o^{\infty} \{p_i h_i(e_i(t), X(t)) - (w_i + \tau_i)\, e_i(t)\} e^{-\delta t} \partial t \qquad i{=}1{,}2 \tag{1}$$

where p, w, h, e, and X denote the unit price of hake, unit cost of effort, total landings, fishing effort, and fish stock, respectively. The parameter δ represents the discount rate.

The usual natural growth function of the marine resource (F) is modified by a new parameter θ, which catches the selectivity of both fleets. The fish stock dynamic is shown as follows:

$$G(X_t, \theta_t) = \theta_t \, \mathrm{F}(X_t) \tag{2}$$

where $F(\cdot)$ is the natural growth function of the resource. The effects that the different technologies have on it are defined as follows [19]:

$$\theta_t = 1 - \sum_i \gamma_i \frac{h_i(X_t, e_{it})}{h(X_t, e_t)} \qquad \text{con } 0<\theta \leq 1; \;\; \mathrm{i}{=}1{,}2 \tag{3}$$

where the parameter $\gamma_i(0\leq;\gamma_i<1, i=1,2)$ shows the level of fishing selectivity of each technology or fleet. If the i-fleet technology has no effects on the fish stock dynamics, the fleet shows a high selectivity level and this fleet can be considered as conservationist with the marine resource. In this case, the parameter γ_i takes on a zero value. In contrast, if technology has effects on the marine stock dynamics in a negative way, the fleet shows a nonselective level and it can be considered as a less conservationist fleet with the fish stock. Therefore, the fishing selectivity parameter will approach the unit value.

From one of the first-order conditions to resolve the problem (1) [20], the following equation is obtained:

$$P_I \frac{\partial H_I(.)}{\partial E_I} = W_I + \tau_I \qquad I=1,2 \quad . \tag{4}$$

In the absence of regulation (i.e., no tax would be implemented), we would obtain the following expression [19]:

$$(p_i - \mu) \frac{\partial h_i(.)}{\partial e_i} + \mu \frac{\partial G(.)}{\partial e_i} = w_i \quad . \tag{5}$$

and if both expressions (4) and (5) are compared, the optimum tax value can be obtained (i=1,2):

$$\tau_i = \mu \left[\frac{\partial h_i(.)}{\partial e_i} - \frac{\partial G(.)}{\partial e_i} \right] = \mu \frac{\partial h_i(.)}{\partial e_i} \left[1 - \frac{(\gamma_j - \gamma_i) h_j(.) F(X)}{(h_i + h_j)^2} \right] . \tag{6}$$

This expression indicates that the tax level depends not only on the social value of the marine resource (μ) and the marginal productivity of the effort ($\partial h_i/\partial e_i$) but also on the effect on the natural growth of the resource ($\partial G(\cdot)/\partial e_i$). On the contrary, the lower (higher) the marginal productivity of the fleet *i*, the lower (higher) the tax level that will have to be paid to fish in the fishery.

On the contrary, and for $\gamma_i \neq \gamma_j$, if fleet *i* shows a high (low) selectivity level and with $\gamma_i < \gamma_j$ ($\gamma_i > \gamma_j$), then $\gamma_i \rightarrow 0$ ($\gamma_i \rightarrow 1$) and the effect of the activity of *i* on the natural growth function will be lower (higher), allowing a greater (smaller) growth of the fish population, that is, ($\gamma_j - \gamma_i$)> 0 (($\gamma_j - \gamma_i$) < 0) and $\partial G(\cdot)/\partial e_i > 0$ ($\partial G(\cdot)/\partial e_i < 0$). Consequently, given that $\partial h_i(\cdot) > 0$, the tax level for this fleet will be higher (lower) than that which corresponds to the other fleet.

4. ESTIMATIONS

Because fishing effort (fishing days) data are not available separately for the trawling and artisanal fleets for the last 10 years, we will use the parameter values estimated by Garza-Gil and Varela-Lafuente [19] for this fishery, who made an econometric estimation through the Ordinary Least Squares (OLS) method with annual observations for 20 years and for different options of the natural resource dynamic and the production functions. These values are summarized in **Table 1**. Substituting those values of the parameters in the above expression (6), the stationary solutions for the tax levels can be estimated.

However, previously, and because the selectivity parameters are unknown, we must assume some value for them. Regarding trawling, this fleet catches mainly smaller-sized individuals, as mentioned in the previous sections, and therefore has a negative impact on the Southern stock hake population by preventing a greater number of young fish from reaching maturity and being able to spawn for next years. On that basis, we will assume a selectivity parameter value for this fleet initially closer to unit value than to zero, in particular γ_1=0.7.

Regarding the artisanal fleet, although it captures mostly mature individuals, it also captures a small amount of young individuals. This figure does not reach 10% of the landings [19]. Therefore, we will assume a selectivity value for artisanal fleet closer to zero (0.1).

TABLE 1. Parameter values for estimations.

	Expression/value	**Unit**
Hake dynamic		
Trawl technology		
Artisanal technology		
Price of trawl landings	4,565	Euros per tonne
Price of artisanal landings	7,482	Euros per tonne
Cost of trawl effort	373	Euros per fishing day
Cost of artisanal effort	419	Euros per fishing day
Discount rate	5	Percentage

"1" indicates trawling and "2" artisanal.

Source: Own compilation from Refs. [19, 21].

On the contrary, the trawling may improve the selectivity of this gear, as the EC [17] and the ICES [18] proposed in its management recommendations with a view to improving the pattern of hake production for this fishery. Accordingly, some options may increase, for example, the size of the mesh and expand the cod-end of the fishing nets (the "cod-end" is the rearmost part of a trawl net, of net of the same mesh size, having either a cylindrical or a tapering shape). If this technology improves its fishing selectivity level, the negative effects of its activity on hake dynamic will decrease. In this case, other possible and lowest values for γ_1 can be posed. The results obtained for different values of parameter $\gamma_1$are shown in **Table 2**.

It can be seen that the tax level on trawling (in euros per fishing day) is higher than that applied to the artisanal fleet in the scenarios contemplated for selectivity parameter due fundamentally to the fact that it shows a greater marginal productivity in the effort and a negative effect on hake biomass. Consequently, it should pay more to fish in the fishery. Furthermore, as the trawling selectivity improves ($\gamma_1 \rightarrow 0$) and therefore the negative effect of the activity of this fleet on the hake population diminishes, the tax per unit of effort applied to this fleet also decreases, whereas, for the artisanal fleet, it increases and its effort level decreases.

TABLE 2. Hake biomass (metric tons) and tax levels (euros/day) for different γ_1 and $\gamma_2 = 0.1$.

γ_1	X		
0.7	31,416	2,389	38
0.6	34,143	2,183	53
0.5	37,116	1,627	83
0.4	37,262	1,075	109
0.3	42,499	1,062	242
0.2	43,014	994	262

"1" indicates trawling and "2" artisanal.

5. DISCUSSION AND CONCLUSIONS

The intensive exploitation of the fishery resources around the world for the last decades has shown the natural limitations of the productivity of fish stocks. In this environment with a depletion of marine resources, economists have been worried about searching for management tools oriented to change the behavior of fishermen to save the resource and also to maintain a positive economic return. From an economic point of view, fish populations are treated as capital assets that can provide flows of income over time. The aim is therefore to determine the path of exploitation of marine resources in a sustainable way and to incorporate the biological conditions of the marine resource and institutional conditions of fishing into the analysis. In this way, the fishing economy has advanced since the first works by Gordon [22] and Scott [23], which includes biological and institutional conditions of basic form, to the development raised by Clark [11] and Clark and Munro [24], who introduced the theory of capital to manage a fishing resource in a dynamic context.

In general, the regulatory mechanisms can be classified into two groups [11]: (1) those that are directed toward the direct control on the fish stocks as well as to maintain high production levels and (2) a group of mechanisms that, in addition to indirectly control the size of the stock, points to sustain activity in economically efficient levels. The methods that have been traditionally implemented, such as production quotas, closed seasons, closed zones, and restrictions on the equipment, correspond to the first group and it has been shown that they have failed to prevent the overexploitation of fish stocks [25–27]. The allocation of property rights and the system of taxes (on production or on the inputs) are in the second group. Among the latter, individual property rights require the creation of markets; the regulator may establish certain rules with respect to fishery exploitation (distribution of the surplus of the marine resource among fishermen involved in the fishery) and allow a rights transaction market to emerge to ensure that fishermen comply with its conduct selling or buying part of that right. Taxes can be defined as mechanisms based on the regulation via prices; the essence of these instruments involves the introduction of a price (cost) linked to the behavior that the regulator wants to promote or discourage.

In this chapter, we have studied the European hake fishery (Southern stock), where two fishing fleets are operating using different technologies. We have shown the way in which effort taxes exercised in this multigear fishery make it possible to reach a socially optimum solution for this marine resource, introducing a variable into the analysis, which includes the effects of fishing activity on the natural growth function of the hake population. The efficient stationary solutions for the hake stock levels, its social value and the effort exercised by the two fleets involved in the fishery (trawling and artisanal), propose different scenarios with regard to the selectivity parameter for the fleet that has a more intensive impact on young individuals and then on marine resource dynamics. If trawling selectivity improves, then the optimum level of the natural resource and its shadow price increases, whereas the global level of effort diminishes, increasing that of the trawling fleet and reducing that of the longline fleet [19].

If the present situation is compared to the optimal estimations obtained in this study, it can be seen that the Southern stock of European hake is being fished in an inefficient way, both from an economic point of view and the conservation of the natural resource point of view. In particular, the amount of hake biomass existing at the end of the period studied is significantly lower than that derived from a socially stationary solution. Even in a few years, landings have exceeded the spawning hake biomass in Iberian-Atlantic waters.

To reach socially stationary solutions, we have incorporated an intervention mechanism based on taxes, particularly a tax based on effort exercised by each fleet. The tax equilibrium level is directly related to the social value of the fishing resource, with the marginal productivity of the effort exercised and with the effect that fishing activity has on the natural growth of the resource. In particular, the tax level on the trawling effort is greater than that applied to the artisanal fleet, as it is more productive and affects the hake population more negatively. Therefore, it will pay more to exercise its effort in the fishery. On the contrary, the equilibrium level obtained for the tax on the effort of the artisanal fleet is lower, as it is less productive and much more selective. However, when the trawling fleet improves its selectivity, its effort equilibrium level increases and the optimum tax decreases, to the detriment of the stationary values that correspond to the artisanal fleet.

In this framework, the proposed regulation involving declines in the level of fishing (reducing the pressure on the stock of fish) is not usually well received by the fishing industry. However, an efficient regulation allows maintaining the marine resources in a sustainable way and it will generate economic income for fishermen. An inefficient situation to an efficient change must be associated with a policy of income distribution suitable based on the criteria of equity. The regulation mechanism based on taxes could offer a solution to the externalities associated with the absence of efficient allocations. Although the analysis shown in this chapter is simple, the results can orient the regulator to achieve a more rational exploitation of the Southern stock of hake.

6. ACKNOWLEDGEMENTS

The authors acknowledge the financial support from FEDER and Xunta de Galicia (GRC2014/022) and the Ministry of Economy and Competitiveness (ECO2014-52412-R and ECO2013-44436-R).

REFERENCES

1. Armstrong C, Clark D. 1997. Just fishing? Equity and efficiency in fisheries management regimes. Marine Resource Economics 12: 203–220.
2. Garza-Gil MD. 1998. ITQ systems in multifleet fisheries. An application for Iberoatlantic hake. Environmental and Resource Economics 15: 93–106.
3. Costa Duarte C, Brasao A, Pintasilgo P. 2000. Management of the Northern Atlantic bluefin tuna: an application of C-Games. Marine Resources Economics 14: 21–36.
4. Bjorndal T, Lindroos M. 2003. International Management of North Sea Herring. Paper presented at the 12th Annual Conference of European Association of Environmental and Resources Economists 2003.
5. Lleonart J, Recasens L. 1996. Fisheries and Environment in the Mediterranean Sea. Resources and Environmental Issues Relevant to Mediterranean Fisheries Management. Studies and Reviews General Fisheries Council for the Mediterranean, No. 66. Rome: FAO.
6. Boyce J. 1996. An economic analysis of the fisheries bycatch problem. Journal of Environmental Economics and Management 31: 314–336.
7. Prellezo R, Gallastegui C. 2003. Gear selectivity based regulation in a mixed fishery. Paper presented at the 12th Annual Conference of European Association of Environmental and Resource Economists 2003.
8. Turner MA. 1997. Quota induced discarding in heterogeneous fisheries. Journal of Environmental Economics and Management 33: 186–195.
9. Espasa M, Prellezo R. 2003. Fishing technology and optimal distribution of harvest rates. Environmental and Resource Economics 25: 377–394.
10. Clark CW. 1980. Towards a predictive model for the economic regulation of commercial fisheries. Canadian Journal of Fisheries and Aquatic Sciences 37: 1111–1129.
11. Clark CW. 1990. Mathematical Bioeconomics. The Optimal Management of Renewable Resources. 2th ed. New York: Wiley-Interscience Publication, 386 pp.
12. Arnason R. 1989. Minimum information management with help of catch quotas. In: Neher P, Arnason R, Mollet N, editors. Right Based Fishing. Netherlands: Kluwer Academic Publishers, pp. 215–240.
13. Surís-Regueiro JC. 1993. Regulation of the Iberoatlantic sardine fishery. Environmental and Resources Economics 3: 457–470.
14. Weitzman M. 2002. Landing fee vs. harvest quotas with uncertain fish stocks. Journal of Environmental Economics and Management 43: 325–338.
15. Jensen F, Vestergaard N. 2002. Moral hazard problems in fisheries regulation: the case of illegal landings and discards. Resource and Energy Economics 24: 281–299.
16. Garza-Gil MD, Varela-Lafuente M, Surís-Regueiro JC. 2003. European hake fishery bioeconomic management (Southern stock) applying an effort tax. Fisheries Research 60: 199–206.
17. European Commission. 2003. Proposal for Recovery Measures for Southern Hake. Brussels: COM 818, Brussels.
18. ICES. 2015. Hake-Southern Stock Report. http://www.ices.dk.
19. Garza-Gil MD, Varela-Lafuente M. 2007. Bioeconomic management and fishing selectivity: an application to the European hake fishery. American Journal of Agricultural and Biological Sciences 2: 69–74.
20. Kamien M, Schwartz N. 1991. Dynamic Optimization. The Calculus of Variations and Optimal Control in Economics and Management. New York: North-Holland Ed., 377 pp.
21. European Union. 2014. Economic Assessment of EU Fisheries Economic Performance of Selected European Fishing Fleets. Annual Reports. https://stecf.jcr.ec.europa.eu.home.

22. Gordon HS. 1954. The economic theory of a common property resource: the fishery. Journal of Political Economy 62: 124–142.
23. Scott A. 1986. Progress in Natural Resource Economics. Clarendon Press.
24. Clark CW, Munro GR. 1975. The economics of fishing and modern capital theory: a simplified approach. Journal of Environmental Economics and Management 2: 92–106.
25. Neher P, Arnason R, Mollet N. 1989. Right Based Fishing. Netherlands: Kluwer Academic Publishers.
26. Townsend R.E. 1990. Entry restrictions in the fishery: a survey of the evidence. Land Economics 66: 359–378.
27. Wilen J. 1988. Limited entry licensing: a retrospective assessment. Marine Research Economics 5: 313–324.

CHAPTER 6

Effects of Astaxanthin on Antioxidant Capacity of Golden Pompano (*Trachinotus Ovatus*) in Vivo and in Vitro

Jia-jun Xie[1,2†], Xu Chen[1†], Jin Niu[3*], Jun Wang[1], Yun Wang[1] and Qiang-qiang Liu[1]

[1]*South China Sea Fisheries Research Institute, Chinese Academy of Fishery Sciences, Guangzhou 510300, China.*

[2]*School of Life Science and Technology, Shanghai Ocean University, Shanghai 201306, China.*

[3]*Institute of Aquatic Economic Animals, School of Life Science, Sun Yat-sen University, NO.135 at Xingang Xi Road, Haizhu District, Guangzhou 510275, Guangdong Province, China*

ABSTRACT

The objective of this research was to study the effect of astaxanthin (AST) on growth performance and antioxidant capacity in golden pompano (*Trachinotus ovatus*) both in vivo and in vitro. In the in vivo study, two diets were formulated with or without astaxanthin supplementation (D1 and D2; 0 and 200 mg/kg) to feed fish for 6 weeks. In the in vitro study, cells from hepatopancreas of golden pompano were isolated and four treatments with or without astaxanthin and H_2O_2 supplementation were applied (control group: without both astaxanthin and H_2O_2 treated; H_2O_2 group: just with

H_2O_2 treated; H_2O_2 + AST group: with both astaxanthin and H_2O_2 treated; AST group: just with AST treated). Results of the in vivo study showed that weight gain (WG) and special growth rate (SGR) significantly increased with astaxanthin supplemented ($P < 0.05$). Feed conversion ratio (FCR) of fish fed D2 diet was significantly lower than that of fish fed D1 diet ($P < 0.05$). Hepatic total antioxidant capacity (T-AOC) and the reduced glutathione (GSH) of golden pompano fed D2 diet were significant higher than those of fish fed D1 diet ($P < 0.05$). Superoxide dismutase (SOD) was significantly declined as astaxanthin was supplemented ($P < 0.05$). Results of the in vitro study showed that the cell viability of H_2O_2 group was 52.37% compared to the control group, and it was significantly elevated to 84.18% by astaxanthin supplementation (H_2O_2 + AST group) ($P < 0.05$). The total antioxidant capacity (T-AOC) and the reduced glutathione (GSH) of cell were significant decreased by oxidative stress from H_2O_2 ($P < 0.05$), but it could be raised by astaxanthin supplementation (H_2O_2 vs H_2O_2 + AST), and the malondialdehyde (MDA) was significant higher in H_2O_2 group ($P < 0.05$) and astaxanthin supplementation could alleviate the cells from lipid peroxidation injury. In conclusion, dietary astaxanthin supplementation can improve the growth performance of golden pompano. Moreover, astaxanthin can improve the golden pompano hepatic antioxidant capacity both in vivo and in vitro study by eliminating the reactive oxygen species.

Keywords: Golden pompano, Astaxanthin, Growth performance, Antioxidant capacity

BACKGROUND

Reactive oxygen species (ROS) are oxidative products, produced continuously in the course of normal aerobic cellular metabolism and respiratory burst (Chew 1995), which participate in a variety of biological processes, including normal cell growth, induction and maintenance of the transformed state, programmed cell death, and cellular senescence (Finkel 2003). However, ROS can, in turn, damage healthy cells if they are not eliminated (Chew 1995). Under normal physiological conditions, the excessive ROS can be removed by internal antioxidants and anti-oxidative systems (Chen et al. 2015), including counter balance such as enzymes (like superoxide dismutase, catalase, and glutathione peroxidase), functionalized large molecules (albumin, ferritin, and ceruloplasmin) and small molecules (ascorbic acid, α-tocopherol, β-carotene, and uric acid) (Martinez-Alvarez et al. 2005). The dietary antioxidants most widely used include vitamin E, vitamin C, carotenoids, flavanoids, zinc, and selenium (Chew and Park 2004). Among those, carotenoids reach the highest plasma and tissue concentrations, despite their lower intake (Olmedilla et al. 2007).

Carotenoids, more than 600 known types, can be classified into two categories, xanthophyll and carotenes. Astaxanthin (3,3′-dihydroxy-β,β-carotene-4,4′dione, AST) is a xanthophyll carotenoid which is found in many microorganisms and marine animals, such as shrimp, crayfish, crustaceans, salmon, trout, krill, microalgae as well as yeast. Its molecule consists of 40 carbon atoms, divided into a portion containing 22 carbon atoms linked with 13 conjugated double bonds and two terminal benzene rings containing hydroxyl and ketone groups, giving rise to the higher polar structure of AST compared with other carotenoids (Britton 1995).The antioxidant activity of astaxanthin was found to be approximately 10 times stronger than β-carotene (Shimidzu et al. 1996).

Except for its antioxidant capacity, AST is also recognized to have growth performance and survival rate promoting in Atlantic salmon (*Salmo salar*) (Christiansen and Torrissen 1996) and red porgy (*Pagrus pagrus*) (Kalinowski et al. 2011), skins coloration enhancing in large yellow croaker (*Larimichthys croceus*) (Yi et al. 2014) and Atlantic salmon(Baker et al. 2002), anti-lipid peroxidation (Leite et al. 2010), and immune response reinforcing in *Astronotus ocellatus* (Alishahi et al. 2015) properties.

Golden pompano (*Trachinotus ovatus*) belongs to family carangidae, genus *Trachinotus*. It is a warm-water species (25–32 °C) and a carnivorous fish that preys mainly on zooplankton, small crustaceans, shellfish, and small fish (Liu and Chen 2009). *T. ovatus*is is widely distributed in China, Japan, Australia, and other countries (Huo-sheng 2006). Pompano is considered one of the most desirable food fishes, and it commands a significantly higher price than many other marine and freshwater species (Tutman et al. 2004). Recently, pompano is widely farmed owing to its high price in the market and resilience to salinity and temperature ranges (Tutman et al. 2004), and annual output was over 100,000 tons. The suitable dietary protein and lipid levels for golden pompano are 46.0 and 6.5% (Wang et al. 2013). The optimum carbohydrate level for juvenile golden pompano could be 11.2–16.8% of the diet (Zhou et al. 2015). The optimal requirements of methionine (Niu et al. 2013), arginine (Lin et al. 2015), and n-6 long-chain polyunsaturated fatty acid arachidonic acid (ARA) (Qi et al. 2016) for golden pompano have been determined as well.

However, fewer studies were conducted to investigate the effects of antioxidants on golden pompano. To date, the effects of AST on various kinds of fishes are mainly reported in vivo and rarely found in vitro. We used in vivo and in vitro models to study and compare the effect of astaxanthin on antioxidant ability of golden pompano, respectively.

METHODS

Diet preparation and dietary treatments

In this study, two isonitrogenous and isoenergetic semi-purified diets were formulated supplementing with or without astaxanthin (D1: 0%; D2: 0.2%; CAROPHYLL Pink, 10% astaxanthin, DSM Nutritional Products France SAS) (Table 1). The method of diet preparation was the same as described by Niu et al. (2015). Briefly, all dry ingredients were finely ground, weighed, mixed manually for 5 min, and then transferred to a Hobart mixer (A-200 T Mixer Bench Model unit; Resell Food Equipment Ltd., Ottawa, ON, Canada) for another 15 min mixing. During the mixing, 6 N NaOH was added to establish a pH level of 7–7.5. The pH of the diet was obtained by homogenizing a 5-g portion of the diet with 50 mL of distilled water with a glass-electrode pH meter on the supernatant (Robinson et al. 1981). Soya lecithin was added to pre-weighed fish oil and mixed until homogenous. The oil mix was then added to the Hobart mixer slowly while mixing was still continuing. All ingredients were mixed for another 10 min. Then, distilled water (about 30–35%, *v*/*w*) was added to the mixture to form dough. Dough of even consistency was passed through a pelletizer with a 2.5-mm-diameter die (Institute of Chemical Engineering, South China University of Technology, Guangzhou, China). The diets were dried until the moisture was reduced to <10%. The dry pellets were placed in plastic bags and stored 20 °C until fed.

Animal rearing and experimental procedures

The feeding trial was conducted at an experimental station of South China Sea Fisheries Research Institute of Chinese Academy of Fishery Sciences (Sanya, Hainan). Prior to the start of the trial, juvenile *T. ovatus* were acclimated to a commercial diet for 2 weeks and were fed twice daily to apparent satiation. At the beginning of the feeding trial, the fish were starved for 24 h, weighed, and then fish with similar size (initial body weight 23.65 ± 0.10 g) were randomly allotted to 8 sea cages (1.0 m × 1.0 m × 1.5 m; four cages per diet treatment); each cage was stocked with 30 fish. Each experimental diet was randomly assigned to four cages. The feeding frequency was once daily at 8:00 and lasted for 6 weeks. To prevent the waste of pellets, fish were slowly hand-fed to satiation based on visual observation of their feeding behavior. Feed consumption was recorded for each cage every day. Water quality parameters were monitored daily.

Table 1 Ingredients and proximate composition of the two experimental diets (%)

Ingredients	Diet 1	Diet 2
Fish meal	32.00	32.00
Soybean meal	30.00	30.00
Wheat flour	20.00	19.80
Krill meal	2.00	2.00
Fish oil	8.00	8.00
Soya lecithin	2.00	2.00
Monocalcium phosphate	2.00	2.00
Pre-vitamin[a]	1.00	1.00
Pre-mineral[b]	1.00	1.00
Choline	0.50	0.50
Vc	0.50	0.50
DL-Met	0.40	0.40
Lys-HCL (78%)	0.60	0.60
AST	0	0.20
Sum	100	100
Nutrient levels		
Moisture	9.00	7.54
Crude protein	40.64	40.55
Crude fat	10.71	10.90
Ash	15.21	15.24

[a]Pre-vitamin (mg or g kg^{-1} diet): thiamin, 25 mg; riboflavin,45 mg; pyridoxine HCl, 20 mg; vitamin B12, 0.1 mg; vitamin K3,10 mg; inositol, 800 mg; pantothenic acid, 60 mg; niacin acid, 200 mg; folic acid, 20 mg; biotin, 1.20 mg; retinal acetate, 32 mg; cholecalciferol, 5 mg; α-tocopherol, 120 mg; ascorbic acid, 2000 mg; choline chloride, 2500 mg; ethoxyquin 150 mg; and wheat middling, 14.012 g (Niu et al. 2013)
[b]Pre-mineral (mg or g kg^{-1} diet): NaF, 2 mg; KI, 0.8 mg; $CoCl_2 6H_2O$ (1%), 50 mg; $CuSO_4 5H_2O$, 10 mg; $FeSO_4 H_2O$, 80 mg; $ZnSO_4 H_2O$, 50 mg; $MnSO_4 H_2O$, 60 mg; $MgSO_4 7H_2O$, 1200 mg; $Ca(H_2PO_4)_2 H_2O$, 3000 mg; NaCl, 100 mg; and zoelite, 15.447 g (Niu et al. 2013)

Sample collection

At the end of the feeding trial, fish were starved for 24 h and then weighed and counted the total number. Ten fish from each cage were randomly collected for sampling: four for analysis of whole-body composition and six were anesthetized to obtain weights of individual whole body, viscera, and liver. The livers were rapidly removed and frozen in the liquid nitrogen separately for analysis of lipid peroxidation and antioxidant status.

Biochemical composition analysis

Chemical composition of diets and fish were determined by standard methods (Latimer 2012). Moisture was determined by oven drying at 105 °C until a constant weight was obtained. Crude protein content ($N \times 6.25$) was determined according to the Kjeldahl method after acid digestion using an Auto Kjeldahl System (1030-Autoanalyzer; Tecator, Höganäs, Sweden). Crude lipid was determined by the ether extraction method using a Soxtec extraction System HT (Soxtec System HT6, Tecator). Ash content was determined after samples were placed in a muffle furnace at 550 °C for 4 h.

Isolation of liver cells

Golden pompano was purchased from a market in Guangzhou, China. Hepatocytes were isolated according to the methods of Wan et al. (2004) with some modifications. In the procedure, a fish was kept in 0.01% potassium permanganate solution for half an hour, after that, its skin was sterilized by alcohol, and its abdomen was dissected with sterile instruments from the anus toward the head. Liver tissue was excised and rinsed three times with phosphate buffer solution. The liver tissue was then minced into pieces of 1 $mm^3$and transferred to a 15 mL tube to which a solution of 0.25% trypsin (1:20 *w*/*v*; Sigma) was added. The mixture was trypsinized on a thermostatic water bath to obtain the cell suspension, which was shaken every 5 min. Then, the mixture was filtered through a 100-mesh sieve. The cell suspensions were pooled and centrifuged at 1000 rpm for 10 min, and the cell pellet was washed and resuspended in a culture medium. The cell number was counted using a haemocytometer, and cell viability was estimated immediately following isolation using the trypan blue exclusion assay.

Cell culture and treatments

A final cell density of hepatocytes was adjusted to 2×10^6 cells mL^{-1} in L-15 culture medium (Jinuo Co, Hangzhou, China) supplemented with 2 mM L-glutamine (Sigma) and 10% foetal bovine serum (Gibco). Cells were seeded into 12-well culture plates with 500 μL cell suspension per well. Cells were cultured in a humidified atmosphere at 28 °C. Once seeded, cells were allowed to attach to culture plates for 24 h. At 24 h, 50% of the culture medium (250 μL) was removed and replaced with fresh medium. Then PBS, 100 mM H_2O_2, 1000 ng/mL astaxanthin dissolved in dimethyl sulfoxide (DMSO) (final concentration 0.01%), and H_2O_2 plus astaxanthin were added in the wells. Every treatment was replicated in three wells. Cell viability was evaluated by trypan blue exclusion test, and cells were harvested for antioxidant capacity analysis after the treatments. As an additional measure of cell viability, lactate dehydrogenase (LDH) activity in the extracellular medium (an indicator of membrane leakage (Misra and Niyogi 2009) was measured.

Antioxidant capacity analysis

Hepatic and cell samples were homogenized in ice-cold phosphate buffer (1:10 dilution) (phosphate buffer; 0.064 M, pH 6.4). The homogenate was then centrifuged for 20 min (4 °C, 3000 g), and aliquots of the supernatant were used to quantify hepatic T-AOC, GSH, SOD, and MDA.

The levels of enzyme activity and lipid peroxidation were measured with commercial ELISA kits (Randox Laboratories Ltd.) in accordance with the instructions of the manufacturer. The assays are briefly described as follows: The T-AOC is the representative of enzyme and nonenzyme original antioxidant in the body; these antioxidants can reduce the ferric ion (Fe^{3+}) to ferrous ion (Fe^{2+}). The latter combines with phenanthroline and produces a stable chelate, which can be measured by spectrophotography at 520 nm (Xiao et al. 2004). The T-AOC was determined in units per milligram of tissue protein.

Total superoxide dismutase (SOD) activity was measured by using a xanthine oxides (Marklund and Marklund 1974). The ratio of autooxidation rates of the samples with or without hepatic homogenate was determined at 550 nm. One unit of SOD activity was calculated using the amount of superoxide dismutase required to inhibit the reduction of nitrobluete trazolium by 50%.

The formation of 5-thio-2-nitrobenzoate (TNB) was followed spectrophotometrically at 412 nm (Vardi et al. 2008). The amount of GSH in the extract was determined as μmol/mg protein utilizing a commercial GSH as the standard. The results are expressed as μmol/mg protein.

Lipid-peroxidation levels were determined based on the malondialdehyde (MDA) level generated by oxidizing fatty acids. In the presence of thiobarbituric acid, malondialdehyde started producing colored thiobarbituric-acid-reacting substances (TBARS) that were measured at 532 nm (Buege and Aust 1978).

Lactate dehydrogenase (LDH) can catalyze lactate into pyruvate, which react with 2,4-dinitrophenylhydrazine and produce a stable compound, which was measured by spectrophotography at 450 nm.

Calculations and statistical analysis

The parameters were calculated as follows:

Weight gain rate (WG, %)=100×(final body weight - initial body weight)/initial body weight

Specific growth rate (SGR, % day^{-1})=100×(Ln final mean weight - Ln initial mean weight)/ number of days

Feed conversion ratio (FCR)=dry diet fed/wet weight gain

Survival rate (%)=100×(final number of fish)/(initial number of fish)

Viscerosomatic index (VSI, %)=100×(viscera weight, g)/(whole body weigh t, g)

Hepatosomatic index (HSI, %)=100×(liverweight, g) /(whole body weight, g)

Condition factor (CF, g/cm3)=100×(bodyweight, g)/(body length , cm^3)

Data from each treatment were subjected to one-way analysis of variance (ANOVA). Homogeneity of variance was verified using Bartlett and Levene's test. When overall differences were significant, Tukey's multiple range tests was used to compare the mean values among individual treatments. The level of significant difference was set at $P < 0.05$. Statistical analysis was performed using the SPSS19.0 (SPSS Inc., Michigan Avenue, Chicago, IL, USA) for Windows, and the results are presented as means ± SEM (standard error of the mean).

Results

Growth performance in vivo

Growth, feed utilization, and biometric parameters of juvenile pompano fed different dietary astaxanthin levels are shown in Table 2. Results showed that final body wet weight (FBW), weight gain (WG), and special growth rate (SGR) significantly increased with astaxanthin supplemented ($P < 0.05$). Feed conversion ratio (FCR) of golden pompano fed the diets supplemented with astaxanthin was significantly lower than that of fish fed the control diet ($P < 0.05$), while no significant differences were found in survival rate between the two diet treatments ($P > 0.05$). Hepatosomatic

somatic indices (HSI), visceral somatic indices (VSI), and condition factor (CF) were significantly decreased in astaxanthin-supplemented diet treatment.

Table 2 Growth performance and survival of golden pompano fed diets with and without supplementation of astaxanthin

Diets (AST mg kg^{-1})	Diet 1 0	Diet 2 200
IBW (g)	32.72 ± 0.21	32.58 ± 0.06
FBW (g)	63.71 ± 0.66a	67.23 ± 0.22b
WG (%)	95.30 ± 0.02a	106.36 ± 0.01b
SGR (% day^{-1})	1.59 ± 0.02a	1.72 ± 0.01b
FCR	1.78 ± 0.02a	1.53 ± 0.02b
SR (%)	97.78 ± 0.01	96.67 ± 0.00
HSI (%)	1.13 ± 0.06a	0.94 ± 0.03b
VSI (%)	5.97 ± 0.11a	5.42 ± 0.15b
CF (g/cm3)	3.4 ± 0.06a	3.04 ± 0.05b

Values are mean ± SEM of four replicates, and values in the same row with different letters are significant different ($P < 0.05$). Diet 1 meant golden pompano groups fed diets without supplementation of astaxanthin (AST). Diet 2 meant groups with supplementation of astaxanthin

IBW initial body weight, *FBW* final body weight, *WG* weight gain, *SGR* special growth rate, *FCR* feed conversion ratio, *SR* survival rate, *HSI* hepatosomatic index, *VSI* viscerosomatic index, *CF* condition factor

Growth performance in vitro

In the in vitro study, the cell viability of H_2O_2 group was 52.37% compared to the control group (PBS group), and it could be significantly elevated to 84.18% with astaxanthin supplementation (H_2O_2 + AST group) ($P < 0.05$) (Fig. 1). The highest lactate dehydrogenase (LDH) activity was found in H_2O_2 group, and it was 159.02% compared to the control group, it could be significantly lessened to 122.96% with astaxanthin supplementation (H_2O_2 + AST group) ($P < 0.05$) (Fig. 2).

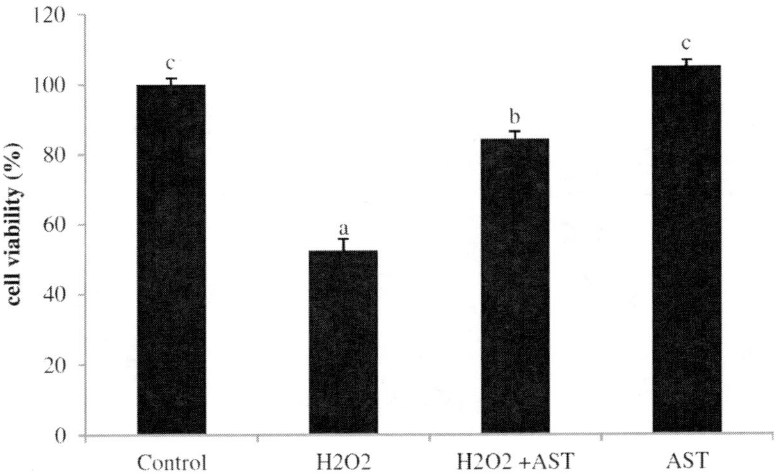

Figure. 1 The relative cell viability in different groups. Control column meant treating with neither H_2O_2 nor AST, H_2O_2column with H_2O_2 only, H_2O_2 + AST column with both H_2O_2 and AST, and AST column with AST only. Data are expressed as mean ± SEM of three replicates; values in the column sharing the same *superscript letter* are not significantly different; however, values in the column with the *different superscript letter* are significantly different

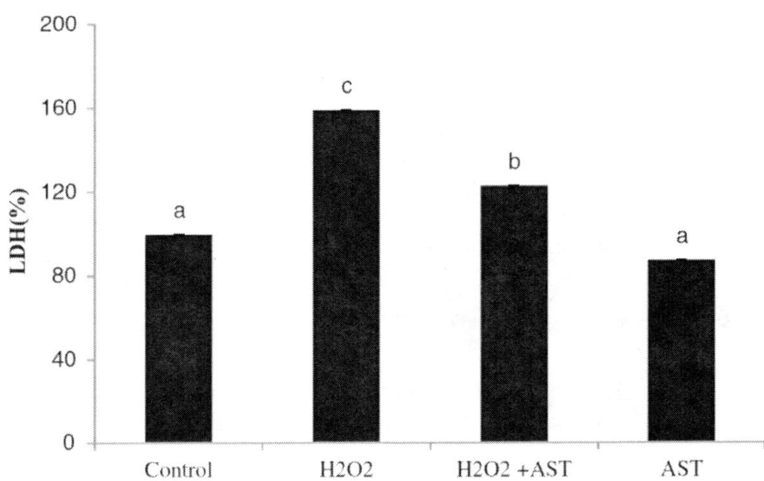

Figure. 2 Lactate dehydrogenase (LDH) activity in the extracellular medium in different groups. Control column meant treating with neither H_2O_2 nor AST, H_2O_2 column with H_2O_2 only, H_2O_2 + AST column with both H_2O_2 and AST, and AST column with AST only. Data are expressed as mean ± SEM of three replicates; values in the column sharing the same *superscript letter* are not significantly different; however, values in the column with the *different superscript letter* are significantly different

Whole-body composition

Whole-body composition of golden pompano fed different dietary astaxanthin levels are shown in Table 3. There were no significant differences in whole-body composition of fish between the two diet treatments ($P > 0.05$).

Table 3 Whole-body compositions (% dry weight) of golden pompano fed diets with and without supplementation of astaxanthin

	Protein	Lipid	Ash	Moist
Diet 1	61.77 ± 1.17	26.68 ± 0.17	17.71 ± 0.23	70.19 ± 1.17
Diet 2	61.68 ± 1.35	26.33 ± 0.46	17.41 ± 0.94	71.68 ± 1.15

Values are mean ± SEM of four replicates, and values in the same column with different letters are significant different ($P < 0.05$). Diet 1 meant golden pompano groups fed diets without supplementation of astaxanthin (AST). Diet 2 meant groups with supplementation of astaxanthin

Antioxidant capacity analysis in vivo

The antioxidant status of juvenile pompano in vivo study are presented in Table 4. The hepatic total antioxidant capacity (T-AOC) and the reduced glutathione (GSH) in fish fed diet supplemented with astaxanthin were significantly higher than that of fish fed the control diet ($P < 0.05$). On the contrary, superoxide dismutase (SOD) declined with astaxanthin supplementation significantly ($P < 0.05$), while hepatic malondialdehyde (MDA) content was not affected by astaxanthin supplementation ($P > 0.05$).

Table 4 Hepatic antioxidant statuses of golden pompano fed diets with and without supplementation of astaxanthin

	T-AOC (U/mg protein)	SOD (U/mg protein)	GSH (μmol/g protein)	MDA (nmol/mg protein)
Diet 1	0.11 ± 0.01a	240.87 ± 5.76a	82.44 ± 4.87a	0.41 ± 0.02
Diet 2	0.15 ± 0.01b	214.24 ± 5.71b	118.52 ± 8.93b	0.41 ± 0.06

Values are mean ± SEM of four replicates, and values in the same column with different letters are significant different ($P < 0.05$). Diet 1 meant golden pompano groups fed diets without supplementation of astaxanthin (AST). Diet 2 meant groups with supplementation of astaxanthin

T-AOC total antioxidant capacity, *GSH* reduced glutathione, *SOD* superoxide dismutase, *MDA* malondialdehyde

Antioxidant capacity analysis in vitro

The antioxidant status of hepatocytes in the vitro study are shown in Table 5. The H_2O_2 as an oxidizing agent can totally damage the healthy cells, and astaxanthin, an antioxidant, can repair the cells from the oxidative stress. The total antioxidant capacity (T-AOC) and the reduced glutathione (GSH) of oxidative stress group (H_2O_2) were the lowest and significantly lower than those of the control group ($P < 0.05$), but it could be significantly improved by astaxanthin supplementation (H_2O_2 + AST group) ($P < 0.05$). The astaxanthin supplemented groups were significant higher than the other ones ($P < 0.05$). The SOD and MDA showed the highest value in H_2O_2 group, which were significantly higher than those in the control group ($P < 0.05$), but it also could be significantly modified by astaxanthin supplementation (H_2O_2 + AST group) ($P < 0.05$). The astaxanthin supplemented groups showed the significantly higher antioxidant capacity than the other groups ($P < 0.05$).

Table 5 The antioxidant statuses of hepatocytes treated with or without astaxanthin and H_2O_2 supplementation

	T-AOC (U/mg protein)	SOD (U/mg protein)	GSH (μmol/g protein)	MDA (nmol/mg protein)
Control	0.35 ± 0.01b	2682.76 ± 127.04b	17.81 ± 0.83b	0.13 ± 0.01a
H_2O_2	0.22 ± 0.02a	3264.92 ± 76.26c	5.92 ± 0.91a	0.40 ± 0.01c
H_2O_2 + AST	0.37 ± 0.01bc	2726.34 ± 74.17b	28.24 ± 1.11c	0.23 ± 0.01bc
AST	0.41 ± 0.01c	2312.19 ± 69.94a	136.51 ± 4.11d	0.12 ± 0.01a

Values are mean ± SEM of three replicates, and values in the same column with different letters are significant different ($P < 0.05$). Control meant golden pompano groups fed diets without supplementation of astaxanthin (AST). Control groups meant treating with neither H_2O_2 nor AST, H_2O_2 groups meant treating with H_2O_2 only, H_2O_2 + AST groups meant treating with both H_2O_2 and AST, and AST groups meant treating with AST only

T-AOC total antioxidant capacity, *GSH* reduced glutathione, *SOD* superoxide dismutase, *MDA* malondialdehyde

DISCUSSION

Growth performance and proximate composition

Carotenoids are reported to improve growth performance of fish with the reason that carotenoids may exert a positive influence on intermediary metabolism in aquatic animals (Segner et al. 1989), which enhance nutrient utilization, ultimately resulting in improving growth (Amar et al. 2001). The other possible mechanism may be to adjust the intestinal flora breaking down indigestible feed components to extract more nutrients and to stimulate the production of enzymes transporting fats for growth instead of storage (James et al. 2006). Kalinowski et al. (2011) believed that astaxanthin could enhance lipid utilization in whole fish and liver, providing more energy and consequently enhancing growth performance. In the present experiment, the growth performance (FBW, WG, and SGR) and feed utilization of fish fed diet with supplemental astaxanthin were significantly higher than that of fish fed the control diet. This result was in agreement with those in previous studies on Atlantic salmon (Christiansen and Torrissen 1996), red porgy (Kalinowski et al. 2011), *Astronotus ocellatus* (Alishahi et al. 2015), and large yellow croaker (Li et al. 2014). However, effect of carotenoids on fish growth is controversial. Many earlier studies have reported that dietary astaxanthin has no significant influence on growth and flesh composition of fish (Tejera et al. 2007; Zhang et al. 2012; Pham et al. 2014; Yi et al. 2014). Kop and Durmaz (2008) believed that the effectiveness of carotenoids in terms of deposition and physiological function is species-specific in fish and not all fish species possess the same pathways for the metabolism of carotenoids. The mechanisms related to these findings have not yet been clearly elucidated. Our latest research results showed that the dietary astaxanthin can increase the apparent digestibility coefficient of the diet and further promote the expression of insulin-like growth factors (IGFs); moreover, as members of the family of transforming growth factors β, myostatin is affected by dietary astaxanthin (unpublished data).

Antioxidant capacity analysis

H_2O_2 is a strong oxidizer, produced in cell metabolism, but the excessive dose may be cytotoxic. As is shown, cell viability was sharply decreased with H_2O_2 supplemented and the increased LDH leakage into the extracellular media by H_2O_2 indicated the occurrence of oxidative stress membrane damage in our present in vitro study. Cellular antioxidative defense mechanisms can intercept the ROS both enzymatically and non-enzymatically. Total antioxidant capacity (T-AOC) is an overall indicator of the antioxidant status of an individual, representing the level of enzyme and nonenzyme original antioxidantin of the body (Xiao et al. 2004). As the value increases, the antioxidant defense against free radical reaction and reactive oxygen intermediates increases (Chien et al. 2003). In both of the in vivo and in vitro study, the T-AOC in the liver of the fish and in the hepatocytes supplemented with astaxanthin were higher, meaning that astaxanthin can improve the antioxidant status whether in vivo or in vitro. Although H_2O_2 may decrease the total antioxidant capacity, the supplementation of astaxanthin can repair it to the same level with the control group.

The stress response might increase free radical contents, which may result in the increase of the lipid peroxidation content and lipid peroxidation injury (Liu et al. 2010). Malondialdehyde (MDA) is a product of lipid peroxidation, through crosslinking with the nucleophilic groups of proteins, nucleic acids, and amino phospholipids, accumulation of MDA leads to cell toxicity, accelerating the damage of cells and tissues (Buege and Aust 1978). The antioxidants and antioxidant enzyme system can play a significant role in resisting lipid oxide damage (Liu et al. 2010). Carotenoids may serve as an antioxidant in systems containing unsaturated fatty acids to quench free radicals (Mansour et al. 2006). The results showed that the MDA were not significant different when no stress appeared in the present in vivo study. However, once the cells suffered from oxidative stress in the present in vitro study, the MDA was increased and the cell viability was decreased, but supplemented astaxanthin could totally decrease the MDA value and save cells from the stress. Increased T-AOC and decreased MDA in the in vitro study reflected that supplemented astaxanthin in media can be totally conducive to eliminate the reactive oxygen species and protect the hepatocytes of golden pompano from free radicals. The MDA in (H_2O_2 + AST) group was lower than that in H_2O_2 group, which indicated that AST can alleviate the lipid oxide damage.

Superoxide dismutase (SOD), a cytosolic enzyme that is specific for scavenging superoxide radicals, is the first enzymes to respond against oxygen radicals and important endogenous antioxidants for protection against oxidative stress (Winston and Di Giulio 1991). Lygren et al. showed that high levels of dietary fat-soluble antioxidants, such as astaxanthin and vitamin E, there was a reduced need for endogenous antioxidant enzymes, such as total SOD (Lygren et al. 1999). The higher the SOD value, the more superoxide radicals need to be reacted (Qingming et al. 2010). It was found that the activities of liver SOD were significantly decreased by dietary astaxanthin supplementation in olive flounder (*Paralichthys olivaceus*) (Pham et al. 2014); large yellow croaker (*Pseudosciaena crocea*) (Li et al. 2014) and rainbow trout (*Oncorhynchus mykiss*) (Zhang et al. 2012). In this present study, SOD was significant lower in vivo and vitro study both supplemented with astaxanthin, implying that astaxanthin can eliminate reactive oxygen species to avoid the cells and tissues to produce more SOD. Once suffering from oxidative stress, the cells may produce much more endogenous SOD, as is shown in the study, to protect the body or cells from being hurt.

Glutathione (GSH), ubiquitous non-enzymatic antioxidants in the cells, is known to play an important role in the scavenging of free radicals and thus protect the important cellular macromolecules and organelles from oxidative damage (Misra and Niyogi 2009). Its role in the detoxification of ROS is important (Mallikarjuna et al. 2009). When suffered from oxidative stress, GSH was significantly lower in the present in vitro study. One mechanism for oxidative stress induced GSH depletion may involve enhanced utilization of GSH for the detoxification of free radicals and other oxidants produced as a result of H_2O_2 exposure (Shaw 1989). Vogt suggested that the increase of lipid peroxidation was not apparent until after GSH levels had been depleted (Vogt and Richie 2007). Astaxanthin can improve the GSH content in both our in vivo and in vitro study.

CONCLUSIONS

In conclusion, dietary astaxanthin supplementation can improve the growth performance of golden pompano. Moreover, astaxanthin can improve their hepatic antioxidant capacity in both vivo and vitro study by eliminating the reactive oxygen species.

Abbreviations

AST: Astaxanthin

CF: Condition factor

FBW: Final body weight

FCR: Feed conversion ratio

GSH: Reduced glutathione

HSI: Hepatosomatic index

IBW: Initial body weight

LDH: Lactate dehydrogenase

MDA: Malondialdehyde

ROS: Reactive oxygen species

SGR: Special growth rate

SOD: Superoxide dismutase

TAOC: Total antioxidant capacity

TBARS: Thiobarbituric-acid-reacting substances

TNB: 5-Thio-2-nitrobenzoate

WG: Weight gain

VSI: Viscerosomatic index

ACKNOWLEDGEMENTS

All authors thank the funding of the Project of Science and Technology of Guangdong Province (2013B090500110, 2013B090600045), the Institutes of Public Welfare Projects (2014A08YQ02), and the Special Project of Marine Fishery Science and Technology of Guangdong Province (A201601C11).

Funding

The design of the study and collection, analysis, interpretation of data, and writing the manuscript were supported by the Project of Science and Technology of Guangdong Province (2013B090500110, 2013B090600045), the Institutes of Public Welfare Projects (2014A08YQ02), and the Special Project of Marine Fishery Science and Technology of Guangdong Province (A201601C11).

Ethics approval and consent to participate

All experimental procedures were conducted in conformity with institutional guidelines for the care and use of laboratory animals in Sun Yat-sen University, Guangzhou, China, and conformed to the National Institutes of Health Guide for Care and Use of Laboratory Animals (publication no. 85–23, revised 1985).

REFERENCE

1. Alishahi M, Karamifar M, Mesbah M. Effects of astaxanthin and Dunaliella salina on skin carotenoids, growth performance and immune response of Astronotus ocellatus. Aquac Int. 2015;23(5):1239–48.
2. Amar E, Kiron V, Satoh S, Watanabe T. Influence of various dietary synthetic carotenoids on bio-defence mechanisms in rainbow trout, Oncorhynchus mykiss (Walbaum). Aquacult Res. 2001;32(s1):162–73.
3. Baker R, Pfeiffer A-M, Schöner F-J, Smith-Lemmon L. Pigmenting efficacy of astaxanthin and canthaxanthin in fresh-water reared Atlantic salmon, Salmo salar. Anim Feed Sci Technol. 2002;99(1):97–106.
4. Britton G. Structure and properties of carotenoids in relation to function. FASEB J. 1995;9(15):1551–8.
5. Buege JA, Aust SD. Microsomal lipid peroxidation. Methods Enzymol. 1978;52:302–10.
6. Chen YY, Lee PC, Wu YL, Liu LY. In vivo effects of free form astaxanthin powder on anti-oxidation and lipid metabolism with high-cholesterol diet. PLoS One. 2015;10(8):e0134733.

7. Chew BP. Antioxidant vitamins affect food animal immunity and health. J Nutr. 1995;125(6 Suppl):1804S–8S.
8. Chew BP, Park JS. Carotenoid action on the immune response. J Nutr. 2004;134(1):257S–61S.
9. Chien Y-H, Pan C-H, Hunter B. The resistance to physical stresses by Penaeus monodon juveniles fed diets supplemented with astaxanthin. Aquaculture. 2003;216(1):177–91.
10. Christiansen R, Torrissen OJ. Growth and survival of Atlantic salmon, Salmo salar L. fed different dietary levels of astaxanthin. Juveniles. Aquacult Nutr. 1996;2(1):55–62.
11. Finkel T. Oxidant signals and oxidative stress. Curr Opin Cell Biol. 2003;15(2):247–54.
12. Huo-sheng Y. Studies on the culture of pompano, Trachinotus ovatus [J]. J Fujian Fish. 2006;1:009.
13. James R., Sampath K., Thangarathinam R. and Vasudevan I. Effect of dietary spirulina level on growth, fertility, coloration and leucocyte count in red swordtail, Xiphophorus helleri. 2006;58(2):97–104.
14. Kalinowski CT, Robaina LE, Izquierdo MS. Effect of dietary astaxanthin on the growth performance, lipid composition and post-mortem skin colouration of red porgy Pagrus pagrus. Aquac Int. 2011;19(5):811–23.
15. Kop A, Durmaz Y. The effect of synthetic and natural pigments on the colour of the cichlids (Cichlasoma severum sp., Heckel 1840). Aquac Int. 2008;16(2):117–22.
16. Latimer GW. Official methods of analysis of AOAC International. Maryland: AOAC International; 2012.
17. Leite M, De Lima A, Massuyama M, Otton R. In vivo astaxanthin treatment partially prevents antioxidant alterations in dental pulp from alloxan-induced diabetic rats. Int Endod J. 2010;43(11):959–67.
18. Li M, Wu W, Zhou P, Xie F, Zhou Q, Mai K. Comparison effect of dietary astaxanthin and Haematococcus pluvialis on growth performance, antioxidant status and immune response of large yellow croaker Pseudosciaena crocea. Aquaculture. 2014;434:227–32.
19. Lin HZ, Tan XH, Zhou CP, Niu J, Xia DM, Huang Z, Wang J, Wang Y. Effect of dietary arginine levels on the growth performance, feed utilization, non-specific immune response and disease resistance of juvenile golden pompano Trachinotus ovatus. Aquaculture. 2015;437:382–9.
20. Liu C, Chen C. The biology and cultured technology of Pompano (Trachinotus ovatus). Shandong Fisheries. 2009;26:32–3.
21. Liu B, Xie J, Ge X, Xu P, Wang A, He Y, Zhou Q, Pan L, Chen R. Effects of anthraquinone extract from Rheum officinale Bail on the growth performance and physiological responses of Macrobrachium rosenbergii under high temperature stress. Fish Shellfish Immunol. 2010;29(1):49–57.
22. Lygren B, Hamre K, Waagbø R. Effects of dietary pro-and antioxidants on some protective mechanisms and health parameters in Atlantic salmon. J Aquat Anim Health. 1999;11(3):211–21.
23. Mallikarjuna K, Nishanth K, Hou CW, Kuo CH, Sathyavelu RK. Effect of exercise training on ethanol-induced oxidative damage in aged rats. Alcohol. 2009;43(1):59–64.
24. Mansour N, McNiven MA, Richardson GF. The effect of dietary supplementation with blueberry, α-tocopherol or astaxanthin on oxidative stability of Arctic char (Salvelinus alpinus) semen. Theriogenology. 2006;66(2):373–82.
25. Marklund S, Marklund G. Involvement of the superoxide anion radical in the autoxidation of pyrogallol and a convenient assay for superoxide dismutase. Eur J Biochem. 1974;47(3):469–74.
26. Martinez-Alvarez RM, Morales AE, Sanz A. Antioxidant defenses in fish: biotic and abiotic factors. Rev Fish Biol Fish. 2005;15(1–2):75–88.

27. Misra S, Niyogi S. Selenite causes cytotoxicity in rainbow trout (Oncorhynchus mykiss) hepatocytes by inducing oxidative stress. Toxicol In Vitro. 2009;23(7):1249–58.
28. Niu J, Du Q, Lin HZ, Cheng YQ, Huang Z, Wang Y, Wang J, Chen YF. Quantitative dietary methionine requirement of juvenile golden pompano Trachinotus ovatus at a constant dietary cystine level. Aquacult Nutr. 2013;19(5):677–86.
29. Niu J, Figueiredo-Silva C, Dong Y, Yue Y, Lin H, Wang J, Wang Y, Huang Z, Xia D, Lu X. Effect of replacing fish meal with soybean meal and of DL-methionine or lysine supplementation in pelleted diets on growth and nutrient utilization of juvenile golden pompano (Trachinotus ovatus). Aquacult Nutr. 2016;22:606–14.
30. Olmedilla B, Granado F, Southon S, Wright AJA, Blanco I, Gil-Martinez E, Berg H v d, Corridan B, Roussel A-M, Chopra M, Thurnham DI. Serum concentrations of carotenoids and vitamins A, E, and C in control subjects from five European countries. Br J Nutr. 2007;85(02):227.
31. Pham MA, Byun HG, Kim KD, Lee SM. Effects of dietary carotenoid source and level on growth, skin pigmentation, antioxidant activity and chemical composition of juvenile olive flounder Paralichthys olivaceus. Aquaculture. 2014;431:65–72.
32. Qi C-L, Lin H-Z, Huang Z, Zhou C-P, Wang Y, Wang J, Niu J, Zhao S-Y. Effects of dietary arachidonic acid levels on growth performance, whole-body proximate composition, digestive enzyme activities and gut morphology of juvenile golden pompano trachinotus. Isr J Aquacult Bamidgeh. 2016;68:1275–84.
33. Qingming Y, Xianhui P, Weibao K, Hong Y, Yidan S, Li Z, Yanan Z, Yuling Y, Lan D, Guoan L. Antioxidant activities of malt extract from barley (Hordeum vulgare L.) toward various oxidative stress in vitro and in vivo. Food Chem. 2010;118(1):84–9.
34. Robinson EH, Wilson RP, Poe WE. Arginine requirement and apparent absence of a lysine-arginine antagonist in fingerling channel catfish. J Nutr. 1981;111(1):46–52.
35. Segner H, Arend P, Von Poeppinghausen K, Schmidt H. The effect of feeding astaxanthin to Oreochromis niloticus and Colisa labiosa on the histology of the liver. Aquaculture. 1989;79(1–4):381–90.
36. Shaw S. Lipid peroxidation, iron mobilization and radical generation induced by alcohol. Free Radic Biol Med. 1989;7(5):541–7.
37. Shimidzu N, Goto M, Miki W. Carotenoids as singlet oxygen quenchers in marine organisms. Fish Sci. 1996;62(1):134–7.
38. Tejera N, Cejas JR, Rodríguez C, Bjerkeng B, Jerez S, Bolaños A, Lorenzo A. Pigmentation, carotenoids, lipid peroxides and lipid composition of skin of red porgy (Pagrus pagrus) fed diets supplemented with different astaxanthin sources. Aquaculture. 2007;270(1):218–30.
39. Tutman P, Glavić N, Kožul V, Skaramuca B, Glamuzina B. Preliminary information on feeding and growth of pompano, Trachinotus ovatus (Linnaeus, 1758)(Pisces; Carangidae) in captivity. Aquac Int. 2004;12(4–5):387–93.
40. Vardi N, Parlakpinar H, Ozturk F, Ates B, Gul M, Cetin A, Erdogan A, Otlu A. Potent protective effect of apricot and beta-carotene on methotrexate-induced intestinal oxidative damage in rats. Food Chem Toxicol. 2008;46(9):3015–22.
41. Vogt BL, Richie Jr JP. Glutathione depletion and recovery after acute ethanol administration in the aging mouse. Biochem Pharmacol. 2007;73(10):1613–21.
42. Wan X, Ma T, Wu W, Wang Z. EROD activities in a primary cell culture of grass carp (Ctenopharyngodon idellus) hepatocytes exposed to polychlorinated aromatic hydrocarbonas. Ecotoxicol Environ Saf. 2004;58(1):84–9.
43. Wang F, Han H, Wang Y, Ma X. Growth, feed utilization and body composition of juvenile golden pompano Trachinotus ovatus fed at different dietary protein and lipid levels. Aquacult Nutr. 2013;19(3):360–7.
44. Winston GW, Di Giulio RT. Prooxidant and antioxidant mechanisms in aquatic organisms. Aquat Toxicol. 1991;19(2):137–61.
45. Xiao N, Wang XC, Diao YF, Liu R, Tian KL. Effect of initial fluid resuscitation on subsequent treatment in uncontrolled hemorrhagic shock in rats. Shock. 2004;21(3):276–80.

46. Yi XW, Xu W, Zhou HH, Zhang YJ, Luo YW, Zhang WB, Mai KS. Effects of dietary astaxanthin and xanthophylls on the growth and skin pigmentation of large yellow croaker Larimichthys croceus. Aquaculture. 2014;433:377–83.
47. Zhang J, Li X, Leng X, Zhang C, Han Z, Zhang F. Effects of dietary astaxanthins on pigmentation of flesh and tissue antioxidation of rainbow trout (Oncorhynchus mykiss). Aquac Int. 2012;21(3):579–89.
48. Zhou CP, Ge XP, Niu J, Lin HZ, Huang Z, Tan XH. Effect of dietary carbohydrate levels on growth performance, body composition, intestinal and hepatic enzyme activities, and growth hormone gene expression of juvenile golden pompano. Trachinotus ovatus Aquaculture. 2015;437:390–7.

CHAPTER 7

Black Spot Infection in Juvenile Steelhead Trout Increases with Stream Temperature in Northern California

Cody J. Schaaf & Suzanne J. Kelson & Sébastien C. Nusslé & Stephanie M. Carlson

Department of Environmental Science, Policy, and Management, University of California, 130 Mulford Hall #3114, Berkeley, CA 94720-3114, USA

ABSTRACT

Climate change will increase water temperature in rivers and streams that provide critical habitat for imperiled species. Warmer water temperatures will influence the intensity and nature of biotic interactions, including parasitism. To better understand the factors influencing a neascus-type parasitic infection known as black spot disease, we examined the relationship between infection rate in juvenile steelhead trout (*Oncorhynchus mykiss*), abundance of another intermediate host (ramshorn snail, *Planorbella trivolvis*), and water temperature. We quantified infection patterns of trout at seven sites within the South Fork Eel River in northern California, visiting each site on three different occasions across the summer, and recording water temperature at each site. We also quantified infection patterns in trout captured from two tributaries to the South Fork Eel River. Overall, trout infection rates were highest in sites with the warmest temperatures. The abundance of ramshorn snails was positively related to both water temperature and black spot infection rates in juvenile trout. Both snail abundance and infection rates increased rapidly above a 23 °C daily maximum, suggesting a threshold effect at this temperature. We suggest that warmer temperatures are associated with environmental and biotic conditions that increase black spot disease prevalence in threatened steelhead trout. A comparison of our results with similar data collected from a more northern latitude suggests that salmonids in California may be warm-adapted in terms of their parasite susceptibility.

Keywords: Trematode Digenea Parasitism Neascus Salmonid *Oncorhynchus Planorbella trivolvis* Cold-water refuge

INTRODUCTION

Climate change is expected to increase aridity, temperature, and drought frequency in many parts of the world over the coming decades (Dai 2011, 2013; Trenberth et al. 2014), including western North America (Cook et al. 2004; Seager et al. 2007; Cook et al. 2015). In fact, the recent drought in California may have been the most severe drought experienced in the region in the past 1200 years (Griffin and Anchukaitis 2014), and several studies predict that drought severity and frequency will continue to increase in this region (Diffenbaugh et al. 2015; Shukla et al. 2015). Decreased stream flow coupled with increased air temperature during periods of drought has the potential to dramatically increase water temperatures in rivers (Mantua et al. 2010; van Vliet et al. 2011), with many potential consequences for riverine taxa and ecosystems.

Beyond the direct effects of temperature on individual performance, elevated stream temperatures can alter and, in some cases, intensify ecological interactions. Increased parasitism and disease with warmer temperatures has been observed across a range of freshwater taxa from invertebrates (e.g., *Biomphalaria glabrata*, Marcogliese 2008) to amphibians (e.g., *Rana boylii*, Kupferberg et al. 2009) to fishes (e.g., *Micropterus salmoides*,Marcogliese 2001). For example, proliferative kidney disease, which is caused by a myxosporean (Hedrick et al. 1993), can increase mortality of brown trout (*Salmo trutta*) in Swiss rivers by 70% when water temperatures exceed 15 °C for over two weeks (Borsuk et al. 2006).

Cold-water taxa including salmonid fishes may be particularly vulnerable to the effects of increased water temperature. Warm temperatures can have direct physiological effects on salmonids with negative fitness consequences. For example, warm water temperatures reduce the ability of Chinook salmon (*Oncorhynchus tshawytscha*) to successfully undergo smoltification (Marine and Cech 2004) and increase chronic hormonal stress in juvenile coho salmon (*O. kisutch*, Thomas et al. 1986). Moreover, salmonids that are physiologically stressed by warm water temperatures may be unable to handle further stressors like pollutants or pathogens (Richter and Kolmes 2005). For example, the dual effects of warmer temperatures – including reduced physiological performance and increased parasite outbreaks – may result in increased mortality (Udey et al. 1975; Ray et al. 2012). In 2002, for example, a bacterial pathogen that proliferated in low flows and elevated water temperature in the Klamath River induced a significant die-off of Chinook salmon (Lynch and Risley 2003).

Trematode flukes are freshwater parasitic flatworms that proliferate in warm temperatures (Poulin 2006; Paull and Johnson 2011). Several digenetic trematode species (e.g., *Uvulifer ambplopitis*, *Crassiphilia bulboglossa*) have metacercarial stages that induce a condition known as black spot disease in fish (Hoffman and Putz 1965; Berra and Au 1978; Harrison and Hadley 1982). These black spot-causing trematodes are often referred to as neascus-type or neascus-like in recent work (Ferguson et al. 2010; Markle et al. 2014). Many families of fish are vulnerable to neascus-induced black spot disease including Cyprinidae (Evans and Mackiewicz 1958; Mahon et al. 1979; Wisenden et al. 2012), Gasterosteidae (Fitzgerald et al. 1994; Karvonen et al. 2013), and Salmonidae (Cairns et al. 2005; Rodnick et al. 2008; Ferguson et al. 2010, 2012), among others. Neascus-type trematodes have a complex life cycle with multiple hosts and life stages. The first intermediate host is often a snail (e.g., *Planorbella* (aka *Helisoma*) *trivolvis*, in California streams, Turgeon et al. 1998). After emerging into the water column from the snail, infectious cercariae penetrate the skin of a fish, the second intermediate host, and encyst as metacercariae that produce a melanin-induced black spot around the raised, fibrous cyst (Berra and Au 1978; Teixeira-De Mello and Eguren 2008; Ferguson et al. 2010). The life cycle of the parasite is completed as infected fish are eaten by a belted kingfisher (*Megaceryle alcyon*), the definitive host, and the trematode carries out reproduction in the bird's intestine (Berra and Au 1978; Lane and Morris 2000). The trematode's eggs are shed with bird feces into the river where larvae hatch and infect a snail, continuing the cycle of infection.

Parasites and pathogens, in combination with increased temperatures, have been considered a potential factor in the decline of wild Pacific salmon populations (Miller et al. 2014). Moreover, previous studies have noted that parasites, including those that cause black spot disease, are more prevalent in coho salmon in warmer stream reaches of the Pacific Northwest (Cairns et al. 2005; Rodnick et al. 2008; Ferguson et al. 2010, 2012). However, the relationship between temperature and neascus-type infection prevalence has yet to be explored in salmonids at their more southern range in California, where salmonids are already on the brink of extinction (Katz et al. 2013).

To better understand the factors influencing black spot prevalence in salmonids near the southern end of their range, we examined the relationship between water temperature and black spot disease infection rates in juvenile steelhead trout (*O. mykiss*) in the South Fork Eel River in northern California. We also explored other variables that may influence infection rates, including intermediate host (i.e., snail) abundance and fish size. We predicted that black spot prevalence in *O. mykiss* would increase with water temperature, leading to spatial variation in infection rates between sites with different temperature histories. We also predicted that infection rates would increase as the summer progressed and water temperature and exposure history increased. We predicted that sites with high infection rates would have high abundance of snails (the first intermediate host). Finally, we predicted that infection rates would be higher in older/larger fish, which have had a longer period of exposure than younger/smaller fish.

MATERIALS AND METHODS

Study system

Our sampling sites were distributed along the mainstem of the South Fork Eel River (SFER) and two of its tributaries in Mendocino County, California, USA (Fig. 1), including sites in and around the UC-Berkeley Angelo Coast Range Reserve (39.733°N, 123.65°W). The SFER experiences Mediterranean seasonality characterized by wet winters and dry, hot summers (with daily summer air temperatures averaging 29 °C in the reserve, Mast and Clow 2000). The dry season extends from May through October and is associated with low river flows and warm water temperatures. The fish community in this region includes our focal organism, steelhead trout (*O. mykiss*), as well as threespine stickleback (*Gasterosteus aculeatus*), Sacramento sucker (*Catostomus occidentalis*), California roach (*Hesperoleucus symmetricus*), and non-native Sacramento pikeminnow (*Ptychocheilus grandis*). Steelhead trout from this region are part of the Northern California Coast Distinct Population Segment (DPS), which is currently listed as threatened under the U.S. Endangered Species Act.

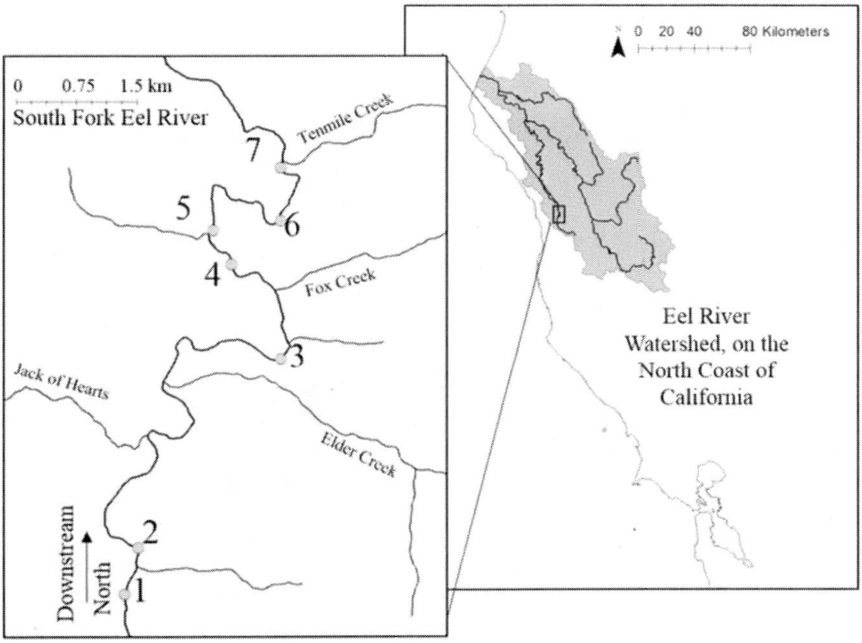

Figure. 1 Map of our study reach on the South Fork Eel River (SFER) in northern California. Numbered circles represent sample sites distributed along the mainstem SFER Our tributary sample sites on Elder and Fox creeks are labelled . The arrow pointing downstream/north represents the direction of water flow

Fish and snail sampling

From June 5 to August 21 2014, we sampled juvenile trout from a 9 km section of the SFER mainstem. We sampled seven sites, spaced from 1 to 2.8 km apart, chosen to represent a gradient of increasing water temperature from upstream to downstream (Fig. 1).

To assess individual fish for black spot infection, we captured young of year (YOY) trout (≤ 80 mm fork length) using a combination of seine and hand nets. At capture, each fish was weighed (±0.01 g), measured (fork length, mm), and assessed for black spot disease infection, always by the same author (CJS). Infection was confirmed via a tactile examination of the black spot to determine whether it was a natural marking or a metacercarial infection (infection spots are raised off the skin, Vaughn 1962; Ferguson et al. 2010). While the vast majority of metacercarial cysts were black, raised nodules that were yellow or brown (Fig. 2) were also considered infectious, since these likely represented new infections that had not yet produced a melanin pigment reaction (Ferguson et al. 2010; Markle et al. 2014).

Figure. 2 Photos showing large (panel **A**) and small (panel **B**) trout with evidence of black spot infection. Cysts on infected trout were small, fibrous, and rounded black spots that were raised off the skin. A few of the heavily infected fish had parasitic cysts that were yellow and brown due to developing infections (e.g., top photo)

We sampled fish at each of the seven mainstem sites on three different dates across the summer, resulting in a total capture of 556 steelhead. To explore infection rate in fish of different size classes, we used a backpack electrofisher (Smith-Root LR-24) to capture 36 additional larger steelhead (i.e., > 80 mm fork length) from our seven mainstem sampling sites during the last sampling effort. Due to length between sample sites (> 1 km), it is highly unlikely that any fish moved between sites during the study period (Edmundson et al. 1968; Keely 2000).

To compare infection rates in the SFER with its reliably cooler tributaries, we report black spot infection rates of 1426 trout that were captured in Elder and Fox creeks (Fig. 1). These fish were captured as part of another ongoing study via a backpack electrofisher between July 20 and Aug 17, 2014 in Elder Creek and July 10–11, 2014 in Fox Creek. Observations from these sites were not included in any statistical models.

To determine the abundance of intermediate snail hosts (*Planorbella trivolvis*), we surveyed snails once (on July 14, 2014) at all seven mainstem sites. We qualitatively observed that snail abundance was highest near the stream bank, where the snails graze algae. Consequently, we surveyed one random site of dimensions 3 m × 2 m at each site, with the long axis paralleling the bank. Within this area, we counted and collected all *P. trivolvis* encountered within a 15-min period, including those visible to the naked eye and those that were found via touch within algae-covered substrates.

Water temperature

At each study site, and in each tributary, we placed one thermistor (HOBO Pendant Temperature/Light Data Logger 64 K - UA-002-64) 10 cm from the substrate, which recorded water temperature at 10-min intervals between the first and last sampling events (6/5/2014–8/21/2014). We use a two-pronged approach to explore temperature variation through space and time.

First, we calculated the daily average temperature at each site to visualize continuous trends in temperature throughout the summer at each sampling site.

Next, we summarized water temperature data for each site and each sampling effort by averaging the maximum daily water temperature that was recorded during a seven-day timeframe including the sampling date (details below). We chose to use the average daily maximum (ADM) water temperature as a metric because it has been used previously in other studies exploring black spot infection rates (Cairns et al. 2005). This resulted in three ADM temperature metrics for each sample site, i.e., one for each of three sampling events that were used in analyses.

For our first sampling effort, we deployed thermistors on the date of the initial sampling. Thus, for the early summer ADM for each site, the seven-day average includes the initial sampling date and the six following days. During our mid-summer sampling event, the seven-day ADM was centered on our sampling date. Because temperature loggers were removed on our last sampling event, the late summer ADM included the six days leading up to and including the final sampling date.

ADM temperature at one site (site three on figures and table) was excluded from analyses because the data logger was exposed to unrealistically warm in-stream conditions due to severe water level drop at that site.

Statistical analyses

To determine the effect of water temperature (measured as the seven-day ADM) on black spot infection rate in YOY trout from our seven mainstem sampling sites, we used a mixed modeling approach, with infection status (infected/not infected) as the response variable (logistic regression). We chose to use mixed models so that temperature and sample date could be included as fixed effects, and site could be included as a random effect on the intercept to avoid pseudo-sampling (Lindstrom and Bates 1990). We tested the fit of three candidate models to determine the effects of temperature and sample date on the proportion of infected fish: 1) temperature as a fixed effect, 2) sample date as a fixed effect, and 3) both as fixed effects. Sample site was included as a random effect for all models. We used Akaike Information Criteria (AIC) (Akaike 1974) to choose between the three candidate models (Sakamoto et al. 1986). All analyses were performed in R (R Core Team 2016). Mixed models were fitted with the *glmer()* function from the package *lme4* (Bates et al. 2015).

To determine the relationship between snail count and water temperature, we performed exponential regressions on snail abundance, i.e., abundance as the response variable, and *e* to the power of temperature as the explanatory variable, this time using the 30-day ADM for each site. We used a longer temperature history for this analysis since there was only a single sampling event for the snails. The 30-day ADM was calculated by averaging the maximum temperature at each site for a 30-day period centered on the date of the snail abundance survey (July 14, 2014). To test the hypothesis that trout infection rates increase with snail abundance, we used a logistic regression with snail count as the independent variable and the proportion of infected steelhead at each site as the response variable, focusing on YOY trout and combining data from all seven mainstem sites from all three sampling events.

Finally, we tested whether infected trout were on average larger than non-infected trout using a Welch two-sample t-test accounting for heteroscedasticity. For this analysis, we only used fish captured in the final sampling event, to include the age-1 and older fish captured during this event.

RESULTS

Water temperature varied among sites and across the summer (Table 1, Fig. 3). In general, downstream sites were consistently warmer than upstream sites (Table 1, Fig. 3), and temperatures increased rapidly from June to July, after which temperature remain elevated and decreased slowly through our final sampling event in mid-August (Table 1, Fig. 3). Water temperatures in the two tributaries were considerably lower than all mainstem sites, ranging from 13 to 18 °C over the course of the summer. The seven-day ADM water temperature in the mainstem ranged from a minimum of 19.0 °C to a maximum 25.3 °C from June 5 to August 21, 2014 (Table 1).

Table 1 Summary of the 7-day average daily maximum temperature (ADM, °C) and black spot infection rates (%) in young of the year steelhead trout at each site and each sampling event (n = sample size). Sites are listed in order from upstream (Site 1) to downstream (Site 7)

Site	Date	ADM	% Infected	n
1	6/5/2014	20.9	0	18
	6/30/2014	20.5	8.6	35
	8/21/2014	21.5	33.3	18
2	6/18/2014	19.0	0	44
	7/7/2014	23.0	7.6	66
	8/21/2014	21.7	36.4	11
3	6/18/2014	20.4	0	28
	7/7/2014	25.2	52	25
	8/21/2014	22.4	100	19
4	6/18/2014	21.5	0	45
	6/30/2014	23.2	11.5	26
	8/21/2014	NA	92.9	28
5	6/23/2014	21.7	6.4	16
	7/13/2014	24.7	100	22
	8/21/2014	24.7	100	42
6	6/25/2014	24.1	76.5	17
	7/13/2014	25.3	100	13
	8/21/2014	24.8	100	11
7	6/30/2014	23.1	96.2	26
	7/13/2014	24.3	87.0	25
	8/21/2014	23.7	100	21

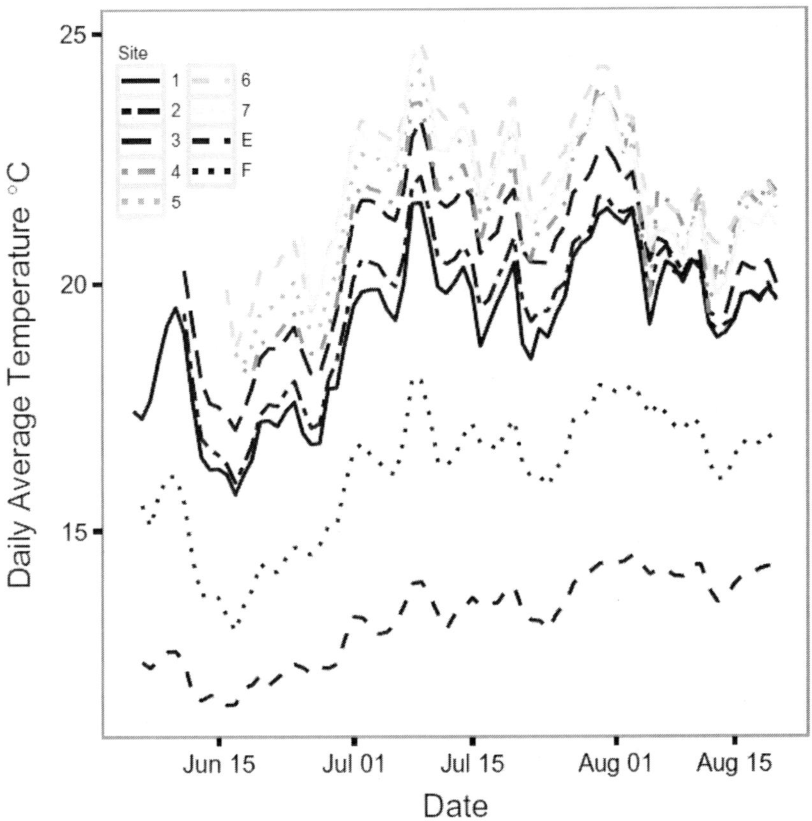

Figure. 3 Average daily water temperature at each site over the course of the summer. E and F refer to Elder Creek and Fox Creek, respectively. Refer to Fig. 1 for the location of sites 1–7 along the mainstem South Fork Eel River

Nearly half (249 of 556, 44.8%) of the juvenile trout that were sampled from the seven mainstem SFER sites showed evidence of black spot infection. In contrast, we observed minimal black spot infection (2 of 1426, < 0.1%) among trout sampled from the cool tributary sites in Elder and Fox creeks.

The mixed models of the seven mainstem sampling sites revealed that both temperature and sample date are important covariates in determining the proportion of infected steelhead at each site. Temperature was significant in the first (i.e., temperature [fixed] and site [random]) model ($z = 6.86$, $p < 0.01$, AIC = 178.4). Sample date was significant in the second (i.e., sample date [fixed] and site [random]) model ($z = 9.88$, $p < 0.01$,

AIC = 124.6). In the final model (temperature [fixed], sample date [fixed], and site), both fixed effects were statistically significant, and this model was associated the lowest AIC value (temperature: $z = 4.35, p < 0.01$, sample date: $z = 7.22, p < 0.01$, AIC = 98.0). From these comparisons, we base our interpretations on the final model, including both sample date and temperature as fixed effects, and site as a random effect.

Overall, the top-supported model suggests that temperature and sample date both influence observed infection rates, with infections accumulating over the course of the summer, and infection rates increasing with temperature. The model suggests that at 23.3 °C, 50% of fish will be infected with black spot disease. In general, trout infection rates tended to be higher at downstream sites (Figs. 4 and 5), reaching infection rates of 95–100%, where water temperatures were the highest (Fig. 3), in comparison to cooler sites, where infection rates never exceeded 40% (Table 1, Fig. 5). In addition, at every site, infection rates increased over the course of the study, peaking at the last sample event on August 21, 2014 (Table 1, Fig. 5).

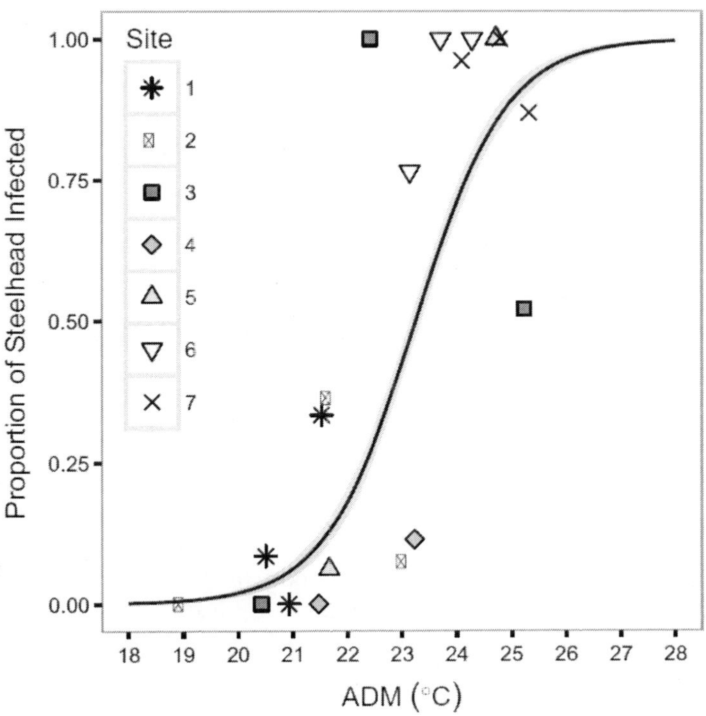

Figure. 4 Relationship between trout infection rates and average daily maximum (ADM) water temperature calculated over 7-day windows

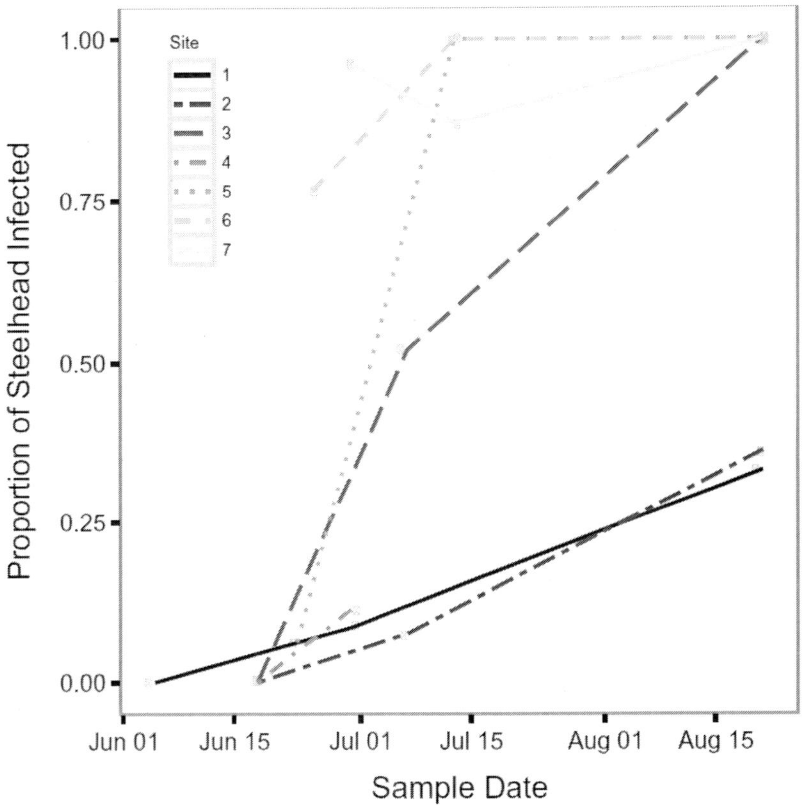

Figure. 5 Temporal trend in trout infection rates at each site over the study period, highlighting differences among early, middle, and late sampling events at each site. Refer to Fig. 1 for the location of sites 1–7 along the mainstem South Fork Eel River

The relationship between snail abundance at each site and water temperature (as measured by the 30-day ADM) was best described by an exponential relationship ($F_{1,5} = 51.6$, $p < 0.001$, r-squared =0.91). We found <10 snails per plot until water temperatures exceeded 23 °C (Fig. 6a). Snail counts were lowest at upstream (cool) mainstem sites ($n = 0$) and increased at the warmer downstream sites (Fig. 6a). The highest number of snails encountered in any given survey ($n = 41$) occurred at the furthest downstream site (i.e., site 7). The relationship between snail abundance and proportion of infected fish was best described by a logistical regression relationship ($z = 12.15$, $p < 0.001$). When there were 11 snails per plot, at least 50% of the steelhead at that site were infected with black spot disease.

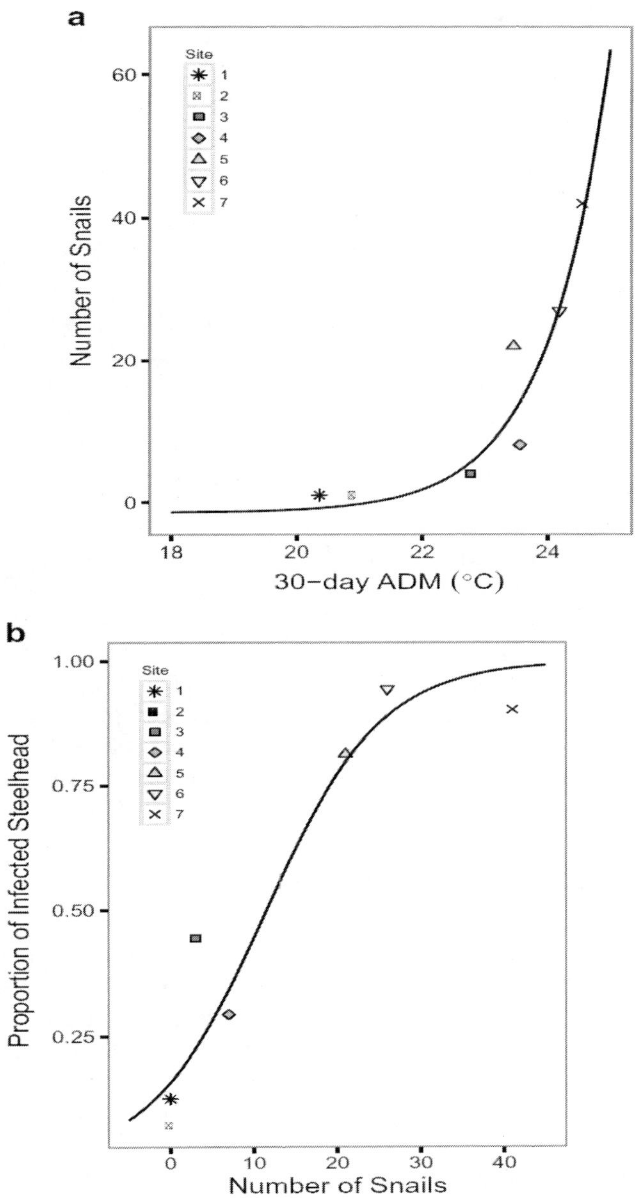

Figure. 6 Relationship between snail counts and average daily maximum temperature (ADM) (panel **A**) calculated over 7-day windows, and between trout infection rate and snail counts (panel **B**)

Finally, we found that the mean size of infected fish was significantly larger than the mean size of not-infected fish (infected: 72.8 ± 23.3 mm, not-infected: 64.6 ± 13.4 mm, $t_{43} = 2.24$, $p = 0.03$).

DISCUSSION

Black spot infection rates in juvenile steelhead trout increased with maximum water temperature in the South Fork Eel River watershed of northern California (Fig. 4). Moreover, spatiotemporal variation in water temperature was associated with spatiotemporal variation in blackspot infection rates. Black spot infection rates were highest in juvenile steelhead trout occupying warmer downstream sites than cooler upstream sites, and infection rates increased over the course of the summer season at every site, possibly due to increased exposure or increased cercarial releases.

We found that infected fish tended to be larger, a pattern that is commonly seen in other parasitized salmonids (e.g., *Salvelinus fontinalis*, Poulin et al. 1991; *O. kisutch*, Barndt and Stone 2003). One possible reason is that larger fish are usually older and, thus, have had a longer cumulative period of exposure to parasites compared to smaller, younger fish (Berra and Au 1978). In fact, metacercarial cysts can persist for up to 4.5 years in some fish species (Hoffman and Putz 1965), suggesting that some of the black spots detected on age-1 and older fish may reflect infections from previous years.

Spatial and temporal variation in infection rates

At the South Fork Eel River, trout captured from downstream sites had the highest infection rates throughout the summer (Table 1, Fig. 5). These same sites were associated with the highest abundance of intermediate-host ramshorn snails (Fig. 6). Since infectious trematode cercariae survive for less than 24 h (Pechenik and Fried 1995), they would likely not influence infection rates at downstream sites by being carried downstream in low flows. Thus, snail host abundance is the most obvious factor that would influence local infection. Increased abundance of intermediate host snails within certain stream reaches may encourage growth of the trematode population, increasing local black spot infection in fish. The proliferation of snails in downstream sites may be linked to features of these sites. Previous work has shown that the abundance of aquatic grazers like snails increases steadily with an increasing biomass of epiphytes and macroalgae (Cattaneo 1983). In the South Fork Eel River, increased solar radiation in wide, downstream channels can cause proliferations of macroalage (Power et al. 2008, 2013), which may explain increased snail abundance at downstream sites.

Infection rates in cool tributaries were considerably lower than in mainstem sites (<1% versus 44.8%). A similar pattern - of lower infection rates in cooler waters - was also observed in other salmonid species (Cairns et al. 2005; Ferguson et al. 2010, 2012) and a cyprinid (Mahon et al. 1979). The tributary sites are shaded and cooler due to groundwater inputs, and lower productivity may limit algal growth (Power et al. 2013) and snail establishment, both of which would impede parasite transmission to trout residing in tributary habitat. Shaded tributary habitats may represent critical refugia for salmonids at the southern end of their range, both because of the cooler temperatures but also by providing escape from warm-water parasites and predators, both of which are more common in the warmer mainstem waters of the South Fork Eel River. Moreover, models predict that such tributary sites are less likely to warm as a result of climate change (Daigle et al. 2015) possibly due to their tendency to be shaded and fed by cool groundwater, which further highlights their importance for conserving salmonid fishes in a warming world.

Black spot infection rates increased with sample date at our South Fork Eel River sampling sites, suggesting that infection increases with exposure history. That both time and temperature are significant in our best-supported model suggests that both are important. However, the large variation in infection rates among sites early in the summer suggests that other (unmeasured) variables could also be important for explaining spatiotemporal variation in infection patterns – e.g., water flow, parasite dispersal, and abundance of other fish hosts.

Water temperature and trematode activity

We found that black spot infection rates increased with water temperature (Fig. 4) in the mainstem South Fork Eel River, as did black spot infection in coho salmon from an Oregon watershed (Cairns et al. 2005; Rodnick et al. 2008). Previous studies have revealed that warm temperatures trigger the growth and release of trematode cercariae from snails (Ataev 1991; Poulin 2006). In one trematode (*Ribeiroia ondatrae*), development was four times faster at 26 °C than 17 °C (Paull and Johnson 2011). In a trematode that commonly infects *O. mykiss*(*Diplostomum spathaceum*), cercariae were shed from snails more than five times faster at 20 °C than 10 °C (Lyholt and Buchmann 1996). We postulate that infection rates increased over the summer because rising temperatures in June and July accelerated trematode development and cercarial releases. Another possibility is that the cumulative period of exposure increases over the summer.

Fitness and health effects of black spot disease

Previous studies on black spot disease have revealed assorted information about the extent to which the disease can negatively impact infected fish. Research by Ferguson et al. (2010) revealed lower overwinter survival in coho salmon heavily infected with *Apophallus sp.* and neascus-type metacercariae. Other freshwater fish species show increased mortality when infected with neascus-type metacercariae, possibly due to increased predation susceptibility (*Esox lucius*, Harrison and Hadley 1982; *Chasmistes brevirostris*, Markle et al. 2014). Reported sub-lethal impacts of neascus-type infection vary greatly from no effect on salmonid metabolism and growth (Ferguson et al. 2012) to significant weight loss at high levels of parasite load in Centrarchids (e.g., *Micropterus dolomieu*, Hunter and Hamilton 1941; *Lepomis macrochirus*, Lemly and Esch 1984). In our study, heavily parasitized steelhead felt rough to the touch (CJS, personal observation), suggesting a possible loss of streamlining and increased swimming burden. Increased swimming burden due to ectoparasites has been observed previously for marine fish (Binning et al. 2012). Future studies on the fitness effects of the parasitic condition in steelhead are warranted given the potential sub-lethal and lethal effects of black spot disease reported in other species. However, Ferguson et al. (2012) stress the importance of examining more than one type of parasite when assessing fitness impacts and mortality.

Temperature thresholds

As in our study, black spot infection rates also increased with stream temperature in coho salmon in the West Fork Smith River, Oregon, USA (Cairns et al. 2005; Rodnick et al. 2008). However, one important difference emerged between the two studies: the threshold temperature at which infection rates increased was *lower* for the Oregon (West Fork Smith River) population compared to the California (South Fork Eel River) population. Specifically, Cairns et al. report over 50% infection in coho salmon from Oregon at 21 °C ADM, while we found that the 50% threshold was not reached until 23.3 °C ADM for steelhead trout in California (Fig. 4).

There are several non-mutually exclusive explanations for the observed geographical differences in temperature thresholds for black spot infection rates in salmonid fishes. One possibility is that the threshold for infection differs for different species of salmonid fishes, due to differences in their physiology or habitat use. Another intriguing possibility is local adaptation to warmer conditions at more southern latitudes. Overall, the South Fork Eel River, California experiences warmer temperatures than the West Fork Smith River, Oregon (average seven-day maximum of 22.8 °C in the SFER (this study) vs 19.3 °C (Rodnick et al. 2008)), suggesting that the more southern salmonids experience warmer temperatures overall may have adjusted to their thermal habitat through local adaptation and increased resistance to parasites. Such local adaptation in salmonids to warm waters is not without precedent (e.g., Verhile et al. 2016). This possibility of local adaptation to thermal regime could provide another reason to conserve salmonids at the southern end of their range as a valuable genetic reservoir for persisting in a changing world (Katz et al. 2013).

Uncertainty regarding the neascus-type trematode in our study

Although we did not collect and taxonomically identify the black spot-causing trematode in our study from any host (fish or snail), we believe that infection was likely caused by a neascus- type trematode. The parasitic cysts were small, fibrous and raised off the skin upon tactile examination (Fig. 2), characteristics of neascus-type trematodes, rather than other black spot-inducing trematodes (e.g., *Apophallus sp.*) (Vaughn 1962; Ferguson et al. 2010; V. Tkach, pers. comm.; J. Ferguson pers. comm.). Two possible trematodes include *Uvulifer amblopitis* and *Crassiphilia bulboglossa*, both of which use *P. trivolvis*, the snail that is very abundant in our study sites, as an intermediate host (Hoffman and Putz 1965; Berra and Au 1978; V. Tkach, pers. comm.).

However, it is possible that non-neascus-type trematodes were responsible for some of the observed infections, since observation of metacercarial cysts under a microscope is the only true way to identify trematode-specific infections (Ferguson et al. 2010). Further investigation into the exact trematode species and corresponding intermediate snail host involved in the infection cycle is needed to confirm linkages between snail and infection data, and would greatly improve our understanding of the causes of black spot infection in fishes, including imperiled salmonid fishes.

Broader implications and conclusions

In summary, we found that the prevalence of black spot disease in juvenile steelhead trout can be high in a large northern California river, but that disease prevalence varies through space and time. Importantly, we found that infection rates increased rapidly once water temperatures exceeded 23 °C ADM, a threshold most likely to be exceeded in sunny, warm mainstem reaches. As climate changes and stream temperatures warm, it is possible that other populations inhabiting sunny river reaches may experience increased vulnerability to black spot disease, highlighting the importance of shaded and cool tributary habitats for sustaining salmonids in a changing climate.

ACKNOWLEDGEMENTS

We thank the National Science Foundation for the Eel River Critical Zone Observatory financial support (CZP EAR-1331940), including an REU that supported CJZ (EAR-1443164), and a Graduate Research Fellowship to SJK (NSF 201517964). Additionally, we thank Tina Mendez, Kurt Spreyer, and Mary Power for feedback on an earlier version of the manuscript, and Katie Kobayashi for assistance in the field. We also thank Vasyl Tkach and Jayde Ferguson for providing expertise in parasite identification. The research activities presented in this manuscript were carried out with approval from the University of California Berkeley Animal Care and Use Committee.

REFERENCES

1. Akaike H (1974) A new look at the statistical model identification. IEEE Trans Autom Control 19:716–723. doi:10.1109/TAC.1974.1100705
2. Ataev G (1991) Temperature influence on the development and biology of rediae and cercariae of *Philophthalmus rhionica* (Trematoda). Parazitologiya 25:349–359
3. Barndt S, Stone J (2003) Infestation of *Salmincola californiensis* (Copepoda: Lernaeopodidae) in wild coho salmon, steelhead, and coastal cutthroat trout juveniles in a small Columbia River tributary. Trans Am Fish Soc 132:1027–1032. doi:10.1577/T02-122
4. Bates D, Maechler M, Bolker B, Walker S (2015) Fitting linear mixed-effect models using lme4. J Stat Softw 67:1–48. doi:10.18637/jss.v067.i01
5. Berra TM, Au R-J (1978) Incidence of black spot disease in fishes in Cedar Fork Creek, Ohio. Ohio J Sci 78:318–322
6. Binning SA, Roche DG, Layton C (2012) Ectoparasites increase swimming costs in a coral reef fish. Biol Lett 9:20120927. doi:10.1098/rsbl.2012.0927
7. Borsuk ME, Reichert P, Peter A et al (2006) Assessing the decline of brown trout (*Salmo trutta*) in Swiss rivers using a Bayesian probability network. Ecol Model 192:224–244. doi:10.1016/j.ecolmodel.2005.07.006
8. Cairns MA, Ebersole JL, Baker JP et al (2005) Influence of summer stream temperatures on black spot infestation of juvenile coho salmon in the Oregon Coast range. Trans Am Fish Soc 134:1471–1479. doi:10.1577/T04-151.1
9. Cattaneo A (1983) Grazing on epiphytes. Limnol Oceanogr 28:124–132. doi:10.4319/lo.1983.28.1.0124
10. Cook ER, Woodhouse CA, Eakin CM et al (2004) Long-term aridity changes in the western United States. Science 306:1015–1018
11. Cook BI, Ault TR, Smerdon JE (2015) Unprecedented 21st century drought risk in the American Southwest and Central Plains. Sci Adv 1:e1400082

12. Dai A (2011) Drought under global warming: a review. Wiley Interdiscip Rev Clim Chang 2:45–65. doi:10.1002/wcc.81
13. Dai A (2013) Increasing drought under global warming in observations and models. Nat Clim Chang 3:52–58
14. Daigle A, Jeong D Il, Lapointe MF (2015) Climate change and resilience of tributary thermal refugia for salmonids in eastern Canadian rivers. Hydrol Sci J 60:1044–1063. doi: 10.1080/02626667.2014.898121
15. Diffenbaugh NS, Swain DL, Touma D (2015) Anthropogenic warming has increased drought risk in California. Proc Natl Acad Sci 112:3931–3936. doi:10.1073/pnas.1422385112
16. Edmundson E, Everest FE, Chapman DW (1968) Permanence of station in juvenile Chinook salmon and steelhead trout. J Fish Res Bd Can 25:1453–1464
17. Evans HE, Mackiewicz JS (1958) The incidence and location of metacercarial cysts (Trematoda: Strigeida) on 35 species of central New York fishes. J Parasitol 44:231–235. doi:10.2307/3274708
18. Ferguson JA, Schreck CB, Chitwood R, Kent ML (2010) Persistence of infection by metacercariae of *Apophallus sp.*, *Neascus sp.*, and *Nanophyetus salmincola* plus two myxozoans (*Myxobolus insidiosus* and *Myxobolus fryeri*) in coho salmon *Oncorhynchus kisutch*. J Parasitol 96:340–347. doi:10.1645/GE-2289.1
19. Ferguson JA, Romer J, Sifneos JC et al (2012) Impacts of multispecies parasitism on juvenile coho salmon (*Oncorhynchus kisutch*) in Oregon. Aquaculture 362–363:184–192. doi:10.1016/j.aquaculture.2011.07.003
20. Fitzgerald GJ, Fournier M, Morrissette J (1994) Sexual selection in an anadromous population of threespine sticklebacks --- no role for parasites. Evol Ecol 8:348–356. doi:10.1007/BF01238187
21. Griffin D, Anchukaitis KJ (2014) How unusual is the 2012–2014 California drought? Geophys Res Lett 41:9017–9023. doi:10.1002/2014GL062433.1
22. Harrison EJ, Hadley WF (1982) Possible effects of black-spot disease on northern pike. Trans Am Fish Soc 111:106–109. doi:10.1577/1548-8659(1982)111<106:PEOBDO>2.0.CO;2
23. Hedrick RP, MacConnell E, de Kinkelin P (1993) Proliferative kidney disease of salmonid fish. Annu Rev Fish Dis 3:277–290. doi:10.1016/0959-8030(93)90039-E
24. Hoffman GL, Putz RE (1965) The black-spot (*Uvulifer ambloplitis*: Trematoda: Strigeoidea) of centrarchid fishes. Trans Am Fish Soc 94:143–151. doi:10.1577/1548-8659(1965)94[143:TBUASO]2.0.CO;2
25. Hunter GW, Hamilton JM (1941) Studies on host-parasite reactions to larval parasites. IV. The cyst of *Uvulifer ambloplitis*. Trans Am Microsc Soc 60:498–507
26. Karvonen A, Kristjansson BK, Skulason S et al (2013) Water temperature, not fish morph, determines parasite infections of sympatric Icelandic threespine sticklebacks (*Gasterosteus aculeatus*). Ecol Evol 3:1507–1517. doi:10.1002/ece3.568
27. Katz J, Moyle PB, Quiñones RM et al (2013) Impending extinction of salmon, steelhead, and trout (Salmonidae) in California. Environ Biol Fish 96:1169–1186. doi:10.1007/s10641-012-9974-8
28. Keely ER (2000) An experimental analysis of territory size in juvenile steelhead trout. Anim Behav 59:477–490
29. Kupferberg SJ, Catenazzi A, Lunde K et al (2009) Parasitic copepod (*Lernaea cyprinacea*) outbreaks in foothill yellow-legged frogs (*Rana boylii*) linked to unusually warm summers and amphibian malformations in northern California. Copeia 2009:529–537. doi:10.1643/CH-08-011
30. Lane RL, Morris JE (2000) Biology, prevention, and effects of common grubs (digenetic trematodes) in freshwater fish. NCRAC Tech Bull Series 115
31. Lemly AD, Esch GW (1984) Effects of the trematode *Uvulifer ambloplitis* on juvenile bluegill sunfish, *Lepomis macrochirus*: ecological implications. J Parasitol 70:475–492. doi:10.2307/3281395

32. Lindstrom MJ, Bates D (1990) Nonlinear mixed effects models for repeated measures data. Biometrics 46:673–687
33. Lyholt HCK, Buchmann K (1996) *Diplostomum spathaceum*: effects of temperature and light on cercarial shedding and infection of rainbow trout. Dis Aquat Org 25:169–173
34. Lynch DD, Risley JC (2003) Klamath River Basin hydrologic conditions prior to the September 2002 die-off of salmon and steelhead: U.S. Geological Survey Water-Resources Investigations Report 03–4099
35. Mahon R, Balon EK, Noakles DLG (1979) Distribution, community structure and production of fishes in the upper speed river, Ontario: a preimpoundment study. Environ Biol Fish 4:219–244. doi:10.1007/BF00005480
36. Mantua N, Tohver I, Hamlet A (2010) Climate change impacts on streamflow extremes and summertime stream temperature and their possible consequences for freshwater salmon habitat in Washington State. Clim Chang 102:187–223. doi:10.1007/s10584-010-9845-2
37. Marcogliese DJ (2001) Implications of climate change for parasitism of animals in the aquatic environment. Can J Zool 79:1331–1352. doi:10.1139/z01-067
38. Marcogliese DJ (2008) The impact of climate change on the parasites and infectious diseases of aquatic animals. Rev Sci Tech 27:467–484
39. Marine KR, Cech JJ (2004) Effects of high water temperature on growth, smoltification, and predator avoidance in juvenile Sacramento River chinook salmon. North Am J Fish Manag 24:198–210. doi:10.1577/M02-142
40. Markle DF, Terwilliger MR, Simon DC (2014) Estimates of daily mortality from a neascus trematode in age-0 shortnose sucker (*Chasmistes brevirostris*) and the potential impact of avian predation. Environ Biol Fish 97:197–207. doi:10.1007/s10641-013-0141-7
41. Mast MA, Clow DW (2000) Environmental characteristics and water quality of hydrologic benchmark network stations in the Western United States, 1963–95. U.S. Geological Survey Circular 1173-D
42. Miller KM, Teffer A, Tucker S et al (2014) Infectious disease, shifting climates, and opportunistic predators: cumulative factors potentially impacting wild salmon declines. Evol Appl 7:812–855. doi:10.1111/eva.12164
43. Paull SH, Johnson PTJ (2011) High temperature enhances host pathology in a snail–trematode system: possible consequences of climate change for the emergence of disease. Freshw Biol 56:767–778. doi:10.1111/j.1365-2427.2010.02547.x
44. Pechenik JA, Fried B (1995) Effect of temperature on survival and infectivity of *Echinostoma trivolvis*: a test of the energy limitation hypothesis. Parasitology 111:373–378
45. Poulin R (2006) Global warming and temperature-mediated increases in cercarial emergence in trematode parasites. Parasitology 132:143–151. doi:10.1017/S0031182005008693
46. Poulin R, Curtis MA, Rau ME (1991) Size, behaviour, and acquisition of ectoparasitic copepods by brook trout, *Salvelinus fontinalis*. Oikos 2:169–174. doi:10.2307/3545334
47. Power ME, Parker MS, Dietrich WE (2008) Seasonal reassembly of a river food web: floods, droughts, and impacts of fish. Ecol Monogr 78:263–282. doi:10.1890/06-0902.1
48. Power ME, Holomuzki JR, Lowe RL (2013) Food webs in Mediterranean rivers. Hydrobiologia 719:119–136. doi:10.1007/s10750-013-1510-0
49. R Core Team (2016) R: A language and environment for statistical computing
50. Ray RA, Holt RA, Bartholomew JL (2012) Relationship between temperature and *Ceratomyxa shasta* -induced mortality in Klamath River salmonids. J Parasitol 98:520–526. doi:10.1645/JP-GE-2737.1
51. Richter A, Kolmes SA (2005) Maximum temperature limits for chinook, coho, and chum salmon, and steelhead trout in the Pacific Northwest. Rev Fish Sci 13:23–49. doi:10.1080/10641260590885861

52. Rodnick KJ, St.-Hilaire S, Battiprolu PK et al (2008) Habitat selection influences sex distribution, morphology, tissue biochemistry, and parasite load of juvenile coho salmon in the West Fork Smith River, Oregon. Trans Am Fish Soc 137:1571–1590. doi:10.1577/T07-138.1
53. Sakamoto Y, Ishiguro M, Kitagawa G (1986) Akaike information criterion statistics. Reidel, Dordrecht
54. Seager R, Ting M, Held I et al (2007) Model projections of an imminent transition to a more arid climate in southwestern North America. Science 316:1181–1184. doi:10.1126/science.1139601
55. Shukla S, Safeeq M, AghaKouchak A et al (2015) Temperature impacts on the water year 2014 drought in California. Geophys Res Lett 42:4384–4393. doi:10.1002/2015GL063666
56. Teixeira-De Mello F, Eguren G (2008) Prevalence and intensity of black-spot disease in fish community from a subtropical stream (Santa Lucía river basin, Uruguay). Limnetica 27:251–258
57. Thomas RE, Gharrett JA, Carls MG et al (1986) Effects of fluctuating temperature on mortality, stress, and energy reserves of juvenile coho salmon. Trans Am Fish Soc 115:52–59. doi:10.1577/1548-8659(1986)115<52:EOFTOM>2.0.CO;2
58. Trenberth KE, Dai A, van der Schrier G et al (2014) Global warming and changes in drought. Nat Clim Chang 4:17–22
59. Turgeon DD, Quinn Jr JF, Bogan AE, et al (1998) Common and scientific names of aquatic invertebrates from the United States and Canada: Mollusks Second edition Am Fish Soc Spec Publ 26
60. Udey LR, Fryer JL, Pilcher KS (1975) Relation of water temperature to ceratomyxosis in rainbow trout (*Salmo gairdneri*) and coho salmon (*Oncorhynchus kisutch*). J Fish Res Board Can 32:1545–1551. doi:10.1139/f75-181
61. van Vliet MTH, Ludwig F, Zwolsman JJG et al (2011) Global river temperatures and sensitivity to atmospheric warming and changes in river flow. Water Resour Res 47:W02544. doi:10.1029/2010WR009198
62. Vaughn LR (1962) The incidence and location of metacercarial cysts (*Neascus spp.*, in "black spot") on species of fishes in Clay County, Missouri. Beta Beta Beta Biol Soc 33:216–220
63. Verhile CE, English KK, Cocherell DE, Farrell AP, Fangue NA (2016) High thermal tolerance of a rainbow trout populaiton near its southern range limit suggests local thermal adjustment. Conserv Physiol 4:1–12. doi:10.1093/conphys/cow057
64. Wisenden BD, Martinez-Marquez JY, Gracia ES, McEwen DC (2012) High intensity and prevalence of two species of trematode metacercariae in the fathead minnow (*Pimephales promelas*) with no compromise of minnow anti-predator competence. J Parasitol 98:722–727. doi:10.1645/GE-2454.1

CHAPTER 8

Trawl Selectivity in the Barents Sea Demersal Fishery

Eduardo Grimaldo, Manu Sistiaga, Bent Herrmann and Roger B. Larsen

[1] *SINTEF Fisheries and Aquaculture, Trondheim, Norway*

[2] *SINTEF Fisheries and Aquaculture, North Sea Science Park, Hirtshals, Denmark*

[3] *Norwegian College of Fisheries and Aquatic Sciences, University of Tromsø, Tromsø, Norway*

ABSTRACT

This chapter provides a general overview of the Barents Sea demersal trawl fishery. First, it reviews historical catch levels and current biomass status of four commercially important demersal species (cod, haddock, Greenland halibut, and redfish) and includes an overview of their management plan that has been carried out by the Joint Norwegian–Russian commission. Then, it presents the evolution of the technical regulations for improving size selectivity in this fishery and describes current challenges in gear selectivity. Later, this chapter describes the concept of size selectivity, introduces the selective parameters that define a selection curve, and progressively introduces different parametric models that describe the selection process. The most common experimental methods and gear used to collect selectivity data are described, and their advantages and disadvantages are discussed. Finally, this chapter describes an alternative, or a complementary method, to the conventional estimation of trawl selectivity—the FISHSELECT method. This method is based on morphology measurements and fish penetration models to estimate the selective properties of different mesh shapes and sizes at different mesh openings, which are later used to provide simulation-based prediction of size selectivity. FISHSELECT has already been applied to four important species of the Barents Sea Demersal Fishery, and the results have in all cases showed to be coherent with the results obtained from sea trial results.

Keywords: selectivity, trawl, bottom trawl, Barents Sea, demersal fishery, FISHSELECT

1. INTRODUCTION

The main target species for the Norwegian bottom trawling fleet north of 62° N are the Northeast Arctic cod (*Gadus morhua*), haddock (*Melanugrammus aeglefinus*), and saithe (*Pollachius virens*); all of these species belong to the gadoid family. During the past 20 years, the catches of cod have averaged 625 649 tonnes per year, with the largest landing being in 2014 (986 449 tonnes) [1]. During the same period, the catches of haddock and saithe averaged, respectively, 165 222 and 162 578 tonnes per year, with the largest landing for haddock registered in 2012 (315 627 tonnes) and for saithe in 2006 (212 822 tonnes) [2, 3]. Of these catches, approximately 83% of cod, 90% of haddock, and 97% of saithe were harvested by Norwegian and Russian vessels; the rest was caught by vessels from Iceland, Greenland, and the EU [1–3]. In 2014, nearly 70% of the total catch of cod, 76% of haddock, and 49% of saithe were fished by bottom trawlers. The rest of the quota for these species was fished by gillnetters, longliners, jiggers, and demersal seiners [1–3]. Other commercially important species, mostly caught as bycatch in this fishery, are redfish (*Sebastes marinus*) and Greenland halibut (*Reinhardtius hippoglossoides*).

A key strategy of the Norwegian–Russian fishing legislation is to minimize the capture of undersized fish. When more than 15% of the catch (by numbers) is undersized, the fishery is closed. Moreover, the discard of dead or dying fish from regulated species is forbidden in this (Norway–Russian) fishery. The aim with size selection is to reduce the capture of undersized fish while also reducing the loss of fish above the minimum legal size (MLS). To achieve this goal, the use of sorting grids with a minimum bar spacing of 55 mm is currently mandatory in all trawls fishing for gadoids in this fishery.

Assessments of the Joint Norwegian–Russian Fisheries Commission show that the Northeast Arctic cod stock is in a fairly good state. The spawning biomass in 2013 was estimated at 2 million tonnes, while the total stock biomass was estimated to be around 3.5 million tonnes [1]. The International Council for the Exploration of the Sea (ICES), which is a global organization that develops science and advice to support the sustainable use of the oceans, advises the Joint Russian–Norwegian Fisheries Commission to set a total allowable catch (TAC) of 805 000 tonnes in 2016 [1].

The Northeast Arctic haddock and saithe stocks are considered to be in good condition. ICES advises that when the Joint Russian–Norwegian Fisheries Commission management plan is applied, landings in 2016 should not exceed 223 000 tonnes for haddock and 140 000 tonnes for saithe [2, 3].

2. TECHNICAL REGULATIONS FOR IMPROVING SIZE SELECTIVITY

In 1979, the Northeast Atlantic Fisheries Commission (NEAFC) agreed to increase the mesh size of codends made of cotton, polyamide, and polyester from 120 to 125 mm. As of 1 January 1980, the minimum legal size of cod and haddock that could be kept was set at 39 and 35 cm, respectively; in addition, the catches of undersized fish could not exceed 15% (by weight) of the catch [4]. In 1982, the NEAFC increased the commercial sizes of cod and haddock to 42 and 39 cm, respectively, and established that the bycatch of undersized fish should not exceed 15% (by numbers) of the catch. Norway also unilaterally increased the minimum mesh size of trawl codends to 135 mm regardless of the trawl material [4]. In 1983, a discard ban was introduced into the fishery [5]; it stated that vessels were obliged to land also the catches of all undersized fish. A surveillance program established during 1983–1984 showed that substantial areas need to be closed to the commercial fishery due to the high proportions of undersized cod and haddock in the catch. Typical catches in these areas could contain 30–50% of juvenile fish (by numbers). Therefore, the temporary closure of certain fishing areas led to a less efficient fishery because the fleet had to search for other areas to fish on [6].

Between 1985 and 1991, researchers in Norway conducted selectivity experiments that aimed at improving size selectivity of bottom trawl codends. These experiments considered different mesh configurations, round straps, different top and bottom panels, codend twine materials, and sorting grid systems [7, 8]. The methods found to be most practical were the shortened lastridge ropes (12–15% reduction) and the escape grids (Sort-X system). The shortened lastridge ropes gave similar size selection estimates as square mesh codends. However, square mesh codends were associated with meshing of fish, and especially redfish, and maneuverability challenges on deck [9].

Even though the Sort-X sorting grid represented a revolutionary method for improving selectivity in trawl codends, in 1990 the Norwegian Directorate of Fisheries, in 1990, increased the minimum size of cod to 47 cm and of haddock to 44 cm. According to this institution, an alternative mean of resource management had to be enforced in the fishery because the development of technical codend modifications had still not given the expected results [10].

Between 1990 and 1996, experiments with sorting grids that aimed at reducing the catch of young fish were conducted in Norway and Russia [11, 12]. In 1995, the Joint Russian–Norwegian selectivity experiments on cod in the Barents Sea proved that the selectivity of the Russian single-steel grid system “Sort-V” and that of the double steel grid system “Sort-X” were similar [12]. Based on the results of these experiments, as of 1 January 1997 both types of grids with 55-mm interbar space began to be used in cod and haddock fisheries in some limited areas of the Barents and Norwegian Seas (Figure 1). Later, the Institute of Marine Research of Norway satisfactorily adapted the Sort-V system to Norwegian trawlers [13, 14]. The new system, which is known as the Single grid and only includes some slight changes from the Sort-V system, was legalized by the Norwegian Directorate of Fisheries in 2000.

Since their introduction, the Sort-X, Sort-V, and Single grid have been associated with crew safety problems, especially when handling the grid sections in bad weather. During the late 1990s, experiments with grids made of other materials, such as plastics, fibreglass, nylon, and rubber, were promoted. The intention behind these experiments was to offer the industry grids that could be more easily handled, were more user-friendly, and were cheaper to purchase [15]. The systems studied were the plastic Sort-X, the Eurogrid, and the Flexigrid. The plastic Sort-X was made of high-density polyethylene (HDPE) materials and weighed considerably less than the steel version. The Eurogrid was made of massive nylon and was developed primarily for the trawl fishery in the North Sea [16]. Although these grids were designed to be more user-friendly alternatives to other grids on the market because of their low weight, flexibility, and the possibility of storing them in net drums, none of them was successfully introduced into the fishery. During three years of experimental work, the Flexigrid (made of Polyamide (PA) bars and rubber frames) (Figure 1) proved to have a selection capacity similar to that of the Sort-X [17–20]. This new grid system was lighter, smaller, more flexible, and therefore easier to handle on deck. Thus, the Flexigrid system was legalized in the beginning of 2002.

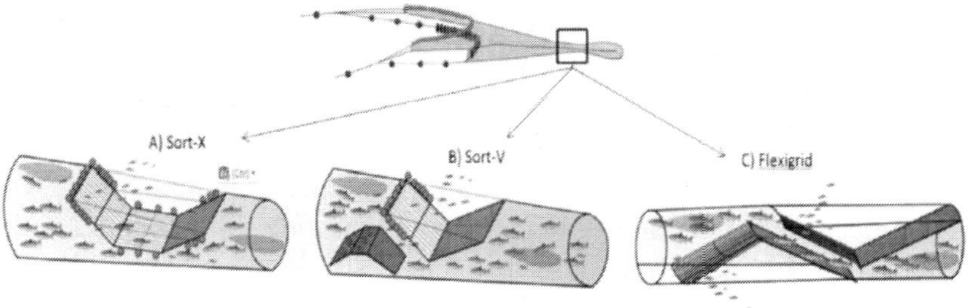

FIGURE 1. Mandatory sorting grids in the Barents Sea Demersal Fishery: (A) Sort-X, (B) Sort-V, and (C) Flexigrid.

Between 2004 and 2008, further experiments with escape panels were performed in the Barents Sea [21]. Despite the results showing similar selective performance to mandatory sorting grids [22, 23], escape panels were not considered as an alternative selection device for the Barents Sea Demersal Fishery.

Today, all four sorting grids described above are legal with a minimum grid bar spacing of 55 mm: the Sort-X, Sort-V, Single grid, and Flexigrid. The Sort-V and the Single grid system are still used by some vessels, but the system employed by the majority of vessels today is the Flexigrid. The two main reasons for fishermen to prefer the flexigrid system are that (a) it is a safer grid to use on deck, especially in bad weather, and (b) the retention of fish above the minimum size for cod and haddock is higher for the flexigrid than for the Sort-V/Single grid [24]. This second argument became especially important when in 2011 the minimum legal size was reduced from 47 to 44 cm for cod and from 44 to 40 cm for haddock. This reduction in minimum size resulted from the Russian–Norwegian joint meetings and the aim of standardizing the regulations for the Barents Sea. This reduction in minimum sizes was also accompanied by a reduction in the minimum codend mesh size. Thus, in addition to the grid, trawlers can today use codends with a minimum diamond mesh size of 130 mm.

The rapid increase in the biomass of Atlantic cod in the period 2008–2014 has brought some unusual challenges for the Barents Sea trawlers, which have often met really high concentrations of cod in the fishing grounds. From the beginning, these high concentrations at sea were quickly reflected in reduced sorting capacity of sorting grids. The high catch rates encountered by the grid sections have led not only to reduced selection performance of the grids but also to more serious problems like breakage of grids and grid sections. Underwater recordings have shown that fish accumulates both in front and behind the grid [25] and does not fall back in the codend. Because the catch sensors are placed in the codend, the skipper does not receive any information on the ongoing catch accumulation in the gear. The consequence of this process is unwanted big catches that exceed the vessel's processing capacity and lead to reduced quality of the catch [25].

The main reason for the accumulation of fish in front and behind the grid is related to the reduction in water flow created by the presence of the grid section and especially the sorting grid(s) [26]. In the past years, there have been efforts to increase the water flow in the grid sections to try to mitigate the fish accumulation problem. If the water flow is too high, grids lose their selective properties and fish can flow through the section without having a chance to contact the grid. Sorting grid sections have traditionally been constructed as two-panel constructions, but in an attempt to improve the flow conditions in the section new four-panel constructions have been tested in the past years. Gjøsund et al. [26] showed that a four-panel Single grid section allows higher water flow through the section than an equivalent two-panel grid section. Because the tests carried out at sea also indicate that at high catch rates four-panel sorting grid sections with larger cross-section areas perform better than two-panel grid sections (Sort-X, Sort-V, and Flexigrid) (own data, unpublished), the use of these types of sections was legalized in 2014 and 2015.

3. DESCRIPTION OF SIZE SELECTIVITY

Selective fishing refers to the ability of a fishing method to target and capture fish by size and/or species. Size selective fishing in the Barents Sea demersal trawl fishery aims at avoiding catching fish below the minimum landing sizes that enters the fishing gear while simultaneously having high probability of retaining fish above the minimum size. In the Barents Sea demersal trawl fishery, the selective devices in the fishing gear consist of a sorting grid followed by a size selective codend. The grid enables releasing fish up to a certain size through the grid, provided that they are able to find their way out between the bars. Fish that do not manage to escape through the grid fall back into the codend where they have an additional chance to escape through the meshes, provided that they are not too big to be able to pass through them. It is obvious that based on this system only fish below a certain size will have a chance to escape through the grid or codend meshes. The smaller the fish the easier they will fit through between the grid bars and through the codend meshes. Therefore, it is to be expected that the retention probability for a fish entering the gear will increase with the size of the fish. For the description of the size selection in a fishing gear, it is convenient to use a parametric mathematical model that is able to describe the retention probability *r* of the fish as a function of its length *l*. Traditionally, as for many other fisheries and trawl constructions the combined size selection of the grid and codend for the species in the Barents Sea demersal trawl fishery has been described by the *logit* size selection model:

$$r(l, L50, SR) = \frac{\exp\left(\frac{\ln(9)}{SR} \times (l - L50)\right)}{1.0 + \exp\left(\frac{\ln(9)}{SR} \times (l - L50)\right)} \tag{1}$$

In a *logit* size selection model, the retention probability will increase monotonously with the length of the fish. The size selection for this model is fully defined based on the value of the two parameters *L50*and *SR. L50* can be interpreted as the length at which a fish will have 50% probability of being retained by the fishing gear conditioned that it enters the gear. *SR* can be interpreted as the difference length between fish with, respectively, 75% and 25% of being retained in the fishing gear conditioned they enter it. Figure 2 illustrates the *logit* size selection model and the meaning of the selection parameters L50 and SR.

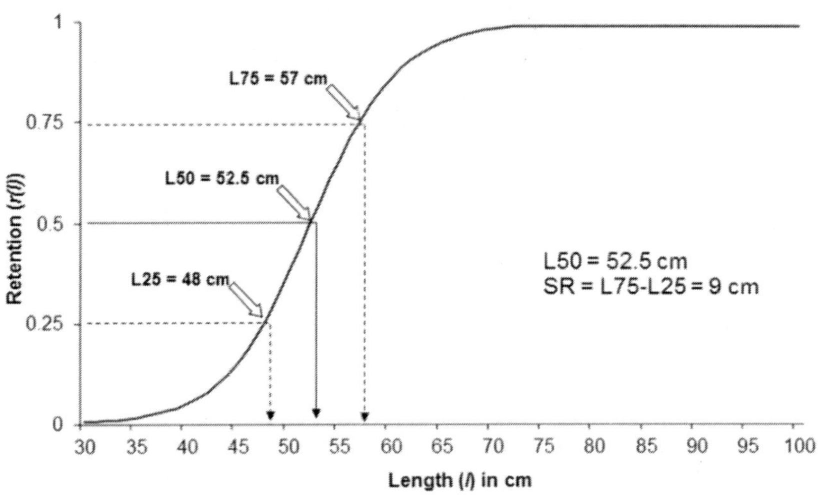

FIGURE2. A selection curve that describes the logit size selection model.

Based on the values of *L50* and *SR* for a *logit* size selection model, the length *Li* of a fish with *i* % of being retained can be calculated by:

$$Li = L50 + \frac{SR}{\ln(9)} \times \ln\left(\frac{0.01 \times i}{1.0 - 0.01 \times i}\right) \tag{2}$$

Several studies have aimed at quantifying size selection for cod and haddock in the selective systems deployed in the Barents Sea demersal trawl fishery in terms of the values for L50 and SR using the *logit* size selection model [11, 13, 14, 16, 21, 22].

However, the *logit* size selection model presents some limitations which are important when considering the size selection systems used in the Barents Sea demersal trawl fishery. First, it does not explicitly reflect the dual nature of the size selection of a system consisting of a size selective grid preceded by a size selective codend. This can lead to a poor description of the size-dependent retention probability in the fishing gear in cases where the size selective properties of the grid and the codend differ considerably. Second, the *logit* model does not enable quantifying the individual contributions of the grid and the codend to the combined size selection in the gear. This is an important limitation when trying to improve size selection based on implementing gear modifications. Third, the *logit* model is not able to account for that not necessary all fish are able to make contact with the grid during their drift towards the codend. Besides potentially leading to a poor overall description of the size selection in the gear, this limitation also disables the quantification effect on the sorting efficiency of the grid by, for example, changing grid area size or making modifications to the lifting panel ahead of a Sort-V/Single grid section.

Based on the above-mentioned limitations, recent studies on the size selection in the Barents Sea demersal trawl fishery [30] have applied a more complex dual size selection model *rdual* that accounts for all of these limitations:

$$r_{dual}(l, C_{grid}, L50_{grid}, SR_{grid}, L50_{codend}, SR_{codend}) = (1.0 - C_{grid} + C_{grid} \times logit(l, L50_{grid}, SR_{grid})) \times logit(l, L50_{codend}, SR_{codend}) \tag{3}$$

The dual size selection model contains five parameters, C_{grid}, $L50_{grid}$, SR_{grid}, $L50_{codend}$, and SR_{codend}. The first three describe the size selection in the grid while the last two describe the size selection in the codend. The parameter C_{grid} quantifies the proportion of fish entering the grid section that makes contact with the grid and is size selected by it. For the fish that makes contact with the grid, their size selection is described by a *logit* model with the parameters $L50_{grid}$ and SR_{grid}. For the codend, the size selection is described by an additional *logit* model with the parameters *L50codend* and SR_{codend}.

The grid contribution to the combined size selection is described by Eq. (3) (the term in the first brackets) and is named as the *clogit* size selection model [31]:

$$r_{clogit}\left(l, C_{grid}, L50_{grid}, SR_{grid}\right) \equiv 1.0 - C_{grid} \times \left(1.0 - r_{logit}\left(l, L50_{grid}, SR_{grid}\right)\right) \quad (4)$$

Model (4) quantifies the length-dependent retention probability for that (conditioned it entered that part of the trawl) a fish is still inside the trawl after it has passed through the section of the trawl where the grid is installed. Based on Eq. (4), Eq. (3) can be simplified to:

$$r_{dual}\left(l, C_{grid}, L50_{grid}, SR_{grid}, L50_{codend}, SR_{codend}\right) = r_{clogit}\left(l, C_{grid}, L50_{grid}, SR_{grid}\right) \times r_{logit}\left(l, L50_{codend}, SR_{codend}\right) \quad (5)$$

4.EXPERIMENTAL METHODS FOR COLLECTING SELECTIVITY DATA

The selectivity of fishing gear should be measured under conditions that mimic commercial fishing. Ideally, the sea trials should be undertaken on commercial fishing grounds, at commercial fishing depths, at commercial catch rates and sizes, and during the commercial fishing season. It is well known that variations in the environmental conditions or fish composition in the fishing grounds can influence the selective performance of fishing gear. This is not only because environmental factors affect fish behaviour towards fishing gears but also because selective fishing gears can have limitations with regard to fish densities entering it for example.

Success in selectivity trials at sea rely first on a good experimental design. It is important to determine when and where the experiments will be carried out so that the trials represent ordinary fishing conditions. Also, the choice of a sampling method to be used is crucial because it will determine what kind of additional gear one will have to use during the trials. Finally, the choice of a sampling method should be done considering the analytical and practical advantages/disadvantages of each of the available methods.

Different methods for measuring selectivity in towed fishing gears are described in reference [27]. These authors divided the methods into two categories: the Paired-gear method and Covered-gear method. In the

Paired-gear method, two gears of equal overall dimensions are towed alternatively or alongside each other. In one of the gears (the test gear), the selectivity device to be tested is installed, whereas the other gear (the control gear) is built in small unselective meshes (Figure 3). Thus, the selective properties of the tested device are calculated, assuming that the small mesh size gear captures a size and species composition that is equal to the one that has entered the test gear.

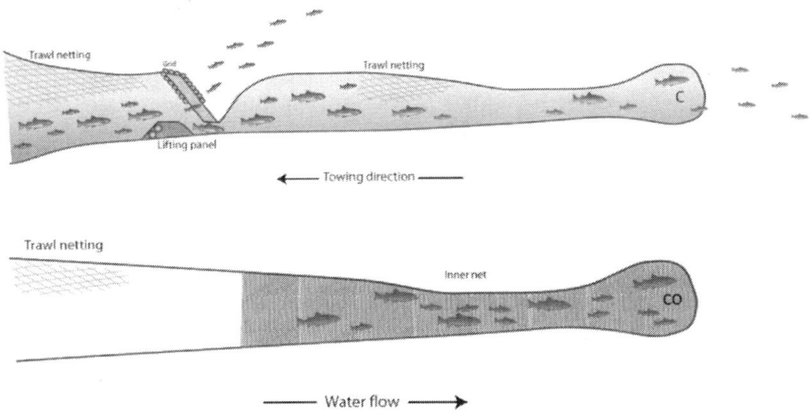

FIGURE 3. Illustration of the Paired-gear method showing the test (above) and the control (below) gear.

The trouser trawl, twin trawl, parallel haul, and alternate haul methods are examples of the Paired-gear method [27]:

- Trouser trawl method: In a single trawl, the belly is divided into two extension pieces and codends in a way that one of them acts as the control gear and the other one acts as the test gear.
- Twin trawl method: In this case, the gear is composed of two trawls. In one of the trawls, the gear to be tested is installed while the other one is used as control.
- Parallel haul method: In this case, two vessels tow one single trawl each alongside each other. In one of the trawls, the gear to be tested is installed while the other one is used as control.
- Alternate haul method: Here one vessel alternates the use of the test and the control gear.

The covers used in the Covered-gear method can vary a lot in size and shape, depending on the gear they need to cover [24]. The most important properties to look for when choosing a cover are that it needs to cover the selective device completely, it should keep the desired geometry and not be an obstacle for fish escaping through the tested device (avoid masking effect), and it should not (or minimally) reduce water flow. Square mesh covers are therefore most often used to avoid geometry problems and masking of the selective device. Still, measurements of the water flow show that the flow inside codends can be substantially reduced even when using these types of square mesh covers [28].

In the Barents Sea, the compulsory selectivity devices fishermen are allowed to use are composed of a grid section followed by a 130-mm codend. Thus, the selectivity process in the gear is a dual selection process [30] where a selection process in the grid section is followed by a selection process in the codend. To measure the selectivity of such dual selection devices, one can either install a single cover over both devices (Figure 4) [11] or install an independent cover over each of the selective devices (Figure 5) [29]. The challenge of using independent covers with respect to using a single cover is that there is an additional compartment to be considered and that the practical operations on board with multiple covers can be more challenging.

If the objective of the experiment is to collect selectivity data in order to assess only the size selection in the grid section, then the covered-grid method can be applied (Figure 6) [25].

In some experiments, the use of covers can be challenging, for example, they cannot be installed in a way that can guarantee that all escapees are collected, there is limited space on the vessel, etc. In such circumstances, the Paired-gear method is applied.

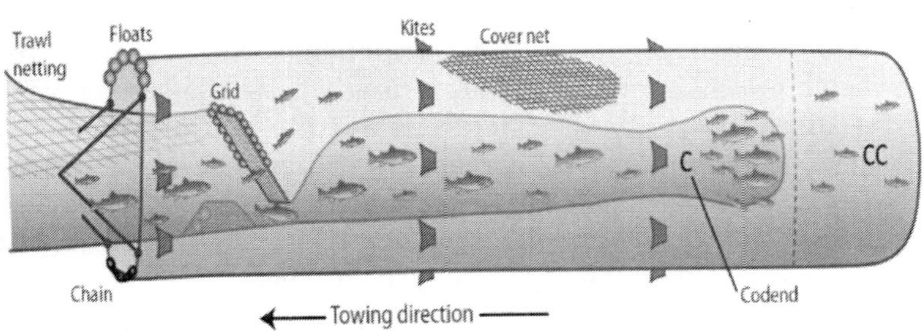

FIGURE 4. Single covered method that covers the grid section and codend.

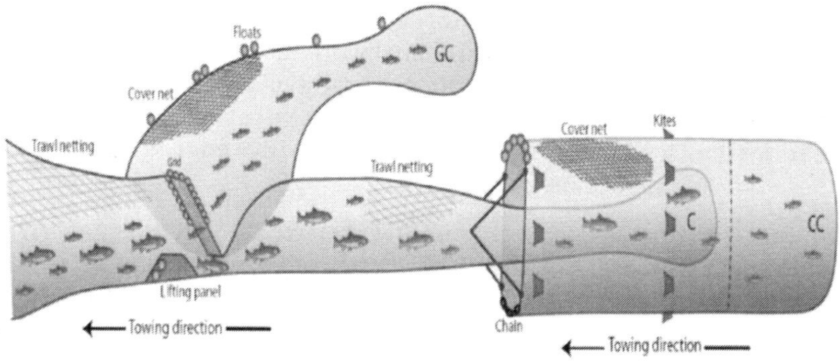

FIGURE 5. Dual cover method.

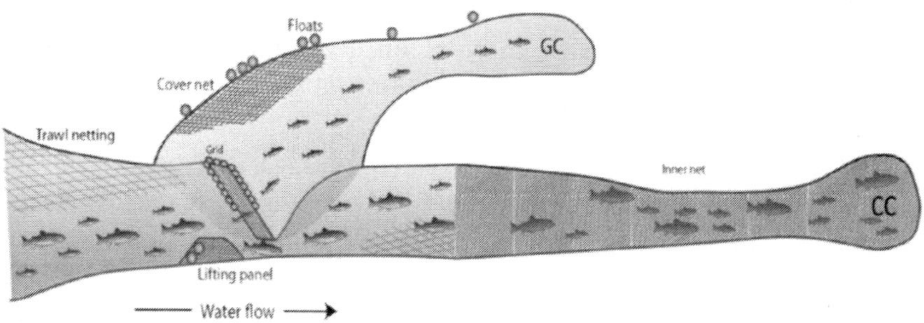

FIGURE6. The Covered-grid method.

When carrying out trawl selectivity experiments, it is important that the trials are carried out within a limited period of time and in an area that holds a fairly similar species and size distribution of fish. Large variations in the availability of fish can create loss of precision in the results, whereas large variations in the size distribution of fish can lead to large between-haul variation in the selectivity parameters L50 and SR. Methods like the Covered-gear method, which provides direct information about the fish escaping from the selective device being tested, are more robust regarding variability in both abundance and size distribution of fish. Other methods, like, for example, the alternate haul method, are more sensible to changes in fishing area as one needs to assume that the size and species distribution entering the gear is the same for the test and control hauls carried out.

Measuring the size selective properties of towed fishing gears requires length measuring the fish collected in the different compartments of the gear (test codend and cover(s) or control codend). The number of fish that needs to be caught and length measured to achieve a certain precision in the selectivity results have for years been an issue among fisheries scientists. Especially in the cases where the catches exceed the number of fish that can be measured, it is important to know how many fish one should measure and how an eventual subsample should be taken. An ideal subsample is a random sample that represents well the size distribution of the fish in the catch. Normally this subsample is taken from the catch in random batches to avoid potential accumulations of specific sizes of fish that can occur, for example, in the collection bins of the vessels. For each species, the fish in the fraction of the catch that is not length measured needs to be counted to calculate the sampling ratio in the catch.

Diverse sampling strategies in experimental trawl selectivity gear were studied by Millar [34], who concluded that sampling the same number of fish from the different compartments was the most efficient sampling method. Also, how the precision in selectivity results varied depending on the amount of fish caught and length measured with both the Paired-gear and the Covered-gear methods was investigated in reference [35]. The results of the investigation showed that the uncertainty in the selection parameters L50 and SR decreased with increasing number of fish measured, and that this relationship could be described by a power model. The results also demonstrated that the sampling effort needed to achieve a specific uncertainty level for the selection parameters was always lower for the Covered-gear method compared to the Paired-gear method (in many cases the number of fish that would need to be measured to maintain a specific uncertainty level was around 10 times higher for the Paired-gear method than for the Covered-gear method). The results of these studies illustrate again the importance of carrying out proper experimental design before starting the sea trials. Both the potential limitations of the operations that need to be carried out on board and the advantages and disadvantages of the different sampling methods need to be always considered.

For each haul conducted with one of the experimental collection methods described above (Figures 3–6), the catch in each compartment (codend, cover(s), control) is length measured species by species. These length measurements are sorted into so-called length classes that in the Barents Sea are typically 1 cm wide. In a dataset, each haul consists of count data that show how many fish of those collected in each of the compartments belong to the same length class. If subsampling is applied only to a proportion of the fish in each compartment, the sampling factor for each compartment is also provided together with the count data. Table 1 shows an example of a data file resulting from one haul. In this case, the data was collected using the single covered method (Figure 4), and 54% of the fish in codend and 62% of the fish in the cover were length measured. The rest of the fish in each compartment was counted. Analysis of such data (Table 1) forms the basis for estimating the selectivity of trawl gear by the methods described in the next section.

TABLE 1. Illustration of a dataset (one single haul) collected using the covered codend method.

Length class (cm)	Number of fish in the codend	Number of fish in the cover
25.5	0	0
26.5	0	0
27.5	0	0
28.5	0	0
29.5	0	0
30.5	2	43
31.5	2	23
32.5	3	64
33.5	1	118
34.5	7	116
35.5	5	279
36.5	7	137
37.5	5	179
38.5	10	180
39.5	13	106
40.5	12	266
41.5	2	61
42.5	7	50
43.5	9	52
44.5	12	26
45.5	16	21
46.5	12	61
47.5	14	9

48.5	21	18
49.5	16	8
50.5	26	3
51.5	13	11
52.5	21	0
53.5	10	5
54.5	22	0
55.5	16	1
56.5	11	2
57.5	10	0
58.5	20	1
59.5	7	0
60.5	9	0
61.5	8	0
62.5	7	0
63.5	5	0
64.5	0	0
65.5	2	0
SAMPLING	0.54	0.62

5. METHODS TO ESTIMATE SIZE SELECTION PARAMETERS

The accepted scientific method for estimation of trawl size selectivity is based on a Maximum Likelihood Estimation (MLE) [27]. The maximization problem this represents is by convenience converted into a minimization problem simply by adding a minus sign in front of the function that else would have been a maximization. Further, to simplify the formulation, the natural logarithm (*ln*) is taken to the function prior to minimization as this leads to a simpler formulation that is easier to minimize. This step is in valid since the *ln* of a function, and the function itself, has the same minimum for the same model parameter values. The function to minimize depends both on the selection system for which the data has been collected and the level of detail intended, and as described in the previous section on which of the experimental data collection has been applied. Conditioned that the size selection model applied in the estimation is able to describe the size selection processes occurring sufficiently well, the MLE estimation provides the model parameters (see Section 3) that make the collected experimental size selection data most likely. If the intention is to assess the combined size selectivity in a single haul *j* based on the covered gear method (Figure 4), based on the *logit* model Eq. (1), the following equation is minimized with respect to $L50_j$ and SR_j:

$$-\sum_{l}\left\{\begin{array}{l} nc_{jl}\times\ln\left(\dfrac{qc_j\times r_{logit}(l,L50_j,SR_j)}{qc_j\times r_{logit}(l,L50_j,SR_j)+qcc_j\times\left(1.0-r_{logit}(l,L50_j,SR_j)\right)}\right)+ncc_{jl} \\ \times\ln\left(\dfrac{qcc_j\times\left(1.0-r_{logit}(l,L50_j,SR_j)\right)}{qc_j\times r_{logit}(l,L50_j,SR_j)+qcc_j\times\left(1.0-r_{logit}(l,L50_j,SR_j)\right)}\right)\end{array}\right\} \tag{6}$$

Where nc_{jl} is the number of fish in length class *l* that is length measured in the haul *j* in the codend; and ncc_{jl} is the number of fish in length class *l* that is length measured in the haul *j* in the cover. qc_j and qcc_j are the corresponding sampling rates. The summation in Eq. (6) is over length classes *l*.

In case the Paired-gear method (Figure 3) is applied for the data collection, the function to minimize would be:

$$-\sum_{l}\left\{\begin{array}{l} nc_{jl}\times\ln\left(\dfrac{qc_j\times SP_j\times r_{logit}(l,L50_j,SR_j)}{qc_j\times SP_j\times r_{logit}(l,L50_j,SR_j)+qco_j\times\left(1.0-SP_j\right)}\right)+nco_{jl}\times \\ \ln\left(\dfrac{qco_j\times\left(1.0-SP_j\right)}{qc_j\times SP_j\times r_{logit}(l,L50_j,SR_j)+qco_j\times\left(1.0-SP_j\right)}\right)\end{array}\right\} \tag{7}$$

Compared to Eq. (6), Eq. (7) includes an additional parameter SP_j that needs to be estimated together with $L50_j$ and SR_j. SP_j is the so-called split parameter that quantifies the proportion of fish entering the test side of the gear. nco_{jl} is the amount of fish in length class *l* that is length measured in the control codend and qco_j is the corresponding sampling rate.

Several of the size selection studies carried out in the Barents Sea demersal trawl fishery have applied Eq. (6) or (7) to estimate the combined size selection for individual hauls. In most cases, a mean size selection for the group hauls is subsequently estimated in a second estimation step using the size selection results obtained in the individual hauls and following an estimation procedure described in references [24, 30, 32]. This procedure accounts for both the uncertainty in the estimated size selectivity in the individual hauls (often named within-haul variation) and the between-haul variation in the size selection.

However, in most of the recent studies the dual nature of the size selection process in the gears has been explicitly accounted for by basing the analysis on the dual selection model described by Eq. (5). Further, in most of these studies the final aim has been to estimate the size selection averaged over a group of hauls. Therefore, the process involves summing data over hauls in the estimation process.

If the single covered data collection method (Figure 4) is applied, this would lead to minimizing the following equation:

$$-\sum_{j=1}^{m}\sum_{l}\left\{\begin{array}{l}\frac{nc_{jl}}{qc_j}\times\ln\left(r_{dual}(l,C_{grid},L50_{grid},SR_{grid},L50_{codend},SR_{codend})\right)+\frac{ncc_{jl}}{qcc_j}\times\\ \ln\left(1-r_{dual}(l,C_{grid},L50_{grid},SR_{grid},L50_{codend},SR_{codend})\right)\end{array}\right\} \quad (8)$$

In the case of Eq. (8), the minimization is carried out in five dimensions to estimate C_{grid}, $L50_{grid}$, SR_{grid}, $L50_{codend}$, and SR_{codend}. The outer summation is over the *m* hauls conducted.

If the Paired-gear data collection method (Figure 3) is applied, then the estimation is conducted in six dimensions because the average split *SP* also needs to be estimated. In this case, the function to minimize becomes:

$$-\sum_{j=1}^{m}\sum_{l}\left\{\begin{array}{l}\frac{nc_{jl}}{qc_j}\times\ln\left(\frac{SP\times r_{dual}(l,C_{grid},L50_{grid},SR_{grid},L50_{codend},SR_{codend})}{SP\times r_{dual}\left(l,C_{grid},L50_{grid},SR_{grid},L50_{codend},SR_{codend}\right)+1-SP}\right)+\frac{nco_{jl}}{qco_j}\\ \times\ln\left(\frac{1-SP}{SP\times r_{dual}\left(l,C_{grid},L50_{grid},SR_{grid},L50_{codend},SR_{codend}\right)+1-SP}\right)\end{array}\right\} \quad (9)$$

If the dual covered data collection method (Figure 5) is applied, the precision in estimating the size selection of the gear can be improved. In this case, the equation to be minimized becomes:

$$\sum_{j=1}^{m}\sum_{l}\left\{\begin{array}{l}\frac{ng_{jl}}{qg_j}\times\ln\left(1-r_{clogit}\left(l,C_{grid},L50_{grid},SR_{grid}\right)\right)+\frac{ncc_{jl}}{qcc_j}\times\\ \ln\left(r_{clogit}\left(l,C_{grid},L50_{grid},SR_{grid}\right)\times\left(1-r_{logit}\left(l,L50_{codend},SR_{codend}\right)\right)\right)+\frac{nc_{jl}}{qc_j}\times\\ \ln\left(r_{clogit}\left(l,C_{grid},L50_{grid},SR_{grid}\right)\times r_{logit}\left(l,L50_{codend},SR_{codend}\right)\right)\end{array}\right\} \quad (10)$$

If the only objective is to assess the average size selection in the grid section, then the Covered-grid data collection method (Figure 6) can be applied. In this case, the function to minimize will be:

$$\sum_{j=1}^{m}\sum_{l}\left\{\begin{array}{l}\frac{ng_{jl}}{qg_j}\times\ln\left(1-r_{clogit}\left(l,C_{grid},L50_{grid},SR_{grid}\right)\right)+\frac{ncc_{jl}}{qcc_j}\times\\\ln\left(r_{clogit}\left(l,C_{grid},L50_{grid},SR_{grid}\right)\right)\end{array}\right\} \quad (11)$$

When estimating the average size selectivity, based on minimizing Eq. (8)–(10) or (11), the uncertainty on the parameters and size selection curve is often estimated using a double bootstrap method as described in references [30, 31, 33].

The ability of a size selection model to describe the experimental size selection data is frequently evaluated based on calculating a *p*-value for model deviance versus the degrees of freedom [31]. This *p*-value quantifies the probability to by coincidence obtain at least as big discrepancy between model and data as observed. Thus, this *p*-value should not be below 0.05 for the model to be able to describe the experimental size selection data well. However, a low *p*-value can also be a result of over-dispersion in the data, which in case of having a low *p*-value should be inspected before rejecting the size selection model.

6. THE FISHSELECT METHODOLOGY

FISHSELECT is a framework of methods, tools, and software developed to determine whether or not a fish is able to penetrate a certain mesh [36]. Through computer simulation, FISHSELECT enables the estimation of the selectivity parameters L50 and SR for a certain species and selection device by comparing the morphological characteristics of the former and the shape and size of the latter. This methodology has been successfully used to estimate mesh selectivity of the most relevant demersal fish species in the Barents Sea: cod (*G morhua*), haddock (*M. aeglefinus*), Greenland halibut (*R. hippoglossoides*), and redfish (*Sebastes ssp.*) . [31, 37–39]. The results obtained with the application of FISHSELECT have proved the reliability of the method as they are coherent with the results registered from earlier sea trials [37–39]. Thus, if we consider the flexibility the method offers compared to the traditional selectivity measuring methods, the value of the method becomes obvious.

By means of the FISHSELECT tools and software, one can predict the selectivity of a specific fishing gear and species. This can also be achieved with the more traditional selectivity study methods; however, FISHSELECT is unique that once the morphology analyses are carried out for a certain species, one can predict the selectivity of this species for endless mesh shapes (including grids), sizes, and opening angles (OAs). Some of the features or advantages of FISHSELECT with respect to the traditional sea trial selectivity studies are that:

I. New estimations do not need additional fieldwork.

II. FISHSELECT provides fast answer on the selectivity parameters that can be expected from an eventual change in the gear. In a similar way, if one wishes to change the selection properties of a gear to achieve certain selectivity for a species or multiple species, FISHSELECT can fast predict the changes necessary in the gear to achieve those selectivity objectives.

III. The method gives an overall estimation of as many species as one wishes. It is interesting to note that most often the gear will be designed as a compromise on the selective properties of several species [37] that different cleaner-fish species with various body sizes and shapes are harvested at the same time.

IV. The results obtained from FISHSELECT can also be used to aid in the interpretation of sea trial selectivity results.

Because FISHSELECT is based on the relationship between fish length and weight, and the shape of the fish's cross-section at different points of its body, factors like the condition of the fish in the different seasons need to be taken into consideration. When collecting the morphological data for a certain species, it is important to cover the whole length spectra for the fish as through its growth not all the body parts grow proportionally. Also, if the fish included in the measurements is captured through different seasons, it would help covering a wider spectra of the different shapes the species can acquire. The wider the spectra of fish covered and the higher the amount of fish included in the measured fish pool, the lower the uncertainty in the predictions.

The application of FISHSELECT is divided into four steps:

Step 1: morphological data

In the first step, the fish length (in mm), weight (in g), and its morphology at different cross sections (CSs) need to be measured. The number and position of the CSs taken are different for each fish species, and are decided based on earlier experiences considering the positions likely to determine if a fish will be able to escape through meshes or grids of different sizes and shapes. For fish, in general, three cross sections have been measured per fish [38]. For cod and haddock, for example, these three cross sections were located at the end of the opercula (maximum girth of the head), at the foremost point of the first dorsal fin, and maximum girth of the fish [37]. The cross sections are measured by means of a mechanical sensing tool named morphometer. The morphometer consists of an aluminum frame and measuring aluminum sticks (2.5 mm wide) that can be shifted horizontally and fixed at a desirable position (Figure 7). The shape formed in the morphometer is later converted into a digital image using a flatbed scanner. The image resulting from the scanner is finally digitized using the image analysis tools implemented in the FISHSELECT software tool [36].

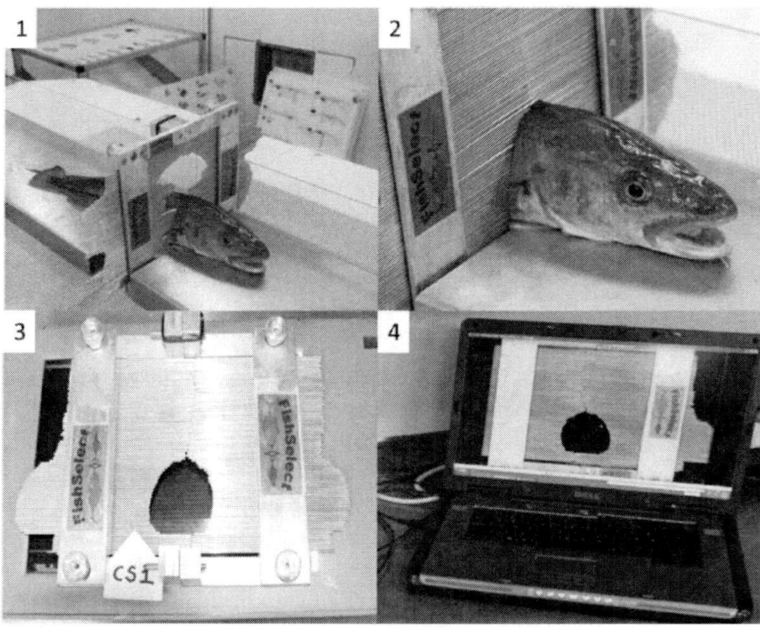

FIGURE 7.Illustration of the use of a morphometer on cod (Gadus morhua).

Step 2: Fall-through experiments

Fall-through experiments are carried out to decide if a fish can in principle physically pass through a certain rigid shape (subjected to the force of gravity only) (Figure 8). These shapes are perforated in 5-mm-thick solid nylon plates [39]. The shapes tested normally include diamonds, hexagons, and rectangles; however, there are no restrictions to the shapes one can test and use in FISHSELECT.

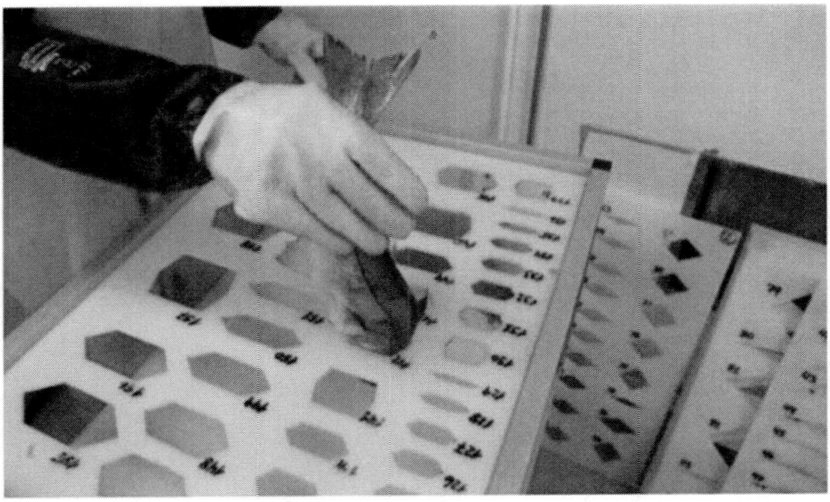

FIGURE 8. Illustration of the fall-through procedure on haddock (Melanugrammus aeglefinus). Each interchangeable plate contains a number of different mesh shapes where each fish is tested. All fish included in the study need to be tested in all meshes.

Step 3: Cross-section modeling

The CS shapes registered with the morphometer need to be modeled so that they can be further analyzed in FISHSELECT. The software has more than 100 different models available to model each of the different cross sections of the fish (Figure 9). One needs to first determine which of the available models seems to represent each of the CSs well enough and later fit all these relevant models to each digitized shape. Each of the models is tested on each of the CSs registered for each fish, and the model with the lowest AIC [40] is chosen for further analysis in FISHSELECT. Once the parameters in the model defining each of the cross sections are the length of the fish, one can create virtual populations with defined CSs.

Step 4: search for penetration model

Fish can be compressed both dorsoventrally and laterally. Thus, different compression models need to be tested for each CS as a first step to establish an optimal penetration model for the species tested (Figure 10). The optimal penetration model is established by comparing the penetration results of each compression model tested with the fall-through results. The degree of agreement (DA) between the simulated and experimental fall-through results is then used to choose an optimal penetration model (see

[36, 37] for the mathematical expression and further information about DA).

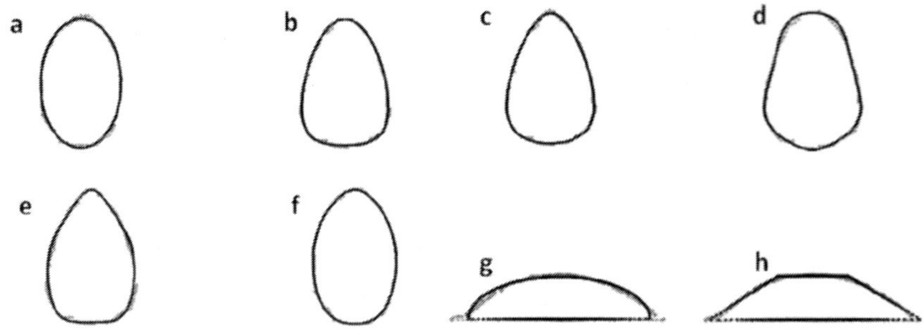

FIGURE 9. Illustration of some of the parametric shapes available in FISHSELECT: Shapes like (a)–(f) are typically used for roundfish while (g)–(h) are normally used on flatfish species.

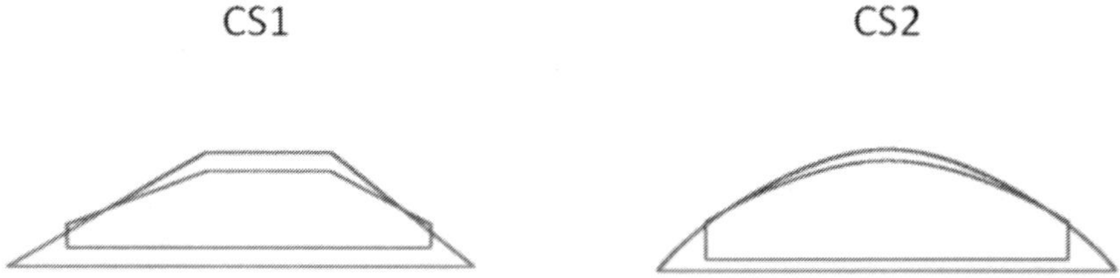

FIGURE 10.Shape of the optimal penetration model (green) overlapped on the original shape modeled from the morphometer (red) for Greenland Halibut (Reinhardtius hippoglossoides).

Given virtual populations with the desired population structure and defined CSs and a defined penetration model, the size selective properties of a range of mesh shapes and sizes can be predicted in FISHSELECT by simulation. The outcome of the method consists of L50 and SR estimations for all of the included mesh sizes, shapes, and OAs.

7. FISHSELECT RESULTS FOR NORTHEAST ARCTIC DEMERSAL SPECIES

Cod, haddock, redfish, and Greenland halibut are the four most important commercial species in the Barents Sea demersal trawl fisheries. FISHSELECT has already been applied to all four species, and the results have in all cases showed to be coherent with the results obtained from earlier sea trials [31, 37–39], . Because the compulsory gear in the Barents Sea demersal trawl fishery is composed of a sorting grid followed by a size selective diamond mesh codend, the FISHSELECT studies present results for both different bar spacing grids and different diamond mesh size codends. Figure 11 shows selectivity results obtained with the different bar spacing grids for both cod and haddock [37]. Figure 12 shows the predicted and observed L50 versus mesh size for *Sebastes* spp. Figure 13 shows historical selectivity results for Greenland halibut compared to FISHSELECT predictions. Finally, Figure 14 shows the variation in L50 for redfish with varying mesh size and OA for diamond meshes.

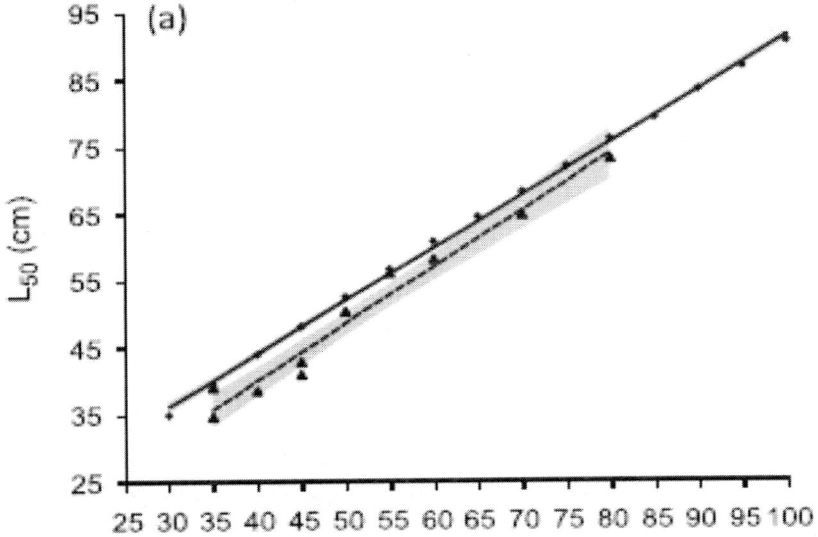

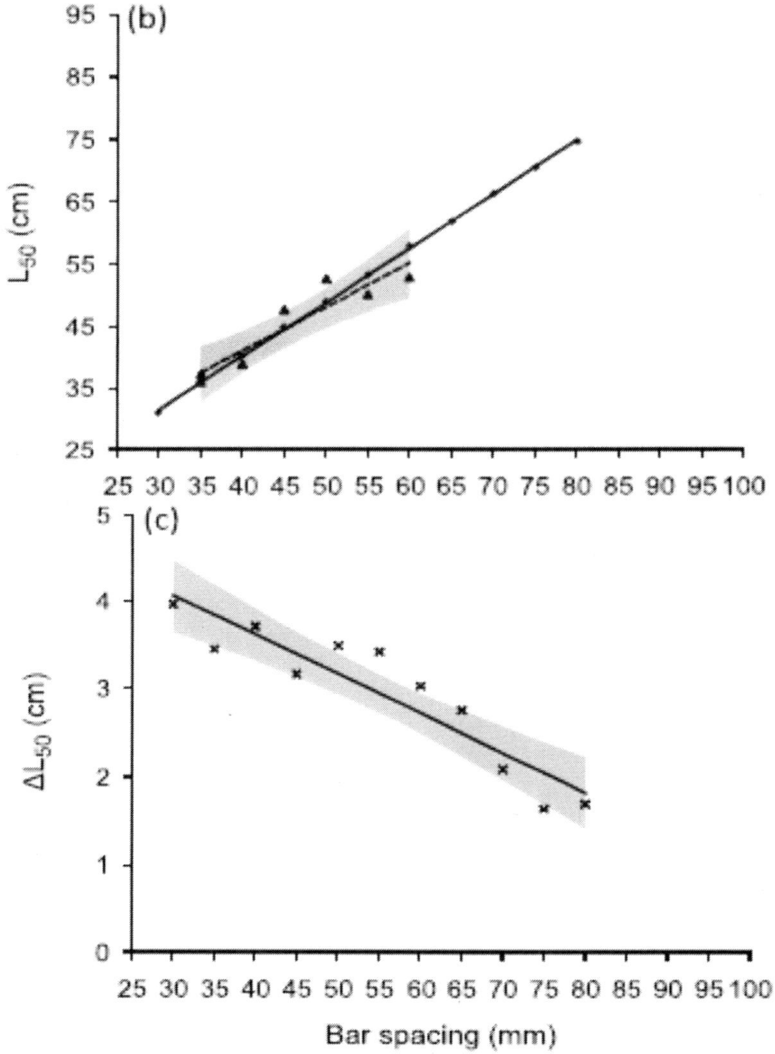

FIGURE 11.Selectivity results obtained with the different bar spacing grids for both cod (G. morhua) (a) and haddock (M. aeglefinus) (b). The solid line is a trend line fitted to the FISHSELECT results (diamonds), which are presented for up to the maximum fish size included in the data collection process. The broken line is a trend line added to the sea trial selectivity results (triangles) obtained for the different bar spacing grids in [41]. Panel (c) shows the L50 differences and a trend line for these differences between cod and haddock. The gray areas in the three panels represent the 95% confidence limits for the added trend lines. Source: [37].

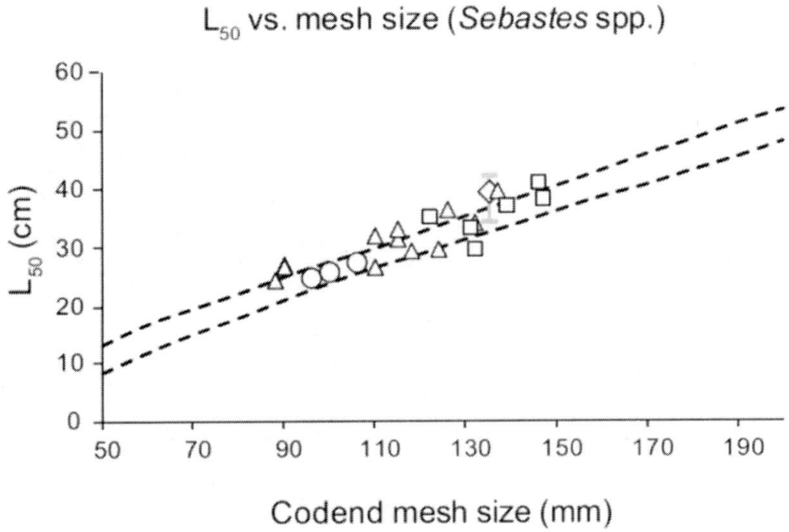

FIGURE 12.Predicted and observed L50 versus mesh size relationships for redfish (Sebastes spp.): Predicted band for codend L50 for different mesh sizes based on the FISHSELECT analysis of the data collected for Sebastes marinus (stippled curves); new results from sea trials for S. marinus (diamonds); previous results for S. marinus (squares); previous results for Sebastes mentella (triangles) and Sebastes mentella/ faciatus (circles). Source: [38].

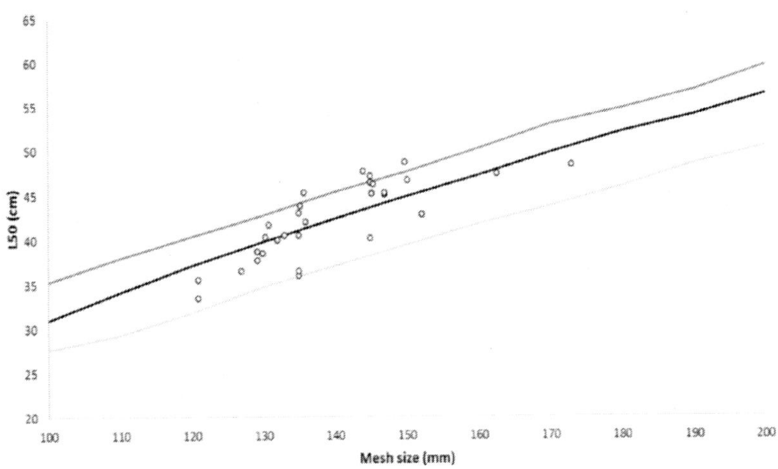

FIGURE 13.Historical data for Greenland halibut (R. hippoglossoides) codend selectivity (circular marks) plotted together with FISHSELECT estimations (lines). Source: [39]

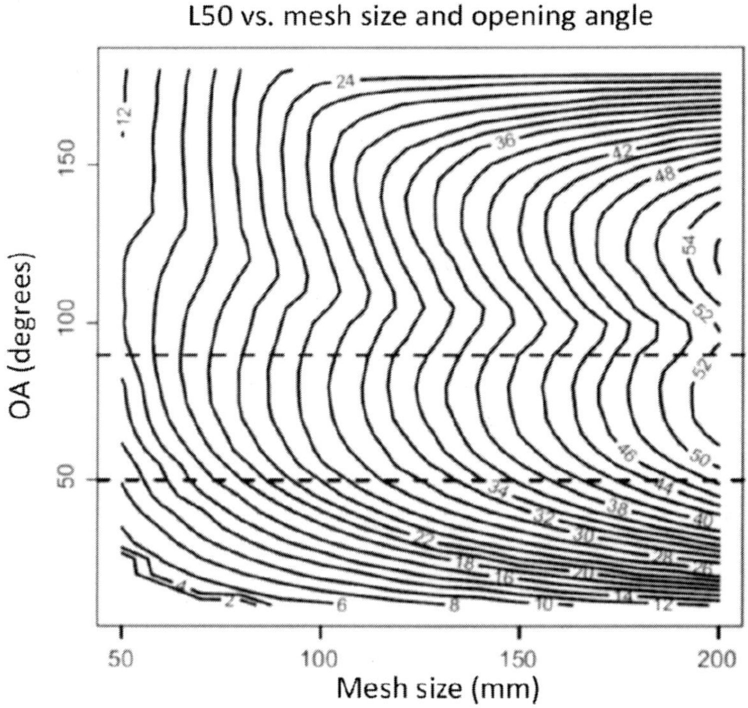

FIGURE 14.Design guide showing the variation in L50 for redfish (S. marinus) with varying mesh size and OA for diamond meshes. Source: [38].

8. ACKNOWLEDGEMENTS

Most of the information presented in this work has been obtained through diverse research projects that have been carried out in the past years. We gratefully acknowledge the financial support from the Norwegian Research Council, the Norwegian industrial research fund (FHF), and the Norwegian Directorate of Fisheries. The initial development of FISHSELECT was financially supported by a project under the development program for sustainable fisheries financed by the European Union and Danish Ministry of Food, Agriculture, and Fisheries.

REFERENCES

1. ICES [2015]ICES Advice: Cod (*Gadus morhua*) in Subareas I and II (Northeast Arctic) [http://www.ices.dk/sites/pub/Publication%20Reports/Advice/2015/2015/codarct. pdf]. 12 June 2015.[Accessed: 26 January 2016]
2. ICES [2015] ICES Advice: Haddock (*Melanogrammus aeglefinus*) in Subareas I and II (Northeast Arctic) [http://www.ices.dk/sites/pub/Publication%20Reports/Advice/ 2015/2015/had-arct.pdf]. 12 June 2015.[Accessed: 26 January 2016]
3. ICES [2015] ICES Advice: Saithe (*Pollachius virens*) in Subareas I and II (Northeast Arctic) [http://www.ices.dk/sites/pub/Publication%20Reports/Advice/2015 /2015/saiarct. pdf]. 12 June 2015.[Accessed: 26 January 2016]
4. Lisovsky, S.F. [2001] Evolution of technical measures regulating fishery on cod and haddock in the Barents Sea and Norwegian Sea. In: Shevelev, M., Lisovsky, S.F. (Eds.) Technical regulations and by-catch criteria in the Barents Sea fisheries. Proceedings of the 9th PINRO-IMR Symposium, Murmansk 14– 15 August, 2001
5. Kvamme, C., Frøysa, K.G. [2008] Assessing the effects on stocks of selectivity changes in a fishery. Fish. Res.; 93: 283-292
6. Huse, I., Ånondsen, S., Ellingsen, H., Engås, A., Furevik, N., Graham, N., Isaksen, B., Jørgensen, T., Løkkeborg, S., Nøttestad, L. & Soldal, A.V. [2002] A desk-study of diverse methods of fishing when considered in perspective of responsible fishing, and the effect on ecosystem caused by fishing activity. Bergen, July 2002. TemaNord 501
7. Isaksen B., Valdemarsen, J.W. [1986] Selectivity experiments with square mesh codends in bottom trawls. ICES CM1986; B:28
8. Larsen, R.B. [1989] A review on the application and selectivity of square mesh netting in trawls and seines. In: Carr, H.A. (Ed.) Proceedings of the square mesh workshop. World symposium on fishing gear and fishing vessel design. St. John's Canada, November 1989; 13–36
9. Isaksen, B. [1991] A brief review of codend selectivity investigations in Norway from 1985-1991. In: Workshop on codend selectivity. June 1991. Marine Laboratory, Aberdeen Scotland
10. Løbach, T. [1991] Gear selectivity: Report of the Nordic conference on gear selectivity. Køge, 28-29 May 1991.TemaNord 513: 115–124
11. Larsen, R.B., Isaksen, B. [1993] Size selectivity of rigid sorting grid in bottom trawls for Atlantic cod (*Gadus morhua*) and haddock (*Melanogrammus aeglefinus*). ICES Mar. Sci. Symp.; 196: 178–182
12. Isaksen, B., Lisovsky, S., Larsen, R.B., Sakhnoe, V., Gamst, K. & Misund, R.[1996] Joint Russian - Norwegian selectivity experiments on cod (*Gadus morhua*) in the Barents Sea 1995. Fisken og Havet.; 1996; 16: 55pp
13. Isaksen, B., Gamst, K., Kvalsvik, K. & Axelsen, B. [1998] Comparison of selectivity and handling characteristics between the Sort-X and the single grid (Sort-V). Experiments performed on board the commercial vessel "Anny Kræmer" in Bear Island, 15-28 August 1997. Norwegian Institute of Marine Research. 1998; 15pp. (In Norwegian)
14. Isaksen, B., Gamst, K., Kvalsvik, K. & Kristiansen, A.[1998] Single sorting grid in to panel trawls for cod. Comparison of selectivity and handling characteristics between Sort-X and the single grid (Sort-V). Norwegian Institute of Marine Research. 1998; 19pp. (In Norwegian)
15. Misund, R. [2001] Development of sorting grids in the Norwegian fishery – A review. In: Shevelev, M. and Lisovsky, S. (Eds.), Technical regulations and by-catch criteria in the Barents Sea fisheries. Proceedings of the 9th PINRO-IMR Symposium, Murmansk, 14-15 August, 2001
16. Jørgensen, T., Galbraith, D., Graham, N., Hansen, K., Holst, R., Larsson, P.O., Mortreux, S., Sangster, G.I., Soldal, A.V. & Tschernij, V.[2002] Design, handling and selectivity

properties of the Eurogrid. In: ICES-FAO Working Group on Fishing Technology and Fish Behaviour, ICES CM 2002/B:01. Sáte, France, 6-8 June 2002

17. Angell, S., Andersen, M., Lilleng, D. & Torsvik, O.K. [2001] Comparison of the selectivity and handling characteristics between a 50 mm Sort-X and 50 mm Flexigrid in the bottom trawl fishery for saithe. Experiments on board"Bliki", 9–20 March 2001. SINTEF Fisheries and Aquaculture. STF80 A013032. 2001; 23pp. (In Norwegian)
18. Angell, S., Lilleng, D., Hassel, K. & Røttingen, Å.[2001] Selectivity experiments with a 55mm Flexigrid in the bottom trawl fishery for saithe. Experiments on board "Bliki", 3–12 April 2000. SINTEF Fisheries and Aquaculture. STF80 A00050.2001; 16pp. (In Norwegian)
19. Angell, S., Lilleng, D., Kolbeinshavn, A., Vollstad, J. & Larsen, R.B. [2001] Selectivity experiments with a 55 mm Flexigrid in the bottom trawl fishery for saithe. Experiments on board "Bliki", 21 September – 2 October 2000.SINTEF Fisheries and Aquaculture. STF80 A0113030. 2001; 18pp. (In Norwegian)
20. Angell, S., Lilleng, D., Vollstad, J. & Grimaldo, E.[2001] Selectivity experiments with 55 and 60 mm Flexigrid in the bottom trawl fishery for cod. Experiments on board "Bliki" 11– 23 June 2001. SINTEF Fisheries and Aquaculture. STF80 A013040. 2001; 19pp. (In Norwegian)
21. Grimaldo, E., Larsen, R.B. & Holst, R. [2007] Exit Windows as an alternative selective system for the Barents Sea Demersal Fishery for cod and haddock. Fish. Res.; 91: 295– 305
22. Grimaldo, E., Sistiaga, M. & Larsen, R.B. [2008] Evaluation of codends with sorting grids exit windows, and diamond meshes: size selection and fish behaviour. Fish. Res.; 91: 271–280
23. Grimaldo, E., Larsen, R.B., Sistiaga, M., Madsen, N. & Breen, M. [2008] Selectivity and escape percentages during three phases of the towing process comparing codends fitted with different selection systems. Fish. Res.; 95: 198–205
24. Sistiaga, M., Herrmann, B. & Larsen, R.B. [2009] Investigation of the paired-gear method in selectivity studies. Fish. Res.; 97: 196–205
25. Grimaldo, E., Sistiaga, M., Herrmann, B., Gjøsund, S.H. & Jørgensen, T. [2015] Effect of lifting panel on selectivity of a compulsory grid section (Sort-V) used by the demersal trawler fleet in the Barents Sea cod fishery. Fish. Res.; 170: 158–165
26. Gjøsund, S.H., Grimaldo, E., Sistiaga, M. & Hansen, K.[2013] Water flow measurements in two and four-panel grid sections. SINTEF Fisheries and Aquaculture. Report A24698. 2013 (In Norwegian)
27. Wileman, D., Ferro, R.S.T., Fonteyne, R. & Millar, R.B.[1996] Manual of methods of measuring the selectivity of towed gears. ICES Coop. Res. Rep. No. 215; 126 pp
28. Breen, M., Sangster, G., O'Neill, B., Kynoch, R., Jones, E. & Soldal, A.[2002] Evidence of sampling induced biases in mortality estimates from experiments investigating mortality in fish escaping from towed fishing gears. ICES CM 2002/V: 25
29. Kvamme, C., Isaksen, B. [2004] Total selectivity of a commercial cod trawl with and without a grid mounted: grid and codend selectivity of North-east Arctic cod. Fish. Res.; 68: 305–318
30. Sistiaga, M., Herrmann, B., Grimaldo, E. & Larsen, R.B.[2010] Assessment of dual selection in grid based selectivity systems. Fish. Res.; 105: 187–199
31. Herrmann, B., Sistiaga, M., Larsen, R.B. & Nielsen, K. [2013] Size selectivity of redfish (Sebastes spp.) in the Northeast Atlantic using grid-based selection systems for trawls. Aquat. Living Resourc.; 26: 109–120
32. Fryer, R.J. [1991] A model of between-haul variation in selectivity. ICES J. Mar. Sci.; 48: 281–290
33. Sistiaga, M., Herrmann, B., Grimaldo, E. & Larsen, R.B. [2015] Effect of lifting the sweeps on bottom trawling catch efficiency: a study based on the Northeast arctic cod (*Gadusmorhua*) trawl fishery. Fish. Res.; 167: 164–173
34. Millar, R.B. [1994] Sampling from trawl gears used in size selectivity experiments. ICES J. Mar. Sci.; 51: 293–298

35. Herrmann, B., Sistiaga, M., Santos, J. & Sala, A. [2015] How many fish need to be measured in trawl selectivity studies? In: ICES. 2015. Second Interim Report of ICESTrawl Selectivity in the Barents Sea Demersal Fishery http://dx.doi .org/10.5772/63019 FAO Working Group on Fishing Technology and Fish Behaviour (WGFTFB), 4-7 May 2015, Lisbon, Portugal. ICES CM 2015/SSGIEOM:22. 183pp
36. Herrmann, B., Krag, L.A. & Frandsen, R.P. [2009] Prediction of selectivity from morphological conditions: methodology and a case study on cod (*Gadus morhua*). Fish. Res.; 97: 59–71
37. Sistiaga, M., Herrmann, B., Nielsen, K.N. & Larsen, R.B. [2011] Understanding limits to cod and haddock separation using size selectivity in a multispecies trawl fishery: an application of FISHSELECT. Canadian J. Fish. Aquat. Sci.; 68: 927–940
38. Herrmann, B., Sistiaga, M., Nielsen, K.N. & Larsen, R.B. [2012] Understanding the size selectivity of redfish (*Sebastes spp.*) in North Atlantic trawl codends. J. Northw. Atlan. Fish. Sci.; 44: 1–13
39. Herrmann, B., Sistiaga, M., Larsen, R.B., Nielsen, K.N. & Grimaldo, E. [2013] Understanding sorting grid and codend size selectivity of Greenland halibut (*Reinhardtiushippoglossoides*). Fish. Res.; 146: 59–73
40. Akaike, H. [1974] A new look at the statistical model identification. IEEE Trans. Autom. Control.; 19(6): 716–723
41. Sistiaga, M., Grimaldo, E., & Larsen, R.B. [2008] Size selectivity patterns in the northeast Arctic cod and haddock fishery with sorting grids of 55, 60, 70 and 80 mm. Fish. Res.; 93: 195–203

CHAPTER 9

Plastic Life-History Variation along The Course of a Steep Mountainous River in Male *Cottus Hangiongensis* (Pisces: Cottidae)

Akira Goto Hideaki Arioka Ryota Yokoyama

Field Science Center for Northern Biosphere, Hokkaido University, 3-1-1 Minato-cho, Hakodate, Hokkaido 041-8611, Japan

Laboratory of Aquatic Animal Breeding, Graduate School of Fisheries Sciences, Hokkaido University, 3-1-1 Minato-cho, Hakodate, Hokkaido 041-8611, Japan

ABSTRACT

To elucidate the causes of clinal life-history variation in male freshwater amphidromous *Cottus hangiongensis* found along the river courses, we examined the genetic composition of upstream, midstream, and downstream groups using microsatellite DNA markers and different growth patterns among juvenile individuals before and after they ascended a small, steep mountainous river in southern Hokkaido, Japan. Microsatellite allele frequencies indicated that the samples collected from three different habitats appeared to represent a panmictic population and no genetic differences were found among the three groups. Ecological field surveys showed that juveniles greatly depended on the timing of their ascent into the river from the sea, and that larger juveniles were able to disperse to more upstream areas and grew larger during the period after they migrated up to the mouth of the river. These results suggested that clinal life-history variations may be induced through phenotypic plasticity and that earlier ascent of juveniles from the mouth of river may be an important source of variation.

Keywords: Life-history traits Juvenile upstream migration Heterogeneous environment Amphidromous sculpin

INTRODUCTION

Phenotypic plasticity, or changes in the phenotypic expression of a single genotype as a function of the environment, has broad significance because it unites perhaps all fields of biology: encompassing genetics, development, ecology, evolution, and even biophysics, physiology, and behavioral science (Scheiner 1993; Dewitt and Scheiner 2004). Within a population, environmentally induced phenotypic traits may be an important source of life-history and genetically induced variation. Genetically based intraspecific variation could arise if the population had a substructure of reproductively isolated subunits (Hendry et al. 1999) and was sufficiently small to yield genetic drift. Three main sources of life-history variation within a fish population can be distinguished: (1) body size (Wootton 1990); (2) environmental conditions (Jonsson et al. 1996; Lobon-Cervia et al. 1997; Morita et al. 1999; Baker and Foster 2002); and (3) breeding dates (Bagenal 1971; Ware 1975; Candolin 1998; Hendry et al. 1999; Poizat et al. 1999). These patterns have been described as morphological constraints, proximal environmental effects, or adaptations (Poizat et al. 2002).

The freshwater amphidromous sculpin, *Cottus hangiongensis* Mori, which is a small bottom-dwelling fish that has a polygynous mating system with eggs guarded by a parental male until hatching, is restricted to northern Japan, Korea and southern Sakhalin, Russia (Goto 1986, 1987a, 1990). In the Daitobetsu River located in the southern part of Hokkaido Island, Japan, male *C. hangiongensis* are known to have clinal differences in population density, sex ratio, age composition and life history along the course of rivers (Goto 1987b, 1989b, 1993a). Sexual maturity occurs earlier in life when fish are a small size in downstream habitats, versus later in life when fish are a larger size in upstream habitats (Goto 1987b, 1989c), although most females are restricted to downstream habitats and reach the first maturity at about 55 mm or more in standard length (Goto 1989b). During the reproductive season, usually in April to May, the larger males inhabiting the upstream area migrate downstream, where their spawning grounds are located, and mate with many females. The smaller males in the downstream habitats stay there and mate with only a few or no females (Goto 1987b, 1993b; Goto and Nakano 1993).

In this study, we sought to elucidate the causes and formation mechanisms of clinal life-history variation in male *C. hangiongensis* along the course of the Daitobetsu River. We examined the genetic heterogeneity of upstream, midstream, and downstream habitat groups using microsatellite markers to infer whether such life-history variation was environmentally induced and surveyed the differential growth patterns among juvenile males of this species before and after ascent to the mouth of the river.

MATERIALS AND METHODS

Sample collection

Adult male *C. hangiongensis* were collected in August 2005 from upstream (St. 1), midstream (St. 2), and downstream (St. 3) sites of the Daitobetsu River. The river is about 21 km in length and has a considerably steep gradient of channels (4/1000–9/1000 in the slope of river bed) in the lower course from the mouth of the river to 4 km from the upstream site (Fig. 1). Thirty-nine (79.5 ± 2.1 mm standard length [SL], mean and SD), 26 (98.8 ± 4.8 mm SL), and 27 (103.6 ± 2.0 mm SL) adult individuals were collected from the upstream, midstream, and downstream sites, respectively (Table 1). All fish specimens were immediately fixed in 99.5 % ethanol and stored at the laboratory of Faculty of Fisheries, Hokkaido University, Hakodate, Japan, until a microsatellite (msDNA) analysis was performed.

Table 1 Number and body length of adult Cottus hangiongensis collected from the Daitobetsu River in 2005 used for microsatellite DNA analysis

			Sex		**Standard length**
Collection site	**Date of collection**	**Number of fish**	**Male**	**Female**	**Mean ± SD (mm)**
Site 1	1 and 15 August	39	1	38	79.5 ± 2.1
Site 2	3 and 13 August	24	24	2	98.8 ± 4.8
Site 3	1 and 10 August	27	27	0	103.6 ± 2.0
Total		92	52	40	

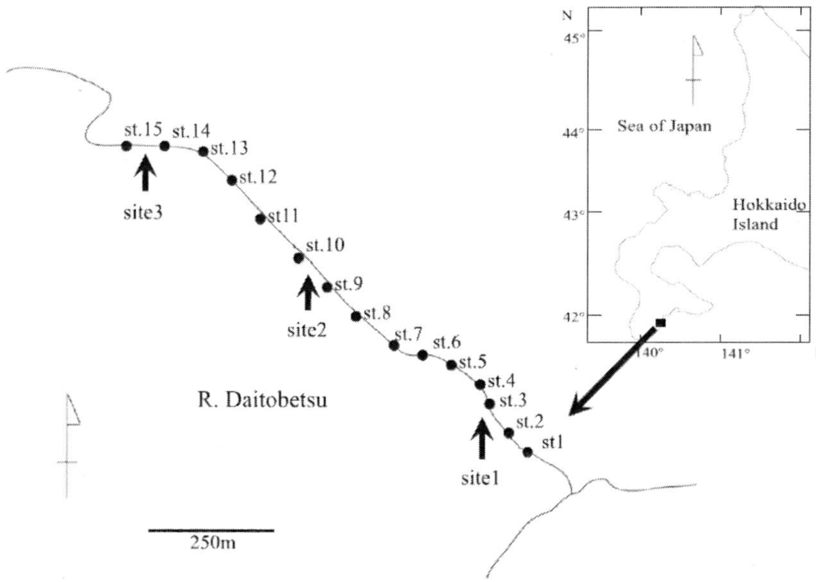

Figure. 1 Map showing the location of the Daitobetsu River and sampling sites for adult males (*black circle*) and upstream migrating juveniles (*black circle*) of *Cottus hangiongensis*

To examine the body size of downstream drifting larvae of *C. hangiongensis* and the duration of their drift, individuals were collected at the mouth of the river using a larva net (net diameter: 50 × 30 cm, net length: 90 cm, net mesh size: 1 mm) at 10-day intervals during the period from early May to mid-June in 2005. The standard length of larvae was measured to the nearest 0.1 mm by using a caliper. To estimate the age (in days) of juvenile male *C. hangiongensis*transitioning from the sea to the mouth of the river, collection of juvenile samples was carried out from late June to mid-July in 2005.

Genetic analysis of heterogeneity among the three different habitat groups

Total DNA extracted from the fin clips was preserved in 99.5 % ethanol using methods previously described by Yokoyama and Goto (2005). Samples were genotyped at five microsatellite loci: Cgo56MEHU, Cgo1016PBBE, Cgo05ZIM, Cgo18ZIM, and Cgo33ZIM, which were derived from *C. gobio* (Englbrecht et al. 1999). Individual amplification volumes (total 10 μL) contained 1 μL template DNA and 2 μM forward and reverse primers in 2× QIAGEN Multiplex PCR Master Mix containing HotStart Taq DNA polymerase. An initial activation step (15 min, 95 °C); followed by

35 cycles of denaturation (30 s, 94 °C), annealing (90 s, see Englbrecht et al. [1999] for annealing temperatures for each primer set), and an extension phase (60 s, 72 °C); followed by a final extension hold (30 min, 60 °C). Amplified products were analysed using an ABI PRISM 3130 Genetic Analyzer (Life Technologies Foster City, CA, USA) with LIZ500 size standards (Life Technologies), and Peak Scanner Software v1.0 (Life Technologies) was used for fragment analysis. msDNA genotype frequencies for each of the three habitat groups were tested for deviations from Hardy-Weinberg equilibrium (HWE) and linkage disequilibrium by an exact test using ARLEQUIN version 3.1 (Excoffier et al. 2005).

Age and size determination for the juveniles migrating upstream from the sea

Sagittal otoliths extracted from juvenile individuals following upstream migration from the sea were used to count the number of daily rings under a binocular microscope. Estimated age (days) after hatching and duration (days) of marine life were measured according to the methods described by Tsukamoto and Kajihara (1987) and Goto and Arai (2006). The estimated age was treated as the duration of marine life of the individual fish because the larvae of *C. hangiongensis* appeared to flow down to the sea less than 3 h after hatching in a small stream, such as the Daitobetsu River (Goto 1981).

Field survey on larval downstream drift and juvenile upstream migration

To elucidate the early life history and upstream migration of juveniles, we examined the ecological aspects of larval drift to the sea and the transition of juveniles from the sea to the river mouth. In the period from late April to early June in 2005, drifting *C. hangiongensis*larvae were collected using a plankton net (net diameter: 50 × 30 cm, net mesh size: 1 mm) at 2-h intervals after sunset. Using the data on larval drift, we estimated the duration of larval drift, the number of larvae collected on each day surveyed, and the body size of larvae collected.

During the period from late May to mid-June, juveniles transitioning from the sea to the mouth of the river were collected using a trap net (length, 120 cm; width, 80 cm; height, 30 cm; mesh size, 2 mm). The net was positioned directly downstream of the mouth of the river on every sampling day. The collected fish were removed from the net twice a day, at 07:00 and 19:00 h, and the number of fish was counted at each sampling time. The SL of the collected fish was measured to the nearest 0.1 mm using a caliper.

To determine the size composition dynamics of ascending juveniles, regression analysis was used to examine the relationship between the SL (mm) of the ascending juveniles and the progress time (days) after the start of ascension from the sea.

To explain the upstream dispersal of juvenile individuals in the lower course (from the mouth of river to 1.5 km upstream) of the Daitobetsu River, 15 sampling sections were established (Fig. 1). Juveniles were collected from each sampling section using a backpack electro fisher (Smith Root Inc., USA) during the period from August to November 2005 and during February 2006. The SL of the collected juvenile fish was measured to the nearest 0.1 mm. To clarify the relationship between the size of juveniles (mm SL) and distance (m) from the mouth of the river, regression analysis was carried out using the SL of juveniles (*Ls*) as the dependent variable and the distance from the mouth of the river (*Dm*) as the independent variable.

Statistic analyses

Differences in genotype frequencies were tested among samples from the three different habitats using the chi-square test. The goodness-of-fit test for the HWE was performed for samples from each of the three habitats and all samples of adult *Cottus hangiongensis* using the chi-square test.

Correlation coefficients (r^2) between the mean total length (mm) of downstream drifting larvae and the sampling date, the SL of upstream migrating juveniles (*Lm*) and the sampling date, the date when the juveniles migrated up to the river mouth (*Dm*) and the *Lm*, the *Dm* and the duration of marine life (*Ls*), the *Dh* and the *Ls*, and the *Lm* and the *Ls* were assessed with single regression analyses. Statistical *p*-values of 0.05 or less were treated as statistically significant. All statistical analyses were performed using the statistical software package R 2.62 (R. Development Core Team 2008).

RESULTS

Genetic heterogeneity among the three different habitat groups

Microsatellite DNA marker analysis revealed no genetic differences among the three habitat groups at three loci, although a significant difference was found at a single locus (Cgo05ZIM, Table 2). When the data for all the fish samples collected from the three different habitats were combined, no loci exhibited a significant departure from the HWE (Table 2).

Table 2 Genotype frequencies and Hardy-Weinberg expectation at 5 loci, and the results of chi-square test for goodness of fit between observed and expected values

Table 2 Genotype frequencies and Hardy-Weinberg expectation at 5 loci, and the results of chi-square test for goodness of fit between observed and expected values

Locus	Genotype	Site 1		Site 2		Site 3		All the samples	
		Observed	Expected	Observed	Expected	Observed	Expected	Observed	Expected
Cgo1016PBBE	*a/a*	39	39	26	26	27	27	92	92
Cgo33ZIM	*a/a*	28	28.6	17	16.08	17	17.04	62	61.74
	a/b	9	7.92	6	8.04	8	7.3	23	23.2
	a/c	1	0.88	1	0.8	1	1.62	3	3.32
	b/b	0	0.48	2	0.88	0	0.68	2	2.09
	b/c	0	0.12	0	0.2	1	0.34	1	0.62
	c/c	0	0	0	0	0	0.02	0	0.03
		(*P*=0.941)		(*P*=0.691)		(*P*=0.808)		(*P*=0.998)	
Cgo56MEHU	*a/a*	36	36.2	26	26	27	27	89	89.01
	a/b	2	1.95	0	0	0	0	2	1.98
	b/b	0	0.03	0	0	0	0	0	0.01
		(*P*=0.98)						(*P*=0.995)	
Cgo05ZIM	*a/a*	14	10.99	9	7.54	7	4.72	30	22.88
	a/b	9	12.96	4	5.38	2	6.16	15	24.4
	a/c	1	2.82	2	2.12	1	2.41	4	8.64
	a/d	2	2.25	4	0.09	6	0.72	12	12.2
	b/b	7	3.56	0	5.92	5	3.7	12	6.3
	b/c	0	1.62	1	1.62	0	3.7	1	4.56
	b/d	0	1.3	5	0.58	3	2.41	8	6.44
	c/c	1	0.14	0	0.96	4	2.01	5	0.76
	c/d	2	0.28	0	0.63	0	1.45	2	2.28
	d/d	0	0.08	1	1.16	0	0.72	1	1.54
		(*P*=0.481)		(*P*=0.005)		(*P*=0.674)		(*P*=0.364)	
Cgo18ZIM	*a/a*	17	15.82	10	9.48	11	9.85	38	34.84
	a/b	2	2.81	6	4.15	4	3.69	12	10.96
	a/c	6	7.73	5	7.11	5	7.38	16	22.58
	a/d	3	2.81	1	1.78	1	1.23	5	5.8
	b/b	0	0.13	0	0.45	0	0.35	0	0.8
	b/c	1	0.69	1	1.56	2	1.38	4	3.52
	b/d	1	0.25	0	0.39	0	0.23	1	0.91
	c/c	2	0.95	2	1.33	2	1.38	6	3.52
	c/d	0	0.69	2	0.67	1	0.46	3	1.86
	d/d	0	0.13	0	0.08	0	0.04	0	0.21
		(*P*=0.814)		(*P*=0.745)		(*P*=0.973)		(*P*=0.746)	

Larval downstream drift and juvenile upstream migration

Downstream drift of larvae of this species was detected mostly during the period from mid-May to early June 2005, with a surge on June 5, 2005. The mean total length of drifting larvae was 7.6 ± 0.4 mm ($n = 347$). No significant correlation was found between the mean total length of larvae at each sampling day and the sampling date (Correlation coefficient $r^2 = 0.002$, $p = 0.608$; Fig. 2).

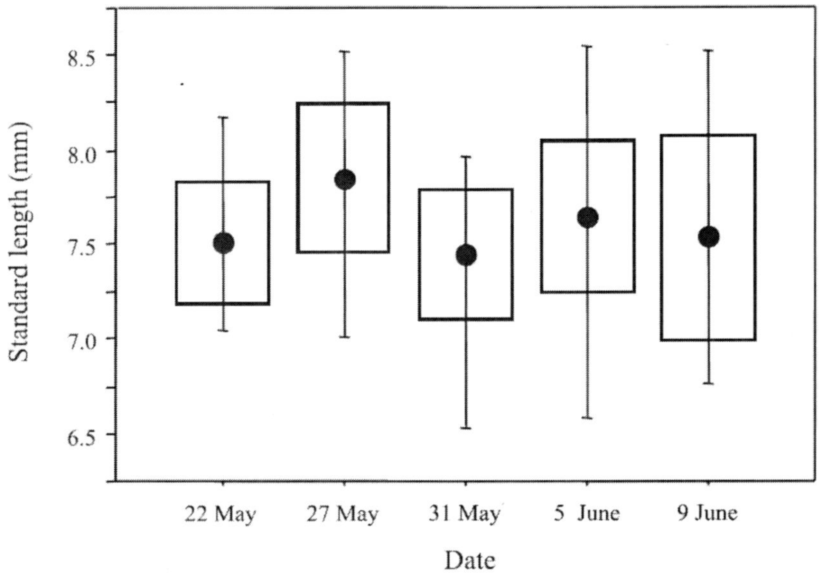

Figure. 2 Standard length of downstream-migrating *Cottus hangiongensis* larvae at each sampling day from 22 May to 9 June in the Daitoubetsu River in 2005. *Dot, square,* and *vertical bar* indicate the mean, the standard deviation and the range, respectively

Upstream migration of juvenile fish having a mean SL of 13.9 ± 0.03 mm ($n = 649$) began on June 10, 2005 and ceased on June 26, 2005 (Fig. 3). The SL of upstream migrating juveniles (*Lm*) tended to decrease over time ($r^2 = 0.028$, $p < 0.001$; Fig. 4). Otolith daily rings and hatching marks were observed in upstream migrating juvenile *C. hangiongensis.* A significant correlation was found between the date when the juveniles migrated to the mouth of the Daitobetsu River from the sea (juvenile migration date: *Dm*) and the SL of juveniles (*Ls*), as estimated by otolith daily rings analysis in 2005 ($r^2 = 0.04$, $p < 0.0001$; Fig. 5). Analysis revealed that the upstream migration of juveniles occurred in 33- to 60-day-old individuals, indicating that the hatching dates (*Dhs*) of upstream migrating juveniles ranged from April 21 to May 17, 2005.

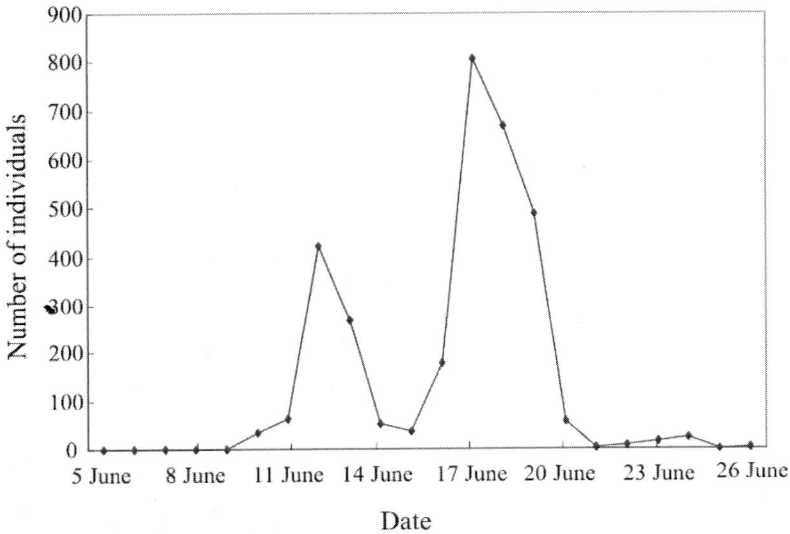

Figure. 3 Chronological changes in the number of upstream migrating juvenile *Cottus hangiongensis* collected in the mouth of the Daitobetsu River in 2005. The surges are found on 12th and 17th June

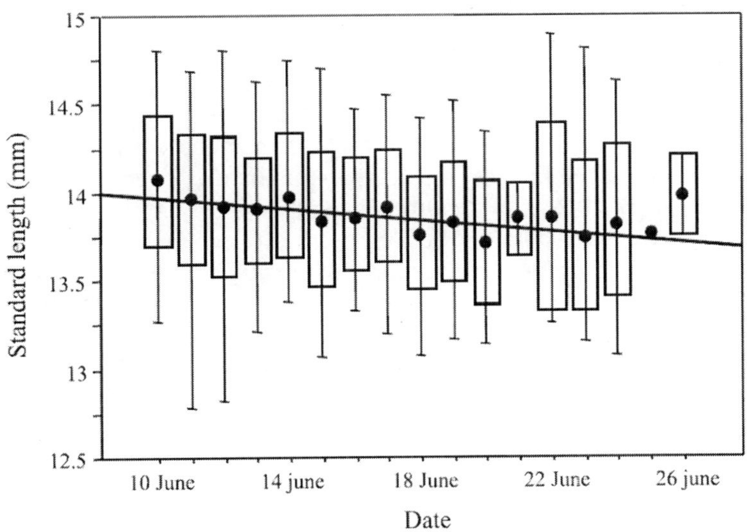

Figure. 4 A relationship between the day migrating to the mouth of the Daitobetsu River from the sea and the mean standard length of *Cottus hangiongensis* juveniles. *Dot, square,* and *vertical bar* indicate the mean, the standard deviation and the range, respectively. The *line* in figure indicates a linear regression showing a significant negative correlation between the two variables

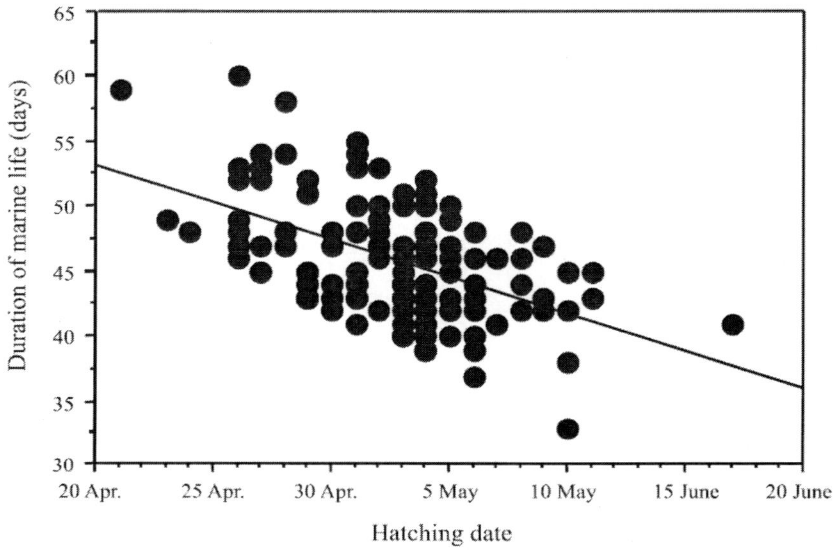

Figure. 5 A relationship between the hatching date (D_hs) and the duration of marine life (D_hs) of upstream-migrating *Cottus hangiongensis* juveniles collected in the mouth of the Daitobetsu River in 2005. The line in figure indicates a linear regression showing a significant negative correlation between the two variables

A significant positive correlation was detected between the *Dm* and the duration of marine life (*Ls*) in each juvenile fish, as estimated from otolith daily rings ($r^2 = 0.349, p < 0.0001$; Fig. 6a). Similarly, a significant negative correlation was found between the *Dh* and *Ls* ($r^2 = 0.249, p < 0.001$; Fig. 6b). On the other hand, significant positive correlations were found between the distance to the collecting site from the sea (*Dn*) and the *Ls* in September ($n = 36, r^2 = 0.720, p < 0.001$), October ($n = 38, r^2 = 0.264, p < 0.001$), and November in 2005 ($n = 24, r^2 = 0.231, p < 0.05$), and February in 2006 ($n = 29, r^2 = 0.324, p < 0.01$; Fig. 7).

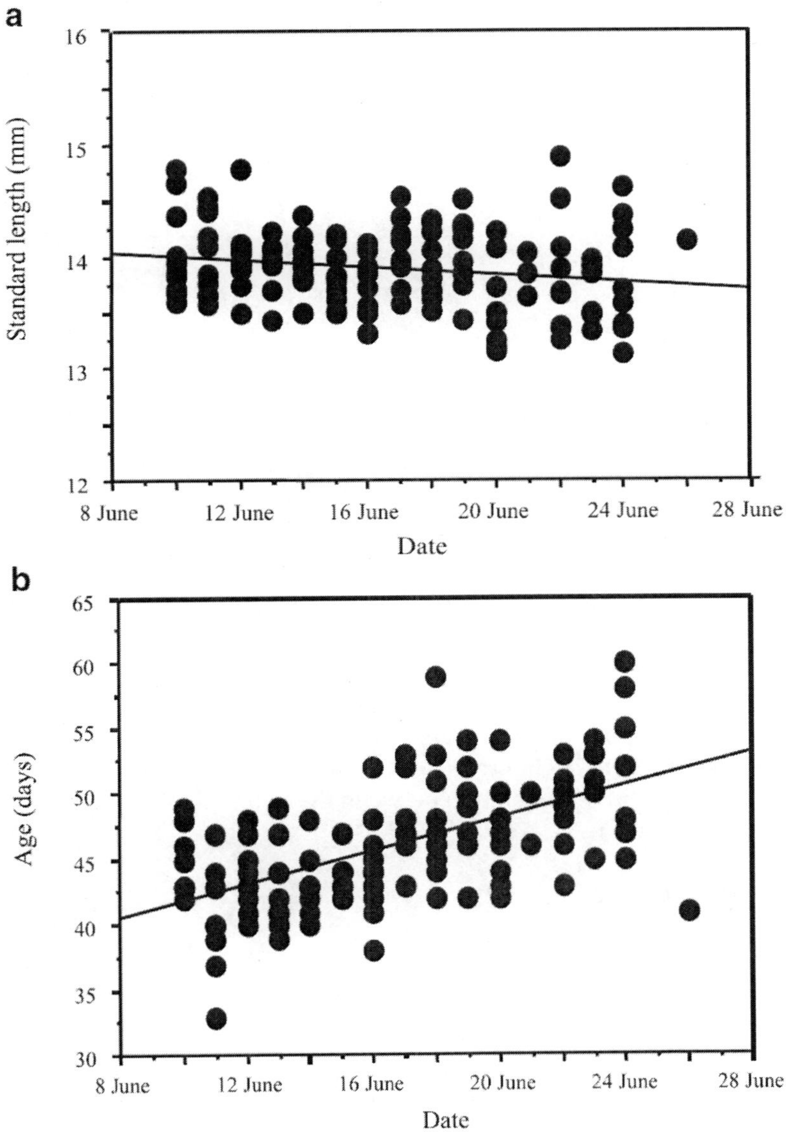

Figure. 6 Relationships between the days after hatching and the number of otolith daily rings (**a**), and between the days after hatching and the standard length (**b**) of downstream-migrating *Cottus hangiongensis* larvae in the Daitobetsu River. The *line* in figure (**a**) indicates a linear regression showing a significant negative correlation between the two variables, and *line* in figure (**b**) indicates that showing a significant positive correlation between them

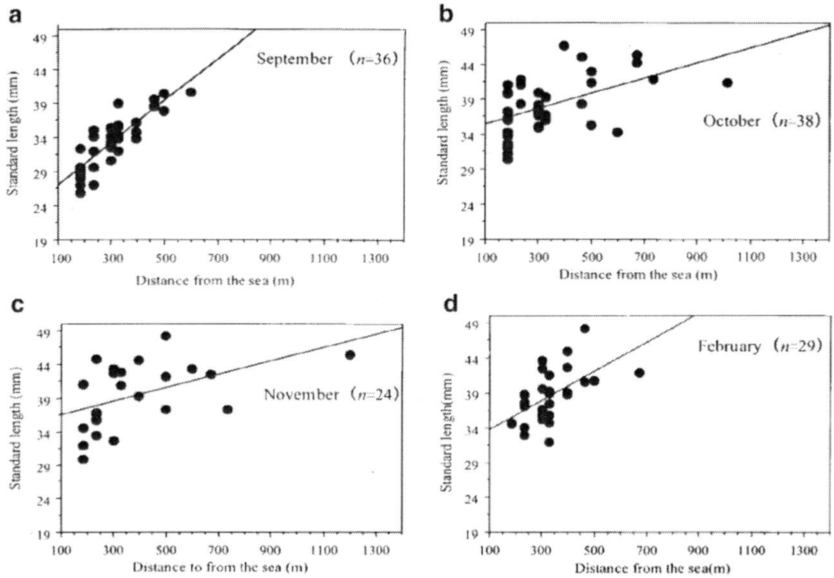

Figure. 7 Relationships between the distance to the collection site from the sea (Dn) and the standard length of juvenile *Cottus hangiongensis* collected in each sampling site of the Daitobetsu River in the period from September 2005 to February 2006. **a** data in September 2005, **b** data in October 2005, **c** data in November 2005 and **d** data in February 2006. The *line* in each of figures (**a–d**) indicates a regression showing a significant positive correlation between the two variables

DISCUSSION

Genetic homogeneity among the three different habitat groups

The limited microsatellite DNA analysis showed no genetic heterogeneity among the three different habitat groups of *C. hangiongensis* along the course of the Daitobetsu River. Additionally, the combined fish samples collected from the three habitats fit the HWE. These results implied that the *C. hangiongensis* population was a single panmictic population. This assumption was supported by the results of genetic composition examination using polymorphic allozyme loci for the upstream and downstream adult samples of the *C. hangiongensis* population (Usui and Goto 1987) and of a common garden experiment on rearing 1+ aged male individuals of this species under different density conditions in which the maturity of fishes was delayed at low density (Goto 1993a). Thus, these results suggest that clinal life-history variation in males of this species (Goto 1988, 1989c, 1993a) may be a result of phenotypic plasticity. To obtain a more reliable result of the panmictic, however, reexamination of the homogeneity among a number of habitat groups of this species using a greater number of polymorphic microsatellite DNA markers is necessary in the near future.

Downstream and upstream migration and growth during early ontogenetic development

For downstream drift of larvae, no significant correlation was found between body size and the date of larval drift. This result indicated that the larvae drifted downstream could reach the mouth of the river while maintaining almost the same body size throughout the period of their downstream drift. For upstream migration of juveniles, our otolith daily rings analysis revealed that juveniles hatching later had a shorter larval duration in the marine pelagic life stage. This suggests that juveniles experiencing marine pelagic life stage later in the season migrated earlier to the mouth of the river from the sea, probably due to their increased growth in the sea under higher water temperature.

In Lake Biwa, Japan, the landlocked ayu *Plecoglossus altivelis altivelis*, which was derived from the freshwater amphidromous population, includes the small "residents," large "spring migrants," and intermediate "summer migrants" (Tsukamoto et al. 1987). These different subpopulations migrate to the mouths of rivers from the lake at different times of the season and have different ages and sizes at their time of migration (Tsukamoto et al. 1987). This study found that juvenile *P. altivelis altivelis* with earlier birth dates and/or large growth rates migrate upstream earlier. However, this was not the case with the amphidromous sculpin *C. hangiongensis*.

In the *C. hangiongensis* population, the date of upstream migration of juveniles was strongly influenced by the growth rates of larvae during the marine pelagic life stage, and significant positive correlations were found between the distance to the collecting site from the sea and the SL of juvenile fishes during upstream migration in the river from September to February in the next year. These results demonstrated that a clinal body-size variation in the juvenile fishes had already been formed along the downstream course of the river after about 3 months since the ascent of juveniles to the mouth of the river and may imply that larger juveniles inhabiting the downstream areas can disperse to more upstream areas and grow larger during the period following their migration upstream to the mouth of the river, most likely due to the abundance of space and food for individual *C. hangiongensis* juveniles in the upstream areas (Goto 1989a, b, 1993a). Therefore, it is possible that such a body-size variation in the juvenile fishes just having recently ascended to the mouth of the river may be one of the proximate causes for forming the clinal life-history variation in this species.

Formation mechanisms of plastic life-history variation along the course of a river

Life history theory predicts that delayed maturation at larger sizes and older ages occurs when juvenile mortality strongly increases with decreasing growth, or when both juvenile and adult mortality increase with decreasing growth rate (Stearns and Koella 1986; Stearns 1989, 1992; Kawecki and Stearns 1993; Jonsson et al. 1996; Taborsky 2006). It is well known that these conditions apply when mortality decreases with body size, as has often been found in fish (Henderson et al. 1988; Wootton 1990; Metcalfe and Monaghan 2001; Taborsky 2006).

In male *C. hangiongensis*, the first maturation of adult individuals that inhabit the upstream area may be delayed until they have reached a larger size and are older than those in the downstream habitat group. This may occur due to the high mortality risk during the juvenile period, when fish migrate from the downstream area to the upstream area. Such high mortality is possibly caused by factors such as predation by the white-spotted charr *Salvelinus leucomaenis* and carnivorous birds, such as the common kingfisher *Alcedo atthis* and the brown dipper *Cinclus pallasii*, and by exhaustion as a result of energy loss from upstream migration. Despite such differences in mortality risks resulting from the life-history variation found in other freshwater fishes, our present empirical study may contribute to verifying the validation of the prediction proposed by Stearns (1992) and Taborsky (2006) that juvenile growth conditions may initiate a feedback loop resulting in individual life-history trajectories located somewhere between 'slow juvenile growth-early reproduction-high reproductive rate' and 'fast juvenile growth-late reproduction-low reproductive rate'. In males of this species, as a result, clinal life-history variation would be formed by the following process: when the upstream migrating juveniles stay in the downstream region, they grow slowly and attain maturity at small body size and younger ages, whereas juveniles that migrate to the upstream area and stay there grow faster and mature at larger body sizes and older ages (Goto 1989a, 1990, 1993a).

Recently, similar life-history variation was reported in the male Japanese fluvial sculpin *C. pollux* by Natsumeda et al. (2013). They found that the variation in life-history traits for males of the congeneric *C. pollux* occurred in relation to nest abundance along a stream course. From this, they concluded that such life history variation was caused by different sexual selection regimes related to differences in nest abundance between the two research sites. However, this may be not the case in amphidromous *C. hangiongensis*, because of its different spawning ecology from that of fluvial *C. pollux* as follows: in the population of *C. hangiongensis*, spawning occurs in a restricted part of the downstream area after the mature females and males gather from the original habitats to the spawning area (Goto 1988), such that different sexual selection regimes found in the *C. pollux* population probably do not function in the population of *C. hangiongensis*.

Therefore, based on our current study, it is possible to consider that this clinal variation of life-history traits within a single gene pool of *C. hangiongensis* population may be adaptive for possible exploitation of broad and heterogeneous habitats along the course of a river, and may stem from the plastic phenotypic responses of juvenile fish to the heterogeneous environments in which they are migrating along the course of a steep mountainous river.

ACKNOWLEDGMENTS

We are grateful to T. Noda, T. Kitamura and M. Uranishi of Graduate School of Fisheries Sciences, Hokkaido University for help in collecting the fish samples, and invaluable technical support for ecological field survey. This work was partly supported by Grant-in-Aid (No. 15405008) for Scientific Research, Japan Society for the Promotion of Science to AG. The field surveys and fish collections carried out in this study comply with the current laws in Japan.

REFERENCES

1. Bagenal TB (1971) The interrelation of the size of fish eggs, the date of spawning and the reproductive cycle. J Fish Biol 3:207–219
2. Baker JA, Foster SA (2002) Phenotypic plasticity for life history traits in a stream population of the threespine stickleback, *Gasterosteus aculeatus* L. Ecol Freshw Fish 11:20–29
3. Candolin U (1998) Reproduction under predation risk and the trade-off between current and future reproduction in the threespine stickleback. Proc R Soc Lond B 265:1171–1176
4. Dewitt TJ, Scheiner SM (2004) Phenotypic Plasticity: Functional and Conceptual Approaches. Oxford University Press, Oxford
5. Englbrecht CC, Largiader CR, Hanfling B, Tautz D (1999) Isolation and characterization of polymorphic microsatellite loci in the European bullhead *Cottus*

gobio L. (Osteichthyes) and their applicability to related taxa. Mol Ecol Notes 8:1957–1969

6. Excoffier L, Laval G, Schneider S (2005) Arlequin (version 3.0): An integrated software package for population genetics data analysis. Evol Bioinform Online 1:47–50
7. Goto A (1981) Life history and distribution of a freshwater sculpin, *Cottus hangiongensis*. Bull Fac Fish Hokkaido Univ 32:10–21
8. Goto A (1986) Movement and population size of the river sculpin *Cottus hangiongensis* in the Daitobetsu River of southern Hokkaido. Jpn J Ichthyol 32:421–430
9. Goto A (1987a) Polygyny in the river sculpin, *Cottus hangiongensis* (Pisces: Cottidae), with special reference to male mating success. Copeia 1987:32–40
10. Goto A (1987b) Life history variation in males of the river sculpin, *Cottus hangiongensis*, along the course of a river. Environ Biol Fish 19:82–92
11. Goto A (1988) Reproductive behavior and homing after downstream spawning migration in the river sculpin, *Cottus hangiongensis*. Jpn J Ichthyol 34:488–496
12. Goto A (1989a) Growth patterns, and age and size at maturity in female *Cottus hangiongensis*, with special reference to their life-history variation. Jpn J Ichthyol 36:90–99
13. Goto A (1989b) Growth differences in males of the river-sculpin *Cottus hangiongensis* along a river course, a correlation of the life-history variation. Environ Biol Fish 24:241–249
14. Goto A (1989c) Reproductive styles and strategies in freshwater sculpins. In: Goto A, Maekawa K (eds) Reproductive Behaviors in Fish: the Modes and Strategies. Tokai University Press, Tokyo, pp 73–84
15. Goto A (1990) Alternative life-history styles of Japanese freshwater sculpins revisited. Environ Biol Fish 28:101–112
16. Goto A (1993a) Clinal life-history variation in the river-sculpin *Cottus hangiongensis*: an example of phenotypic plasticity. Jpn J Ichthyol 39:363–370
17. Goto A (1993b) Duration of male mating activity and male mate choice in the river-sculpin, *Cottus hangiongensis*. Jpn J Ichthyol 40:269–272
18. Goto A, Arai T (2006) Diverse migratory histories of Japanese *Trachidermus* and *Cottus*species (Cottidae) as inferred from otolith microchemistry. J Fish Biol 68:1731–1741
19. Goto A, Nakano S (1993) Distribution and ecology of freshwater fishes in Hokkaido. In: Higashi S, Osawa A, Kanagawa K (eds) Biodiversity and ecology in the northernmost Japan. Hokkaido Univ. Press, Sapporo, pp 113–126
20. Henderson PA, Holmes RHA, Bamber RN (1988) Size selective overwintering mortality in the sand smelt, *Atherina boyeri* Risso, and its role in population regulation. J Fish Biol 33:221–234
21. Hendry AP, Berg OK, Quinn TP (1999) Condition-dependence and adaptation-by-time: breeding date, life history, and energy allocation within a population of salmon. Oikos 85:499–514
22. Jonsson N, Jonsson B, Fleming IA (1996) Does early growth cause phenotypically plastic response in egg production of Atlantic Salmon. Funct Ecol 10:89–96
23. Kawecki TJ, Stearns SC (1993) The evolution of life histories in spatially heterogeneous environments: optimal reaction norms revisited. Evol Ecol 7:155–174
24. Lobon-Cervia J, Utrilla CG, Rincon PA, Amezcua F (1997) Environmentally induced spatio-temporal variations in the fecundity of brown trout *Salmo trutta* L.: trade-offs between egg size and number. Freshw Biol 38:277–288
25. Metcalfe N, Monaghan P (2001) Compensation for a bad start: grow now and pay later? Trends Ecol Evol 16:254–260PubMed
26. Morita K, Yamamoto S, Takeshima Y, Matsuishi T, Kanno Y, Nishimura K (1999) Effect of maternal growth history on egg number and size in wild white-spottrd char (*Salvelinus leucomaenis*). Can J Fish Aquat Sci 56:1585–1589

27. Natsumeda T, Yuma M, Hori M (2013) Variation in life-history traits of male Japanese fluvial sculpin *Cottus pollux* in relation to nest abundance along a stream course. Environ Biol Fish. doi:10.1007/s10641-013-0108-8
28. Poizat G, Rosecchi E, Crivelii AJ (1999) Emprical evidence of a trade-off between reproductive effort and expectation of future reproduction in female three-spined sticklebacks. Proc R Soc Lond B 266:1543–1548
29. Poizat G, Rosecchi E, Crivelii AJ (2002) Life-history variation within a three-spined stickleback population in the Camargue. J Fish Biol 60:1296–1307
30. R Development Core Team (2008) R: A language and Environment for Statistical Computing. R Foundation for Statistical Computing, Vienna, Austria. ISBN 3-900051-07-0. http://www.R-project.org
31. Scheiner SM (1993) Genetics and evolution of phenotypic plasticity. Ann Rev Ecol Syst 24:35–68
32. Stearns SC (1989) The evolutionary significance of phenotypic plasticity. Bioscience 39:436–445
33. Stearns SC (1992) The evolution of life histories. Oxford University Press, Oxford
34. Stearns SC, Koella J (1986) The evolution of phenotypic plasticity in life-history traits, predictions for norm of reaction for age- and size-at-maturity. Evolution 40:893–913
35. Taborsky B (2006) The influence of juvenile and adult environments on life-history trajectories. Proc R Soc B 273:741–750PubMedCentralPubMed
36. Tsukamoto K, Kajihara T (1987) Age determination of ayu with otolith. Nippon Suisan Gakkaishi 53:1985–1997
37. Tsukamoto K, Ishida R, Nara K, Kajihara T (1987) Switching of size and migratory patterns in successive generations of landlocked ayu. Am Fish Soc Symp 1:492–506
38. Usui K, Goto A (1987) Characteristics of the large-sized altricial males in the river sculpin, *Cottus hangiongensis*. Bull Fac Fish Hokkaido University 38:38–49
39. Ware DM (1975) Relation between egg size, growth, and natural mortality of larval fish. J Fish Res Board Can 32:2503–2512
40. Wootton RJ (1990) Ecology of Teleost Fishes. Croom London, London
41. Yokoyama R, Goto A (2005) Evolutionary history of freshwater sculpins, genus *Cottus*(Teleostei: Cottidae) and related taxa, as inferred from mitochondrial DNA phylogeny. Mol Phylogenet Evol 36:654–668

CHAPTER 10

Fishery Status and Taxonomy of the Carangids (Pisces) in the Northern Arabian Sea Coast of Pakistan

Nazia Qamar[1], Sher Khan Panhwar[1*] and Ghazala Siddiqui[1]

[1] Center of Excellence in Marine Biology, University of Karachi, Sindh, Pakistan

ABSTRACT

The objectives of this study were i) to evaluate number of existing members of the family Carangidae in the area ii) to establish a distinguishable and lucid key based on the taxonomic characteristics, meristic count and otolith description. In this study, thirty-six species were collected from the main fish landing facilities between 2012~2015. Fish body colour, taxonomic characteristics, fin rays and otolith shape description were used to identify each species. Otolith description comprises of shape of ostium, sulcus and margins of anterior and posterior surface along with distinct definite shape possess by each species make it easier for identification.

This species catalogue would treasure and latest information on the existing species of the family Carangidae in the area, and provide an ease to identification especially for closely resembling species. In addition, this list includes seven unreported species from the area.

Keywords: Carangidae, taxonomy, otolith, Fishery, Northern Arabian Sea Coast, Pakistan

1. BACKGROUND INFORMATION

Family Carangidae comprises 140 species and is widely distributed throughout the Western Indian Ocean and Western Pacific, from Japan to Australia and eastward to Fiji [1–3] (**Table 1**). The carangids inhabit coastal, estuarine, and marine waters of tropical, subtropical, and temperate regions. This family has attracted many researchers around the world due to its recognition for sport fishing, recreational, and substantial marketable value particularly species of *Scomberoides, Megalaspis, Parastromateus,* and *Caranx*. Genus *Scomberoides* comprises of four species: *S. commersonnianus, S. lysan, S. tala*, and *S. tol* found in the Indian Ocean. In terms of biomass, many species of this family are well known for instance [4] recorded 27.2 kg *Caranx hippos*, this fish species is considered as a recreational and commercial throughout its wide distribution. Popularity of jacks in game fishing highlighted [5, 6] and added that because of huge demand and human consumption indicated that stocks of such fishes have been gradually depleting throughout the world. The carangids has specific role in food web and ecosystem such as these are large predatory species exclusively feeds on clupeid, engraulid, trichurid, cuttlefish [7]. They are active swimmer fishes, pelagic, and gregarious, live in turbid water by forming school (small to large) or solitary, in clear water [8]. Most of the species rather common in the inshore reef, coastal or in shore waters within the depth range of 20–100 m. Some species such as *Carangoides ferdau* and *C. hedlandensis* are benthic confined to 100–200 m depth [8]. Recently, Panhwar et al. and Qamar et al. [9, 10] documented the population dynamics of *Megalaspis cordyla, Scomberoides commersonnianus*, and *S. tol* in Pakistan. Skeletal anomalies found in *M. cordyla* collected from Karachi fish harbor and Ormara based on x-ray photograph was also reported [11].

TABLE 1. List of genera and species of the family Carangidae reported around world.

Author	Year	Location	Genera	Species
Present study	2015	Northern Arabian Sea Coast of Pakistan	19	36
Abdussamad et al. [14]	2013	Tuticorin, India.	20	60
Matsunuma et al. [15]	2011	East coast of Malay peninsula, Malaysia	15	27

Quigley [16]	2007	North-Eastern Atlantic & Mediterranean	14	25
Kuiter [17]	2000	South-eastern Australia	13	23
Gunn [18]	1990	Australia	18	63
Huda [19]	1988	Pakistan	16	34
Farooq and Niazi [20]	1988	Pakistan	4	9
Bianchi (FAO) [1]	1985	Pakistan	20	42
Laroche et al. [21]	1984	World wide	30	140
Nelson [22]	1984	World wide	32	140
Fischer & Bianchi [8]	1983	Indian ocean	22	65
Jalil & Khaliluddin [23]	1972	Pakistan	N/A	24

[i] - *Courtesy by Ahmed M, Marine Fisheries of Pakistan, 1985.

Torpedo trevally, *Megalaspis cordyla* is monotypic schooling species and comprises approximately 25% in terms of biomass to the family Carangidae in Pakistan. Marine fisheries statistics indicated highest landing (9722 mt) in 2001 and the lowest (3559 mt) in 2008 [11]. However, no separate data for queenfishes are recorded by MFD, though the members of *Scomberoides* genus have largely contributing in the capture fishery of the country, but recent trends indicated that the landing of genus *Scomberoides* has been reduced from 17,779 MT in 1999 to 9073 MT in 2009 [12]. The major portion of the catch belongs to *S. commersonnianus* whereas torpedo trevally, *Megalaspis cordyla* is also landed in sufficient quantity recorded at the major fish landing sites. The FAO, 2012 has figured out landing records of *S. commersonnianus* in the Western Indian Ocean, which have increased from 4994 in 2001 to 11,374 in 2010 [13]. In the local market, talang queen fish is mainly sold at approximately US$ 2–3/kg, whereas needle scaled queenfish, and torpedo trevally at approximately US$ 1–2 and 0.5 and 1.5 kg, respectively, depends upon the premium to the lowest quality. Carangids mainly caught with gill nets, seines on hook and line and trawl nets. The seasonal variations in the landings of *S. commersonnianus* showed higher quantities in between the period of August to October (Katti) and smaller quantities in April to May (Cheeta) following in June to July (Unaro) due to the excessive monsoon currents and closed fishing season in the area [1].

This family encompasses a diverse group of fishes known variously by common names such as jacks, scads, pompanos, queen fishes, king fishes, and trevallies, among others, which are characterized by variable body shapes elongated, fusiform to extremely ovate and strongly compressed [22, 24, 25]. The characteristics features of carangids are the presence of elongated Pterygiophores and one or two anal spines often embedded in adults [26], deeply forked caudal fin with equal lobes, slender caudal peduncle [27], and a distinctive swimming mode known as "carangiform" responsible for the name of this family [28]. A detailed description, comparison, ecological distribution, and key characters of the carangids documented from the Australian waters [18]. Abdussamad et al. [14] have reported occurrence of 60 species belonging to 20 genera of the family Carangidae based on macrotaxonomic characters from Indian waters. Further, they added a new morphological feature based on the first dorsal fin for the differentiation of carangids: group I, first dorsal fin is absent such as genus *Parastromateus*and *Alectis*; group II, described as modification of first dorsal fins into spines such as Genus *Scomberoides, Trachinotus, and Naucrates*; group III, well-developed first dorsal fin with spiny rays such as genus *Megalaspis, Elagatis, Decapterus, Seriolina, Seriola, Atropus, Ulua, Uraspis, Gnathanodon, Carangoides, Alepes, Caranx, Selaroides, Atule,* and *Selar*.

In Pakistani waters, the number of species is not clearly known, since Bianchi [1] identified 42 and Huda [19] 34 species. Nevertheless, the family Carangidae is a dominant finfish group in these waters contributing a major contribution of the commercial fisheries resources. According to Bianchi [1], species representing four subfamilies (tribes) of carangids in Pakistani waters: *Alectis ciliaris A. indicus, Alepes djedaba, A. melanoptera, A. vari, Atropus atropos, Atule mate, Carangoides caeruleopinnatus, C. chrysophrys, C. ferdau, C. fulvoguttatus, C. gymnostethus, C. malabaricus, C. bajad, C. hedlandensis, C. oblongus, C. plagiotaenia, C. praeustus, C. talamparoides, Caranx ignobilis, C. melampygus, C. sexfasciatus, C. para, Decapterus macrosoma, D. russelli, Elagatis bipinnulata, Gnathanodon speciosus, Megalaspis cordyla, Naucrates ductor, Parastromateus niger, Scomberoides commersonniansus, S. lysan, S. tol, Selar crumenophthalmus, Selaroides leptolepis, Seriolina nigrofasciata, Trachinotus africanus, T. baillonii, T. blochii, T. mookalee, Trachurus indicus, Uraspis secunda*.

In the present study, various surveys of major fish landing sites were made along the Sindh and Baluchistan coast during 2012–2015 to evaluate number of existing species of the family Carangidae in Pakistani waters. In these surveys, thirty-six species were identified with seven new from the area, namely *Alepes kleinii, Carangoides armatus, Caranx sem, Scomberoides tala, Seriola dumerili, Trachinotus russelii, Uraspis uraspis.* The morphometric and meristic count was taken for each specimen following the protocols of Smith-Vaniz [27]: Standard length(S L), head length (HL), curve and straight lateral line length (CLL, SLL), number of gill rakers (upper and lower), number of spines and fin rays, number of lateral line scales and scutes, number of anal spine, pattern of breast squamition, type of adipose eyelid, and all possible body measures. The parameters of each sagittal otolith such as length, height, weight, and shapes of cauda, ostium, excisura, and sulcus acusticus were recorded. Finally, the sex identification was established from macroscopic observation based on color pattern and size of the gonad [29].

The study was aimed to prepare a species catalogue with otolith descriptions for the available species of the family Carangidae to provide an ease in taxonomic identification of closely resembling species on the basis of characteristics derived from [18, 27] (**Table 2**), (**Figure 1**).

TABLE 2. Summary of the length and otolith data of the 36 species and sex.

Serial	Species (N)	Author name	TL	FL	GIRTH	BW	OL/TL	OH/OL	OW	SEX
1	*Alectis ciliaris* (2)	Bloch, 1788	43	39.5	26.2	895	50	2.2	0.0037	M
2	*Alectis indicus* (4)	Rüppell, 1830	23.5	20.5	24	108.4	50	2.1	0.0048	F
3	*Alepes djedaba*(18)	Forsskål, 1775	24	21.8	19.5	270	27	2.6	0.0048	M
4	*Alepes kleinii* (16)	Bloch, 1793	17.6	14.8	12.8	50	50	2.4	0.0034	M
5	*Alepes melanoptera* (5)	Swainson, 1839	17.5	15	11.5	52	44	2.1	0.0029	M
6	*Alepes vari* (9)	Cuvier, 1833	22.1	19	14.2	101.75	24	2.3	0.0047	F
7	*Atropus atropos*(25)	Schneider, 1801	15.8	13.5	16	60	57	2.1	0.0009	M
8	*Atule mate* (15)	Cuvier, 1833	27	23.4	16	61.08	45	2.1	0.0054	M
9	*Carangoides armatus* (4)	Rüppell, 1830	37	32	29	85	43	2.5	0.011	F
10	*Carangoides chrysophrys* (7)	Cuvier, 1833	21.4	18.5	18.5	136.47	44	2.4	0.0022	M
11	*Carangoides*	Whitley	18.	16	15	51	40	2.9	0.01	M

	hedlandensis (12)	, 1933	1						0	
12	*Carangides malabaricus* (20)	Bloch & Schneider	23	19.8	17	113	56	2.5	0.0042	F
13	*Carangoides praeustus*(18)	Bennett, 1830	20.8	18.2	14	94	44	2	0.0039	F
14	*Caranx ignobilis*(3)	Forsskål, 1775	28.5	24.5	21	265.4	36	1.9	0.0054	M
15	*Caranx para* (2)	Cuvier, 1833	16.5	15	13	53.5	44	2.7	0.0039	M
16	*Caranx sem* (2)	Cuvier, 1833	76	69.5	49	4000	36	0.7	0.0055	M
17	*Caranx sexfasciatus* (17)	Quoy & Gaimard, 1824	19	16.3	13	83	38	2.5	0.0069	M
18	*Decapterus russelli*(22)	Rüppell, 1830	19.5	18.5	9.5	60.8	58	3.6	0.0041	F
19	*Elagatis bipinnulata* (03)	Quoy & Gaimard, 1824	35	28.5	16	271	40	1	0.0024	M
20	*Gnathanodon speciosus* (02)	Forsskäl, 1775	20.8	18.5	14	94	56	1.8	0.0027	F
21	*Megalaspis cordyla* (298)	Linnaeus, 1758	42.7	38.7	22	710	50	2.5	0.0015	M
22	*Parastromateus niger* (56)	Bloch, 1975	25.6	27.2	24	320	56	1.8	0.0027	F
23	*Scomberoides commersonnianus* (1044)	Lacepêde, 1802	64.3	56.1	34.5	1900	33	2	0.0016	F
24	*Scomberoides lysan* (18)	Forsskäl, 1775	73	63	32.5	2200	38	1.1	0.0078	M
25	*Scomberoides tala*(9)	Cuvier, 1832	51	43.5	28	213.9	38	1.6	0.0083	F
26	*Scomberoides tol*(255)	Cuvier, 1832	68	59	24.5	1700	38	1.4	0.0079	M
27	*Selar crumenophthalmus*(23)	Bloch, 1793	21	19.5	12	93.2	60	2.4	0.0081	M
28	*Selaroides leptolepis* (8)	Cuvier, 1833	17	15	12	62.18	63	2.6	0.0046	F
29	*Seriola dumerili*(10)	Risso,1810	21.3	19	12.5	95	40	1.7	0.003	M
30	*Seriolina nigrofasciata* (14)	Rüppell, 1829	35.1	31.6	19.6	405.5	56	1.3	0.003	F
31	*Trachinotus*	Lacepêde,	15.	14	15	55.8	44	23	0.00	M

	baillonii (4)	1801	8						21	
32	*Trachinotus blochii* (6)	Lacepê de, 1801	26	22	23	226	49	2.5	0.00 17	M
33	*Trachinotus mookalee* (6)	Cuvier, 1832)	27. 8	22. 8	22	238	50	1.5	0.00 29	M
34	*Trachinotus russelii* (4)	Cuvier, 1832	54	40	32	1200	27	23	0.00 58	M
35	*Trachurus indicus*(3)	Nekras ov, 1966	27. 6	24	14	168	50	2.9	0.02 29	F
36	*Uraspis uraspis* (8)	Günther , 1860	25. 6	22. 5	21.4	257. 5	27	2.3	0.00 15	F

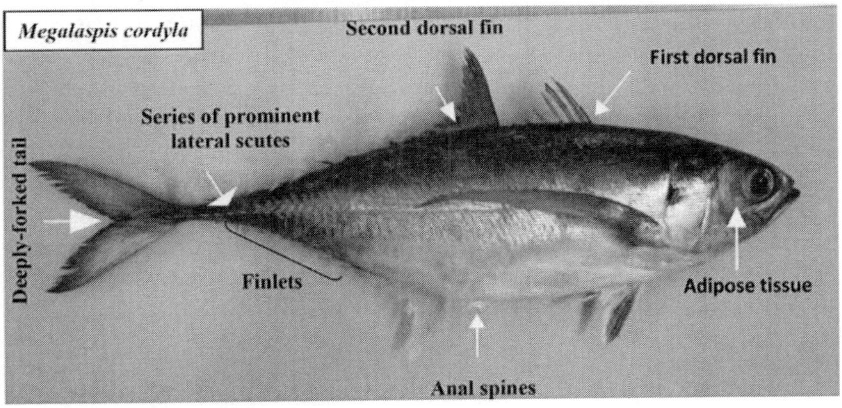

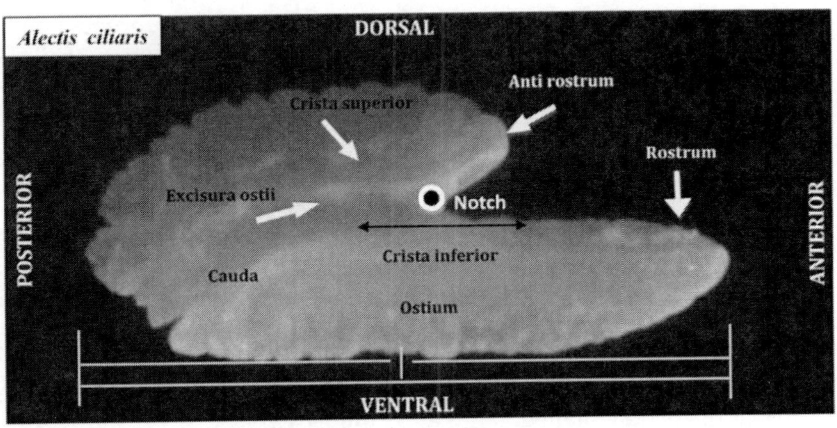

FIGURE 1. General Description of Fish and Otolith.

2. MORPHOLOGICAL AND OTOLITH DESCRIPTIONS 36 SPECIES

2.1. African Pompano, *Alectis Ciliaris*

Body has ovate (to pentagonal) shape, strongly compressed, tapering evenly from midpoint posteriorly and elongated with growth. Dorsal and ventral profile is equally convex. Eye is large and greater than snout length. Mouth is low with protractile jaws. Villiform tooth are arranged in bands in lower jaw, villiform teeth also present in tongue, palatines, and vomer. Maxilla extends two-thirds behind of eye. Head large comprised 30% of FL and nape is rounded. Gill rakers 5 upper and 13 on the lower limb of first gill arch. Dorsal fin with 7 short-free embedded spines. Second soft dorsal fin with 1 spine followed by 19 rays. First 7–8 rays of soft dorsal and anal fin are elongated and filamentous in juveniles. Pectoral fin long, falcate, extended beyond the straight lateral line and greater than pelvic fin. In juveniles, pelvic fin is elongated. Anal fin with 1 spine followed by 19 rays, and caudal fin is deeply forked. Lateral line is strongly concave above pectoral fin, deeply arched and junction of curved and straight lateral line is below 12–14 dorsal soft rays. Curve lateral line is greater than straight lateral line, straight lateral line with 22 scutes. Scales are minute and deeply embedded in the skin, in juveniles, 5–7 crosses extensive bands throughout the body present (**Figure A1**).

Otolith shape: lanceolated. Margins: dorsoventrally crenate. Sulcus acusticus: heterosulcoid, ostiocaudal, median in position. Ostium: funnel-like shorter than cauda. Cauda: tubular and elliptic in shape, slightly curved ending closed to the postero ventral region. Anterior region: lanceolated; rostrum narrow, elongated, and pointed; antirostrum short, broad, pointed; excisura wide “v” shaped with a deep notch. Posterior region: oblique - round (**Figure A2**).

2.2. Indian Threadfish, *Alectis Indicus*

Body has ovate to pentagonal shape, strongly compressed, tapering evenly with midpoint posteriorly and strongly compressed laterally. Dorsal and ventral profile is evenly convex. Eyes are small and about half the length of snout. Mouth is low with protractile jaws, villiform teeth are arranged in bands in lower jaw and also present on tongue, palatines, and vomer. Teeth are absent in adults. Maxilla extends slightly before the anterior margin of eye. Head is angled-shaped, large and comprises 33% of FL. Gill rakers: 8 upper and 21 in the lower limbs of first gill arch. Dorsal fin with 7 short-free embedded spines. Second soft dorsal fin with 1 spine followed by 20 rays.

First five rays of soft dorsal and anal fin are elongated and filamentous in young. Pectoral fin is long, falcate, reaching the straight lateral line, and equal than HL. In juveniles, pelvic fin is elongated. Anal fin with 1 spine followed by 16 rays. Caudal fin is deeply forked and 28% of FL. Lateral line is strongly concave above pectoral fin and deeply arch and junction of curved and straight lateral line is below sixth dorsal soft rays. Curve lateral line is greater than straight lateral line. Straight lateral line with 4 scales and 14 weak scutes. Scales are minute and deeply embedded in the skin (**Figure B1**).

Otolith shape: oblong. Margins: dorsal sinuate, ventral crenate. Sulcus acusticus: heterosulcoid, ostial, median in position. Ostium: funnel-like. Cauda: elliptic, straight ending close to the ventral margin. Anterior region: peaked; rostrum broad, long, rounded; antirostrum absent; excisura wide without notch. Posterior region: oblique (**Figure B2**).

2.3. Shrimp Scad, *Alepes Djedaba*

Body has fusiform shape, strongly compressed. Dorsal and ventral profile is evenly convex. Eye is covered with adipose eyelid posteriorly. Snout is pointed and equal to eye diameter about 25% of HL. Superior maxilla is large, slightly concave posteriorly. A single row of minute comb-like teeth present in row on both jaws. Minute teeth also present in tongue, vomer, and palatines. Shoulder girdle margin is smooth and without papilla. Head length is about 25% of FL. Gill rakers 10 upper and 29 lower, in first gill arch.

Otolith shape: lanceolated. Margins: dorsal margine crenate, ventral margin irregular. Sulcus acusticus: heterosulcoid, ostial, median. Ostium: funnel-like. Cauda: tubular, strongly curved ending close to the ventral margin. Anterior region: peaked; rostrum broad, long, pointed; antirostrum short, narrow, pointed; excisura wide a shallow notch. Posterior region: oblique (**Figure C2**).

2.4. Razorbelly Scad, *Alepes Kleinii* (New Record)

Body has oblong shape, laterally compressed. Ventral profile is more convex than dorsal profile. Eye diameter is equal to snout length. Posterior half of the eye is covered with adipose eyelid. Maxilla is slightly concave posteriorly. Tiny comb-like teeth arranged in band is present in upper jaw. Opercular spot is very distinct. Head length comprises of 24% FL. Gill rakers 10 in upper limb and 24–28 in lower limb.

Fish and otoliths

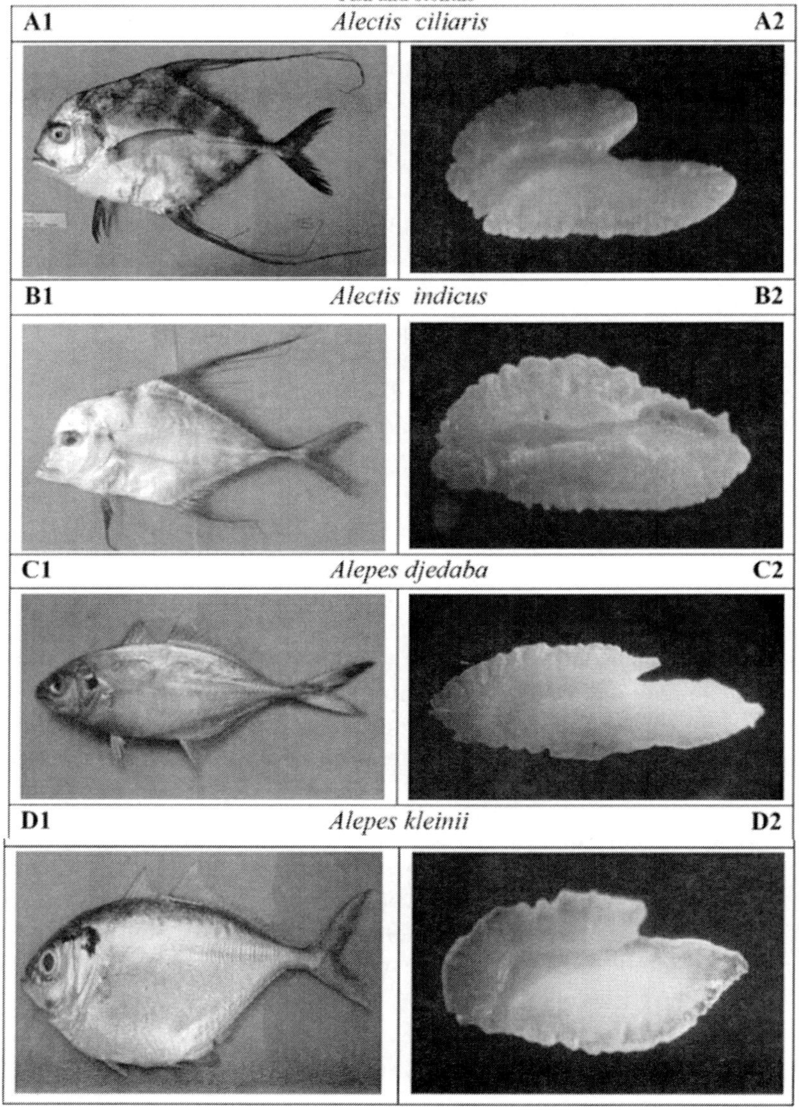

First dorsal fin is membranous, with 8 spines (third spine is moderately high), second dorsal fin with 1 spine followed by 25 rays. Two detached anal spines are present. Anal fin with 1 spine and 19 soft rays. Pectoral fin is long and falcate about 33% of FL. Dorsal fin is low, anal fin slightly falcate. Caudal fin is deeply forked about 26% FL. The distance from snout to anal fin origin is 51% FL. Curved lateral line is shorter than straight line and junction is below second soft dorsal fin rays. Curve lateral line with 33 scales and straight lateral line with 2 scales and 46 scutes (**Figure C1**).

Two separate dorsal fin. First dorsal fin is membranous with 7–8 spines. Second dorsal fin with 1 spine followed by 20–23 soft rays. Pectoral fin is long and falcate. Pelvic fin is short and hyaline. Anal fin with 2 detached anal spines and 1 spine followed by 18–20 soft rays. Caudal fin is deeply forked about 33% of the FL. Vertical 9–10 brownish band equal in width are present above the lateral line. Curve lateral line is shorter than straight lateral line. Scales on curve lateral line are 28–31, and scutes on straight lateral line are 32–40. Junction of CLL and SLL is below the seventh–eighth ray of second dorsal fin (**Figure D1**).

Otolith shape: lanceolated. Margins: dorsal margin irregular, ventral margin crenate. Sulcus acusticus: heterosulcoid, ostial, median. Ostium: funnel-like. Cauda: tubular, strongly curved ending close to the posterior margin. Anterior region: peaked; rostrum broad, medium, pointed; antirostrum short, broad, round; excisura wide with a shallow notch. Posterior region: round (**Figure D2**).

2.5. Blackfin Scad, *Alepes Melanoptera*

Body has oblong shape, moderately compressed. Dorsal and ventral profile evenly convex. Eye covered with adipose eyelid posteriorly. Snout is pointed and equal to eye. A single row of minute uniserate, comb-like teeth present in row on both jaws. Biserate dentition on premaxilla. Maxilla ends at the center of eye. Head length about 24% forked length. Gill rakers 9 upper and 23 lower, in first gill arch. Smooth Shoulder girdle margin without papillae.

First dorsal fin is membranous, with 8 spines, second dorsal fin with 1 spine followed by 24 rays. Two detached anal spines are present. Anal fin with 1 spine and 18 soft rays. Pectoral fin long and falcate about 30% forked length. Longest spine of first dorsal fin about equals to second dorsal fin. Soft dorsal fin low, anal fin slightly falcate. Caudal fin deeply forked about 28% forked length. The distance from Snout to anal fin origin is 57% forked length. Curved lateral line shorter than straight line and junction below second soft dorsal fin rays. Chord of curved line 2.2 times shorter than straight lateral line. Curve lateral line with 30 scales and straight lateral line with 4 scales and 50 scutes (**Figure E1**).

Otolith shape: lanceolated. Margins: crenate. Sulcus acusticus: heterosulcoid, ostial, median. Ostium: funnel-like. Cauda: tubular, slightly curved at the end, ending far from the ventral margin. Anterior region: peaked; rostrum narrow, long, pointed; antirostrum absent or few developed; excisura wide without notch. Posterior region: oblique (**Figure E2**).

2.6. Herring Scad, *Alepes Vari*

Body has fusiform shape, moderately compressed. Dorsal and ventral profile evenly convex. Eye is covered with adipose eyelid posteriorly. Snout is pointed and greater than eye. A single row of minute uniserrate, comb-like teeth present in row on both jaws. Bi-serrate dentition in premaxilla. Maxilla extends to one-third of eye. Head length is about 24% of FL. Gill rakers 10–12 upper and 26–29 lower, in first gill arch. Smooth shoulder girdle margin without papillae.

First dorsal fin is membranous, with 8 spines, second dorsal fin with 1 spine followed by 20–25 rays. Two detached anal spines are present. Anal fin with1 spine and 19 soft rays. Pectoral fin is long and falcate about 27% of FL. Longest spine of first dorsal fin is shorter than second dorsal fin. Soft dorsal fin is low, anal fin slightly falcate. Caudal fin is deeply forked about 26% of FL. The distance from snout to anal fin origin is 50% of FL. Curved lateral line is shorter than straight line, and junction is below third soft dorsal fin rays. Chord of curved line is 2.1–2.5 times to straight lateral line. Curve lateral line with 38–42 scales and straight lateral line with 2 scales and 50–55 scutes (**Figure F1**).

Otolith shape: fusiform. Margins: dorsal margin sinuate, ventral margin crenate. Sulcus acusticus: heterosulcoid, ostial, median in position. Ostium: funnel-like longer than cauda. Cauda: tubular, markedly flexed posteriorly from the ventral margin. Anterior region: peaked; rostrum broad, elongated, pointed; antirostrum short, peaked; excisura wide with a small shallow notch. Posterior region: oblique (**Figure F2**).

2.7. Cleftbelly Trevally, *Atropus Atropos*

Body is ovate, strongly compressed. Nape profile is convex. Eye is equal or slightly larger to snout length. Narrow band of teeth is present on upper jaw. Lower jaw with a single series of teeth. Head is about 26% of FL. Gill rakers upper 9, lower 18 in first gill arch. Ventral side of breast is naked, and diagonally extends to the base of pectoral fin. Prominent opercular spot is present.

First dorsal fin with 8 spines, soft dorsal fin with 21 soft rays. Pectoral fin is falcate and greater than pelvic fin. Pelvic fin is large and blackish. Lobe of second dorsal fin length is 15% of FL. Anal fin with 1 spine followed by 20 soft rays. Adult shows sexual dimorphism between the second and twelfth ray of soft dorsal fin, filamentous in male. Belly has a deep median groove covering pelvic fin and anal spine. Two detached anal spines are present. Tip of pelvic fin extends to the origin of anal fin. Caudal fin is deeply forked and constituting 27% of FL. Lateral line is concave above pectoral fin and junction is below 4–6 rays of soft dorsal fin. Straight lateral line is greater than curved lateral line, having 3 scales and 38 strong scutes (**Figure G1**).

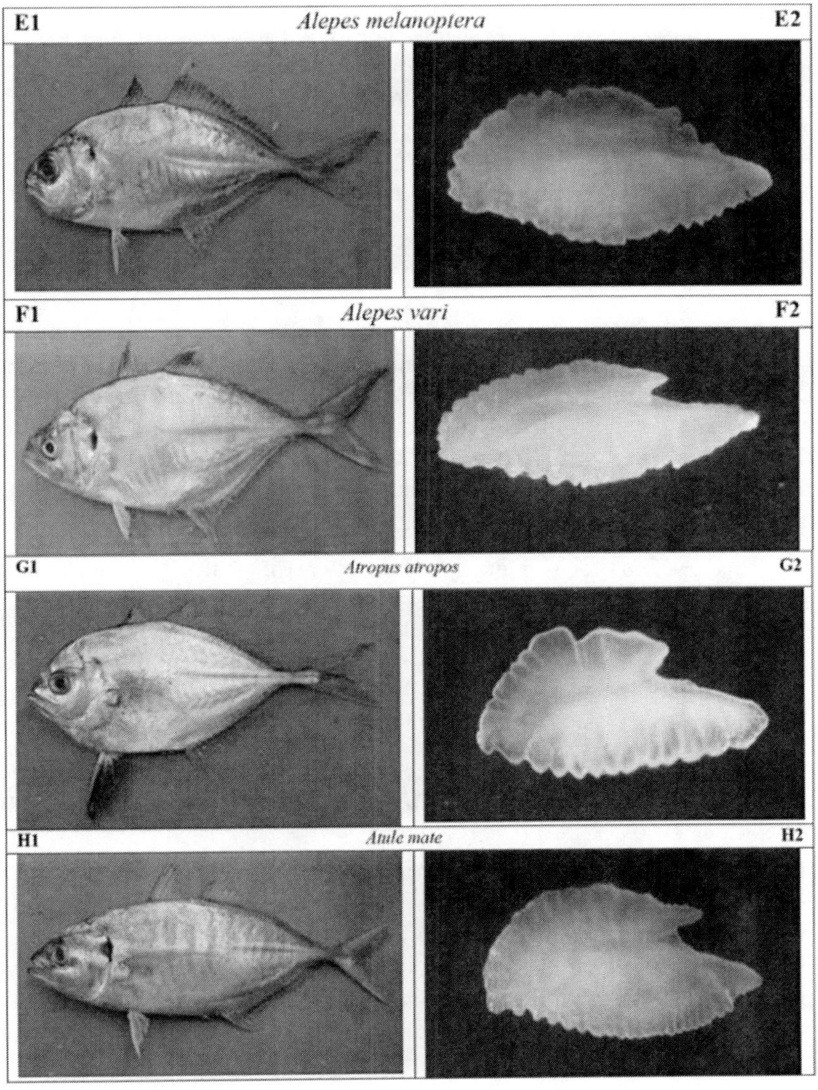

Otolith shape: elliptic-lanceloted. Margins: dorsal irregular, ventral serrate. Sulcus acusticus: heterosulcoid, ostial, median in position. Ostium: funnel-like shorter than cauda. Cauda: tubular, strongly flexed ending close to the ventral margin. Anterior region: peaked; rostrum broad, long, pointed; antirostrum short, narrow, pointed; excisura wide with an acute v-shaped medium notch. Posterior region: oblique (**Figure G2**).

2.8. Yellowtail Scad, *Atule Mate*

Body has fusiform shape, moderately compressed. Dorsal and ventral profile is evenly convex. Eye is covered with adipose eyelid except for vertical slit centered on pupil. Snout is pointed and greater than eye. A single row of minute uniserate, comb-like teeth present on both jaws. Biserate dentition in premaxilla. Maxilla extends to one-third of eye anteriorly. Head length is about 26% of FL. Gill rakers 10–12 upper and 24–26 lower in first gill arch. Smooth shoulder girdle margin without papillae.

First dorsal fin is membranous, with 8 spines, second dorsal fin with 1 spine followed by 20–24 rays. Two detached anal spines are present: anal fin with 1 spine and 19 soft rays. Pectoral fin is long and falcate about 31% of FL. Soft dorsal fin is low, anal fin slightly falcate. Caudal fin is deeply forked about 33% of FL. The distance from snout to anal fin origin is 50% of FL. Curved lateral line is shorter than straight line, and junction is below sixth soft dorsal fin rays. Chord of curved line is 2.1 times to straight lateral line. Curve lateral line with 40 scales and straight lateral line with 2 scales and 52 scutes. Last soft dorsal and anal fin finlet are joined by inter radial membrane and about twice the length of previous ray (**Figure H1**).

Otolith shape: fusiform. Margins: dorso-ventral posterior margin dentate. Sulcus acusticus: heterosulcoid, ostial, median. Ostium: funnel-like. Cauda: tubular, strongly curved ending close to the posterior margin. Anterior region: lanceolated; rostrum broad, medium, pointed; antirostrum short, narrow, pointed; excisura wide with a shallow notch. Posterior region: oblique (**Figure H2**).

2.9. Longfin Trevally, *Carangoides Armatus* (New Record)

Body has ovate (to hexagonal) shape, deep, and compressed; eye diameter is less than snout length. Head is steep and straight from snout to nape without any break in contour (bump). Villiforms teeth are arranged in bands in both jaws, which expended anteriorly. Vomerine tooth is wedge patch shaped. Breast is ventrally naked to behind the origin of pelvic fin up to naked base of pectoral fin. Head length comprises of 30% of FL. Gill rakers are 10 lower and 23 upper in first gill arch.

First dorsal fin is modified into 8 free spines, and soft dorsal fin consists of 1 spine followed by 22 soft rays. Dorsal and anal fin 3–12 middle rays elongate and filamentous, longer than head length in male. Pectoral fin is long and falcate. Pelvic fin is long in juveniles. Anal fin with two detached anal spines with 1 spines followed by 20 rays. Caudal fin is deeply forked and 27% of FL. Lateral line is strongly arched, and junction is below 12–14 soft dorsal rays. Curved lateral line is longer than straight line. Straight lateral line consist of 20 scales and 23 scutes (**Figure I1**).

Otolith shape: lanceolated. Margins: dorsal margin sinuate, ventral margin crenate. Sulcus acusticus: heterosulcoid, ostial, median in position. Ostium: funnel-like longer than cauda. Cauda: tubular, strongly flexed posteriorly. Anterior region: peaked; rostrum broad, elongated, pointed; antirostrum short, roundly peaked; excisura wide with a shallow notch. Posterior region: oblique (**Figure I2**).

2.10. Longnose Trevally, *Carangoides Chrysophrys*

Body has oval shape, strongly compressed. Dorsal profile is more convex. Snout is greater than eye diameter. Dorsal outline of snout to nape is straight and then convex to the origin of dorsal fin. Small villiform teeth are arranged in bands anteriorly widened. Head length comprises of 30% FL. Gill rakers: 5 in upper limb and 15 in lower limb. Two separate dorsal fin. First dorsal fin is membranous with 7–8 spines, second dorsal fin with 1 spine followed by 21 soft rays. Pectoral fin is long and falcate. Pelvic fin is short and hyaline. Anal fin with 2 detached anal spines and 1 spine followed by 16 soft rays. Caudal fin is deeply forked about 27% of FL. Curve lateral line is shorter than straight lateral line. Scales on curve lateral line is 28–31 and scutes on straight lateral line are 32–40. Junction of CLL and SLL is below seventh–eighth ray of second dorsal fin (**Figure J1**).

2.11. Bumpnose Trevally, *Carangoides Hedlandensis*

Body has ovate shape and deep, laterally compressed. Ventral profile is more convex than dorsal side. Snout length is equal to eye diameter. Head profile is steep, dorsal contour of forehead convex, with "bump" on interorbital space. Villiform teeth are arranged in bands in both jaws which widens anteriorly. Patchy wedge-shaped vomerine teeth. Breast is naked and confined below pectoral fin. Head constituted 27% of FL. Gill rakers 8 upper and 25 lower limbs of first gill arch.

First dorsal fin membraneous with 8 spines and second dorsal fin with 1 spine followed by 21 rays (3–8) producing filamentous and elongate, longer than head length in male. Pectoral fin is long and falcate. Pelvic fin short. Anal fin with two detached spines with 1 spine followed by 18 soft rays. Caudal fin is deeply forked and 26% of FL. Lateral line is strongly arched, and junction is below 12–14 soft dorsal rays. Curved lateral line is longer than the straight line. Straight laterally consists of 12 scales and 28 scutes (**Figure K1**).

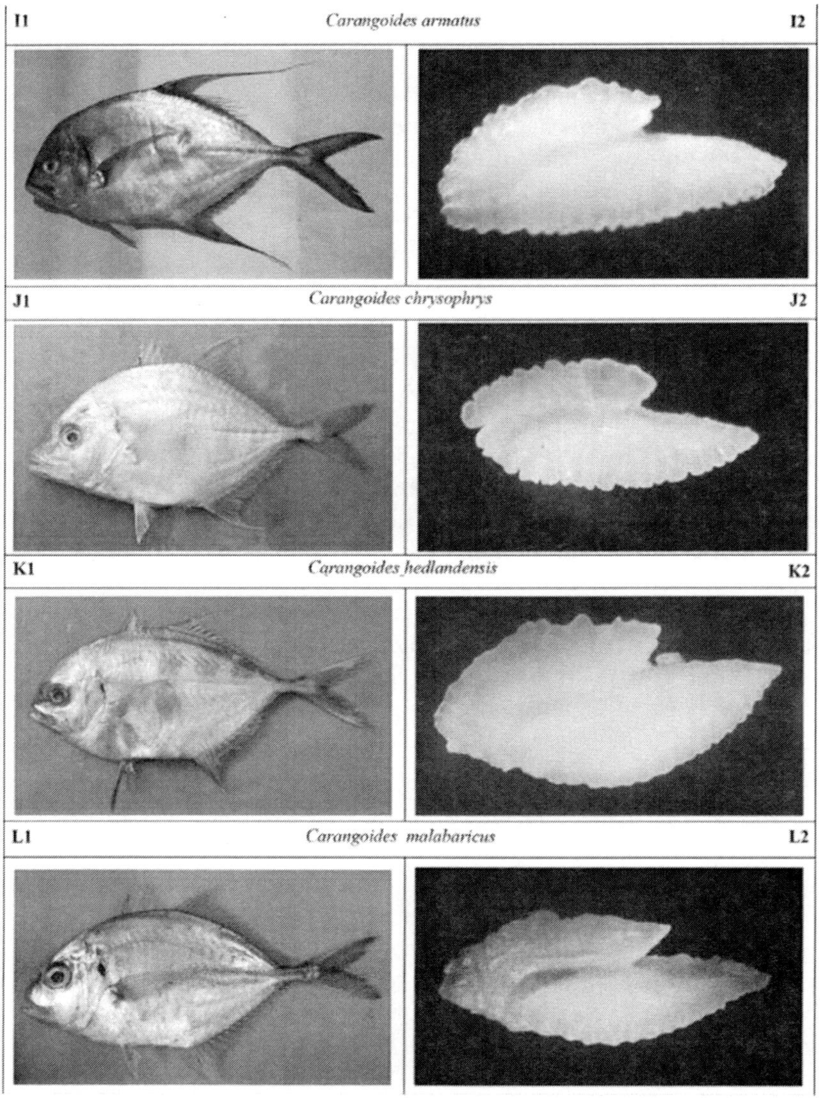

Otolith shape: lanceolated. Margins: dorso-ventrally crenate. Sulcus acusticus: heterosulcoid, ostial, median in position. Ostium: funnel-like. Cauda: tubular, strongly flexed posteriorly. Anterior region: lanceolated; rostrum narrow, elongated, peaked; antirostrum short, broad, round; excisura wide with a shallow notch. Posterior region: oblique-irregular (**Figure J2**).

Otolith shape: lanceolated. Margins: crenate. Sulcus acusticus: heterosulcoid, ostial, median. Ostium: funnel-like. Cauda: tubular, strongly curved ending close to the ventral margin. Anterior region: peaked; rostrum narrow, long, pointed; antirostrum short, broad, round; excisura wide with a shallow notch. Posterior region: oblique (**Figure K2**).

2.12. Malabar Trevally, *Carangides Malabaricus*

Body has oval shape, deep, and compressed; dorsal profile more convex is prominent to nape. Eye diameter is less than snout length. Head is steep and straight from snout to nape without any break in contour (bump). Villiform teeth are arranged in bands in both jaws, which expended anteriorly. A conical outer tooth is present. Vomerine tooth is roughly triangular. Breast is ventrally naked and behind the origin of pelvic fin upto origin of second dorsal fin. Head length comprises of 24–28% of FL. Gill rakers are 12 lower and 25 upper in first gill arch.

First dorsal fin is modified into 8 free spines and second dorsal fin consists of 1 spine followed by 23 soft rays. Pectoral fin is long and falcate. Pelvic fin is short. Anal fin with two detached anal spines with 1 spines followed by 18–19 rays. Caudal fin is deeply forked and 27% of FL. Lateral line is strongly arched, and junction is below 12–14 soft dorsal rays. Curved lateral line is longer than straight line. Straight lateral line consists of 19 scales and 38 scutes (**Figure L1**).

Otolith shape: fusiform. Margins: dorsal, lobed, ventral, sinuate. Sulcus acusticus: heterosulcoid, ostial, median. Ostium: funnel-like. Cauda: tubular, strongly curved posteriorly ending close to the ventral margin. Anterior region: peaked; rostrum broad, short, pointed; antirostrum short, broad, pointed; excisura wide with a shallow v-shaped notch. Posterior region: peaked (**Figure L2**).

2.13. Brownback Trevally, *Carangoides Praeustus*

Body is elliptic, elongated, and laterally compressed. Dorsal and ventral profile is identical. Snout is greater than eye. Straight head profile dorsally. Mouth is pointed and terminal. Teeth are conical, arrange in small irregular row in both jaws. Narrow bands of anterior tooth are arranged in upper jaw. Head length is 26% of FL. Breast ventrally naked, small patch of prepatch scale, remained separated from naked base of pectoral. Gill rakers upper 12 and 30 in lower limb of first gill arch.

First dorsal fin is membranous with 8 spines, and second dorsal fin consists of 1 spine followed by 23 soft rays. Height of third longest spinous dorsal fin is equal to soft dorsal fin lobe. Pectoral fin is long and falcate. Pelvic fin is short. Anal fin with two detached anal spines with 1 spines followed by 18–19 rays. Caudal fin is deeply forked and 28% of FL. Lateral line is strongly arched, and junction is below 11–12 soft dorsal rays. Curved lateral line is longer than straight line. Straight lateral line consists of 10 scales and 30 scutes (**Figure M1**).

Otolith shape: Trapezoidal-lanceolated. Margins: dorsal margin sinuate, ventral margin crenate. Sulcus acusticus: heterosulcoid, ostial, median. Ostium: funnel-like. Cauda: tubular, strongly curved ending close to the posterior margin. Anterior region: peaked; rostrum broad, medium, pointed; antirostrum short, narrow, pointed; excisura wide with an acute medium notch. Posterior region: oblique (**Figure M2**).

2.14. Giant Trevally, *Caranx Ignobilis*

Body has oblong shape, laterally compressed. Profile of dorsal body is more convex from snout to soft dorsal fin. Snout is pointed and greater than eye. Adipose eyelid covered the eye, extending beyond the posterior border of the eye. Maxilla slightly extends to posterior margin of eye. Strong canines in outer row tooth of both jaws. Small villiform tooth is present in inner bands of upper jaw. In upper jaw, canines are widely spaced in adult. Ventral profile of breast is completely naked with a patch of prepelvic scale. Head length comprises 29% of FL. Gill rakers 4 upper and 14 lower in first gill arch.

First dorsal fin is membranous, with 8 spines and second dorsal fin with 1 spine followed by 18 rays. Two detached anal spines are present, anal fin with 1 spine followed by 16 soft rays. Pectoral fin is long and falcate about 37% of FL and longer than HL. Pelvic fin is short. Soft dorsal fin and anal fin lobe is slightly falcate. Caudal fin is deeply forked about 23% of FL. The distance from snout to anal fin origin is 62% of FL. Curved lateral line is longer than straight line, and junction is below fourth–fifth soft dorsal fin rays. Curved lateral line with 46 scales and straight lateral line with 4 scales and 29 scutes. Scale in body is cycloid and small, entirely covered the body except small portion behind pectoral fin (**Figure N1**).

Otolith shape: fusiform. Margins: crenate. Sulcus acusticus: heterosulcoid, ostial, median. Ostium: funnel-like. Cauda: tubular, strongly curved ending close to the ventral margin. Anterior region: slightly curved at tip, peaked; rostrum broad, long, pointed; antirostrum absent; excisura wide without notch. Posterior region: oblique (**Figure N2**).

2.15. Green Jack, *Caranx Para*

Body oval, laterally compressed and ventral profile more convex. Eye equal to snout length and covered with adipose eyelid which extends posterior half of eye. Maxilla broad and concave posteriorly. Teeth of lower jaw are arrange in irregular rows. Lower jaw with a single row of conical teeth. Head profile steep from snout to origin of second dorsal fin and ovate posteriorly. Head length is 28% of forked length. Shoulder girdle margin smooth and without papillae. Gill rakers upper 12 and 26 lower limbs of the first gill arch.

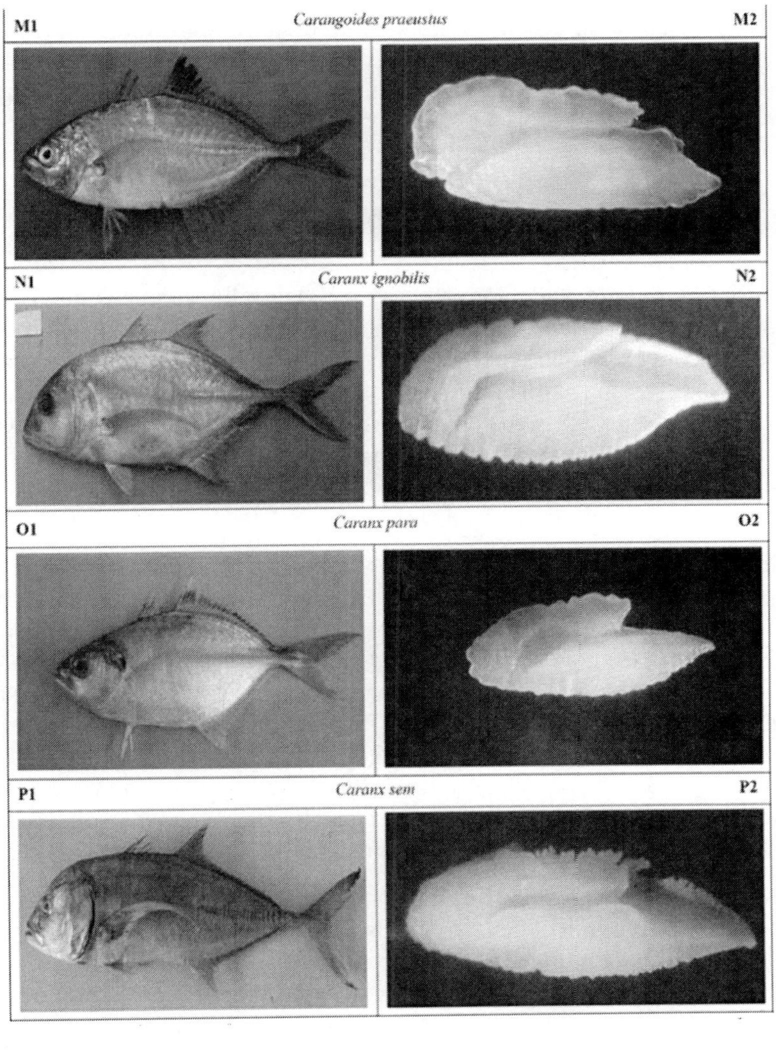

First dorsal fin membranous with 8 spines, soft dorsal fin with 1 spine followed by 25 soft rays and 1.75 times shorter than head length. Dorsal and anal fin lobe falcate. Pectoral fin long 34% of forked length, falcate, tip end reaches beyond the junction of CLL and SLL. Anal fin with two detached anal spines with 1 spines followed by 20 rays. Anal fin behind the origin of soft dorsal fin. Caudal fin deeply forked, dorsal lobe longer, and 31% of forked length. Lateral line strongly arched and junction below fourth–sixth soft dorsal rays. Straight lateral line is longer than curved lateral line. Straight lateral line consists of 45 scutes and scales small and cycloid (**Figure O1**).

Otolith shape: lanceolated. Margins: dorso-ventral posteriorly sinuate. Sulcus acustics, heterosulcoid, ostial, median. Ostium: funnel-like equal to the length of cauda. Cauda: tubular, markedtly curved posteriorly from the middle region. Anterior region: peaked; rostrum elongated, pointed; antirostrum short, broad, peaked; excisura wide with a V-shaped notch. Posterior region: round-oblique (**Figure O2**).

2.16. Blacktip Trevally, *Caranx Sem* (New Record)

Body has oblong and elongated shape, laterally compressed. Profile of dorsal body is more convex from snout to soft dorsal fin. Ventral profile is slightly convex. Snout is pointed and greater than eye. Eye is moderate and covered anteriorly with small adipose eyelid and posteriorly extends to the pupil. Maxilla extends to the posterior margin of the eye. Strong canines present in outer row teeth of both jaws. Small villiform teeth are present in inner bands of upper jaw. Head length comprises 30% of FL. Gill rakers: 4 upper and 15 lower in first gill arch.

First dorsal fin is membranous, with 8 spines, second dorsal fin with 1 spine followed by 20 rays. Two detached anal spines are present. Anal fin with 1 spine followed by16 soft rays. Pectoral fin is long and falcate longer than HL. Pelvic fin is short. Soft dorsal and anal fin is slightly falcate. Caudal fin is deeply forked about 30% of FL. Curved lateral line is longer than straight line, and junction is below fourth–fifth soft dorsal fin rays. Straight part of lateral line is 2.4 times of forked length with 4 scales and 30 scutes. Scale in body is cycloid and small, entirely covered the body and in breast, small-to-large patch of prepelvic scale. All fins are brightly yellow to olive yellow in color, and distal half of upper lobe of caudal fin is dark black (**Figure P1**).

Otolith shape: fusiform. Margins: dorsal dentate, ventral crenate. Sulcus acusticus: heterosulcoid, ostial, median. Ostium: funnel-like. Cauda: tubular, strongly curved ending close to the ventral margin. Anterior region: slightly curved at tip, peaked; rostrum broad, long, pointed; antirostrum absent; excisura wide without notch. Posterior region: oblique-round (**Figure P2**).

2.17. Bigeye Trevally, *Caranx Sexfasciatis*

Body has fusiform shape, laterally compressed. Profile of dorsal body is more convex from snout to soft dorsal fin. Snout is pointed and greater than eye. Adipose eyelid completely covered the eye, which extends moderate anteriorly and posteriorly to the margin of pupil. Maxilla extends to posterior margin of eye. Strong canines in outer row tooth of both jaws. Small villiforms teeth are present in inner bands of upper jaw. Breast is completely covered with scales. Head length comprises 27–30% of forked length. Gill rakers 6 upper and 16–18 lower in first gill arch. First dorsal fin is membranous, with 8 spines, second dorsal fin with 1 spine followed by 20–21 rays. Two detached anal spines are present. Anal fin with 1 spine followed by16–19 soft rays. Pectoral fin is long and falcate about 31% of FL and longer than HL. Pelvic fin is short. Soft dorsal fin is low, anal fin slightly falcate. Caudal fin is deeply forked about 30% of FL. The distance from snout to anal fin origin is 58% of FL. Curved lateral line is longer than straight line, and junction is below fourth–fifth soft dorsal fin rays. Curve lateral line with 46 scales and straight lateral line with 6 scales and 32–37 scutes. Scale in body is cycloid and small, which is entirely covered the body except small portion behind pectoral fin (**Figure Q1**).

Otolith shape: fusiform. Margins: dorsal crenate, ventral serrate. Sulcus acusticus: heterosulcoid, ostial, median in position. Ostium: funnel-like equal to ostial length with perforated colliculum. Cauda: tubular, strongly curved posteriorly ending close to the ventral margin. Anterior region: slightly curved at tip, peaked; rostrum broad, elongated, pointed; antirostrum absent; excisura wide without notch. Posterior region: oblique (**Figure Q2**).

2.18. Indian Scad, *Decapterus Russelli*

The body has fusiform shape and elongated, quite slender, and considerably compressed. Prominent moderate eye covered entirely with adipose except vertical slit centered on the pupil. Eye diameter is smaller than snout length. Eye is moderate and covered by adipose eyelid in a vertical slit. The upper jaw is slightly concave, ending straight above with a narrow band of minute teeth. Lower jaw has an irregular series of minute teeth. Body is bright yellow with transparent fins. Head length comprises of 28% of FL length. Gill rakers: upper 10, lower 33–34 on first gill arch. Shoulder girdle margin with two papillae: The lower papilla is larger.

First dorsal fin is separate and membranous with 7 spines and second dorsal fin with1 spine followed by 26–28 rays. Terminal dorsal and anal fin with distinct single-detached finlet. Pectoral fin is long, and end tip reaches to the second ray of second dorsal fin. Anal fin with 2 detached spines and 1 spine followed by 24 soft rays. Caudal fin is deeply forked with 18% of FL. Anal fin origin is slightly beyond the vertical line of second dorsal fin. Curve lateral line shorter than straight lateral line, junction of curve and straight lateral line below eleventh–twelfth ray of second dorsal fin. The scale in CLL is 38 and 2 in SLL followed by 35–39 scutes (**Figure R1**).

Otolith shape: elliptical. Margins: dorsal sinuate, ventral crenate. Sulcus acusticus: heterosulcoid, ostial, median. Ostium: funnel-like. Cauda: tubular, strongly curved ending close to the ventral margin. Anterior region: peaked; rostrum broadly pointes, medium; antirostrum short, broadly pointed; excisura wide with a medium notch. Posterior region: peaked (**Figure R2**).

2.19. Rainbow Runner, *Elagatis Bipinnulata*

Body has elongated shape, subcylindrical, and fusiform. Snout diameter is greater than eyes. Mouth is small, and head is pointed. Eye is small, moderately developed adipose. Maxilla does not extend to the anterior margin of the eye. Villiform teeth are arranged in both jaws, vomer, and palatines in bands, tongue centrally arranged in bands. Head length comprised of 24% FL. Gill rakers: upper 11 and 23, lower on first gill arch.

First dorsal fin is modified into 7 spines. Second dorsal fin with 1 spine followed by 26 rays excluding finlet. Terminal detached dorsal and anal fin with distinct double rayed finlet. Pectoral fin length is equal to pelvic fin length and double in head length. Anal fin without detached spines with 1 spine attached followed by 19 soft rays. Caudal fin is deeply forked with 31% FL. Anal fin origin is beyond the soft dorsal fin. Anal fin base is shorter than about 1.5 times lower than soft dorsal fin base. Lateral line is slightly arch without scutes. (**Figure S1**).

Otolith shape: lanceolated, clearly curved. Margins: dorsal irregular, ventral serrate-dentate. Sulcus acusticus: heterosulcoid, ostial, median. Ostium: funnel-like. Cauda: tubular slightly curved ending posteriorly very close to the ventral margin. Anterior region: lanceolated-curved; rostrum narrow, long, pointed; antirostrum short, broad, pointed; excisura wide with a U-shallow notch. Posterior region: oblique-irregular (**Figure S2**)

2.20. Golden Trevally, *Gnathanodon Speciosus*

Body has oblong shape, laterally compressed. Snout length is greater than eye. Maxilla is protractile and extends to below posterior one-third of an eye. Lips are papillose. Upper jaw without teeth and lower jaw with small teeth (absent in adult). Head is 26% of FL. Gill rakers: 6 upper, 20 lower in first gill arch.

First dorsal fin is membranous with 7 spines and second dorsal fin with 1 spine followed by 22 soft rays. Soft dorsal is low. Anal fin is falcate. Pectoral fin is long, falcate and 35% of FL. Anal fin with 2 detached spines. Anal fin with 1 spine and 18 soft rays. Anal fin is slightly behind the second dorsal fin. Snout to anal fin distance is 49% of FL. Caudal fin deeply forked 30% of FL. Body color is golden with 7–11 alternating broad black bands. Lateral line is moderately arching anteriorly, junction of curved and straight lateral line below twelfth ray of second dorsal fin. Curve lateral line is longer than straight lateral line contained 0.85 times in a straight line. Straight part of lateral line containing 10 scale followed by 24 scutes. Breast is completely covered with scale (**Figure T1**).

Otolith shape: elliptic-fusiform. Margins: crenate. Sulcus acusticus: heterosulcoid, ostial, median. Ostium: funnel-like shorter than cauda. Cauda: tubular, strongly flexed posteriorly from the middle margin. Anterior region: peaked; rostrum broad, medium, pointed; antirostrum short, narrow, pointed; excisura wide with an acute medium notch. Posterior region: oblique (**Figure T2**).

2.21. Torpedo Scad, *Megalaspis Cordyla*

Body has fusiform and elongate shape, posteriorly compressed and subcylindrical. Eye is large and greater than snout length with well-developed adipose eyelid covered entirely except vertical slit, snout is blunt, and maxilla extends posteriorly to the center of eye. Small villiform teeth is in upper jaw, anteriorly narrow bands and single row of teeth in lower jaw. Head length comprises of 28% of FL. Gill rakers are 10–12 lower and 24–30 upper in first gill arch.

First dorsal fin is membranous with 8 spines, and second dorsal fin consists of 1 spine followed by 17 soft rays posteriorly 8–10 consisting of detached finlets. Second dorsal fin and anal fin lobe is falcate. Pectoral fin is long falcate and tip end reaches to the straight lateral line. Anal fin with two detached anal spines with 1 spines followed by 18 soft rays, posteriorly 8–10 consisting of detached finlets. Anal fin is behind the origin of soft dorsal fin. Caudal peduncle with a marked median keel. Caudal fin is deeply forked and 28% of FL. Curved lateral line is shorter than straight line. Straight lateral line consists of 8 scales and 22 scutes. Scales are small and cycloid (**Figure U1**).

Otolith shape: lanceolated. Margins: dorsally sinuate, ventrally dentate. Sulcus acusticus: heterosulcoid, ostial, median in position. Ostium: funnel-like. Cauda: tubular, markedly flexed posteriorly from the middle region ending close to the ventral margin. Anterior region: lanceolated with dentate protuberances; rostrum narrow, elongated, pointed; antirostrum absent; excisura wide with a square-shaped medium notch. Posterior region: oblique (**Figure U2**).

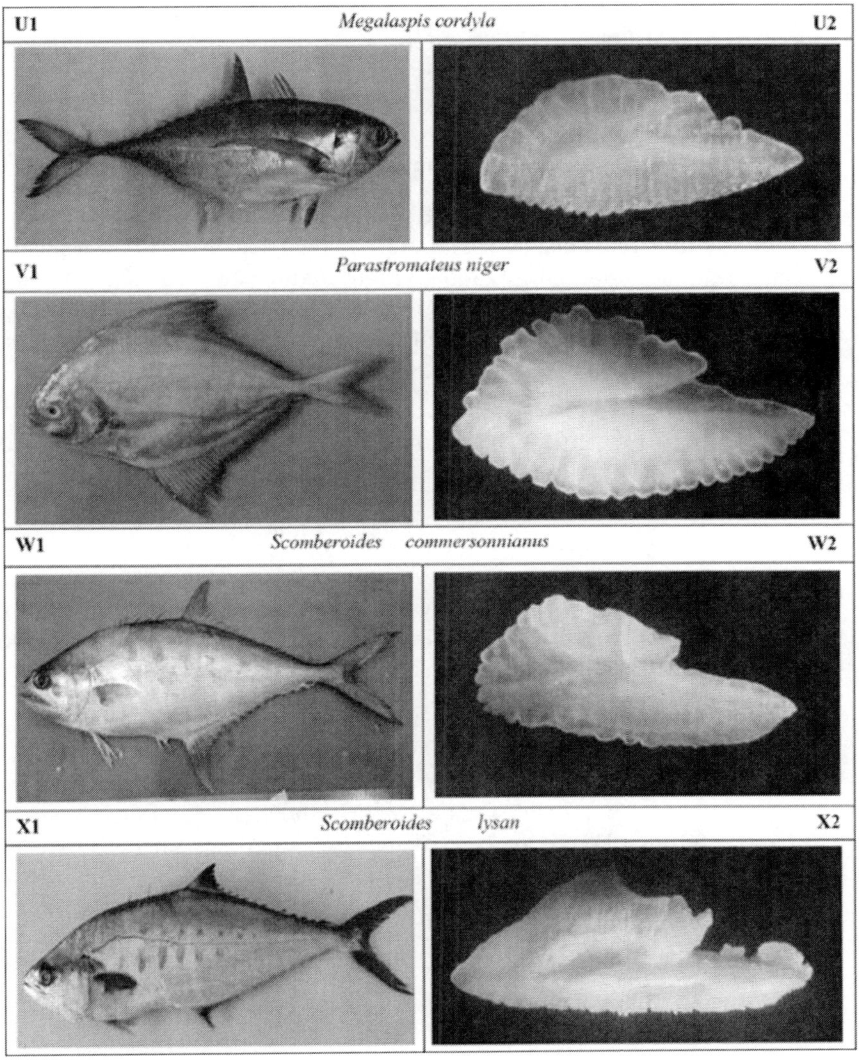

2.22. Black Pomfrets, *Parastromateus Niger*

Body has deep and ovate shape, laterally compressed, dorsal and ventral profile is convex. Mouth is terminated with enlarged teeth that bridge the epibranchial 4-ceratobranchial 4 joints. Form of teeth on the pharyngeal teeth plates is elongate and filamentous. Color is dark brown in juvenile and silvery gray to bluish brown in adult. Gill rakers: upper 8–10 and lower 15–24 on first gill arch. The first dorsal fin is superficially absent. The first dorsal fin with small, short 4–5 embedded spines followed by one spine and

38–44 soft rays. Pectoral fin is long and falcate. Pelvic fin absent in adult. Dorsal and anal fin entirely rounded, broad lobes, and identical. Anal fin with 2 anal spines followed by 1 spine and 33–42 soft rays. The lateral line is visible with slight curved entirely become straight at 29–31 soft dorsal fin rays. Straight part of lateral line with 15- to 17-week scutes forming straight keel on caudal peduncle (**Figure V1**).

Otolith shape: elliptical-lanceolated. Margins: dorsal irregular, ventral crenate-dentate. Sulcus acusticus: heterosulcoid, ostial, median in position. Ostium: funnel-like equal to caudal length. Cauda: tubular, slightly curved ending far to the ventral margin. Anterior region: peaked; rostrum broad, elongated, pointed; antirostrum short, broad, round; excisura wide with an acute shallow notch. Posterior region: oblique (**Figure V2**).

2.23. Talang Queenfish, Scomberoides Commersonnianus

The body has fusiform shaped and elongated, laterally compresses. Dorsal and ventral profiles are identical. Adipose eyelid is developed. Snout length is greater than eye diameter. Snout blunt, head, and nape are slightly concave. Maxilla extends beyond the posterior margin of the eye. Lower jaw has two rows of teeth. Large conical tooth presents in outer row. Small villiform tooth is present in inner row. Head length comprises 19% of FL. Gill rakers are large and slender, 2–5 upper, and 11–14 lower.

First dorsal fin with 1 embedded spine and 6–7 free short spines, and second dorsal fin with 1 spine followed by 18–20 soft rays. Pectoral fin is short and greater than pelvic fin. Pelvic fin is short, depressible into shallow groove. Soft dorsal and anal fin is falcate, equal in length. Posteriorly consisting of semi-detached finlet. Two detached anal spine are present. Anal fin with 1 spine followed by 17–20 soft rays. Distal half of dorsal and anal fin is dusky. Caudal fin is deeply forked and comprises 25% of FL. Dorsal and anal fin is originated in a line. Scales are lanceolate below the lateral line and deeply embedded. Lateral line is little angulated just above the pectoral fin, slightly irregular and without scutes. A series of plumbaeous blotches (6–8) round to oval is present just above or touching the lateral line. Caudal peduncle groove is absent (**Figure W1**).

Otolith shape: lanceolated-triangular. Margins: dorsal sinuate, ventral crenate. Sulcus acusticus: heterosulcoid, ostial, median. Ostium: funnel-like. Cauda: tubular slightly curved posteriorly ending very close to the ventral margin. Anterior region: lanceolated; rostrum narrow, elongated, pointed; antirostrum short, broad, pointed; excisura wide with a shallow notch. Posterior region: round-irregular (**Figure W2**).

2.24. Doublespotted Queenfish, *Scomberoides Lysan*

Body has fusiform and elongated shape, laterally compressed. Dorsal and ventral profile is identical. Adipose eyelid is developed. Snout length is greater than eye diameter. Snout blunt, head, and nape are slightly concave. Maxilla extends beyond the posterior margin of the eye. Lower jaw has two rows of teeth. Large conical tooth presents in outer row. Small villiform teeth is present in inner row. Dentary is subequal in length. Head length comprises 19% of FL. Gill rakers are large and slender, 4–9 upper, and 14–21 lower.

First dorsal fin with 1 embedded spine and 6–7 free short spines, and second dorsal fin with 1 spine followed by 18–21 soft rays. Pectoral fin is not falcate and equal in length of pelvic fin. Pelvic fin is short, depressible into shallow groove. Soft dorsal and anal fin is falcate, equal in length, posteriorly consisting of semidetached finlet. Two detached anal spine are present, anal fin with 1 spine followed by 18–20 soft rays. Soft dorsal and anal fin comprises 11–12% of FL, and distal half of soft dorsal fin is pigmented black. Caudal fin is deeply forked and comprises 26% of FL. Anal fin is originated slightly just behind the second dorsal fin. Scales are lanceolate below the lateral line and mid-body, deeply embedded. Lateral line is little angulated just above the pectoral fin, slightly irregular and straight posteriorly without scutes. Two series of (6–8) blotches rounded to vertically oblong is present just above and below the lateral line. Caudal peduncle groove is absent (**Figure X1**).

Otolith shape: approximately triangular. Margins: dorsal sinuate, ventral crenate. Sulcus acusticus: heterosulcoid, ostial, median. Ostium: funnel-like. Cauda: tubular, strongly curved posteriorly from the middle region ending far from the ventral margin. Anterior region: lanceolated with dentate protuberances; rostrum narrow, elongated, pointed; antirostrum very short, broad, pointed upward; excisura wide with a shallow notch. Posterior region: oblique (**Figure X2**).

2.25. Barred Queenfish, *Scomberoides Tala* (New Record)

Body shaped is elongated, dorsal, and ventral profile strongly compressed. Adipose eyelid is well developed. Snout length is greater than eye diameter. Snout blunt with midline joint to the upper lip by a frenum. Maxilla extends beyond the posterior margin of the eye. Lower jaw has two rows of teeth, large conical teeth present in inner row. Small villiform teeth are present in outer row. Head length comprises 19% of the forked length. Gill rakers are large and slender, 2–3 upper, and 8–10 lower.

First dorsal fin with 1 embedded spine and 6–7 free short spines. Second dorsal fin with 1 spine followed by 18–20 soft rays. Pectoral fin is slightly falcate and shorter in pelvic fin length. Pelvic fin is short, depressible into shallow groove. Soft dorsal and anal fin is falcate, equal in length, posteriorly consisting of semidetached finlet. Two detached anal spine are present, anal fin with 1 spine followed by 17–19 soft rays. Soft dorsal and anal fin comprises 14–15% of the forked length in a specimen greater than 40 cm FL. Caudal fin is deeply forked and comprises 27% of the forked length, anal fin origin in a line with second dorsal fin. Scales are lanceolate below the lateral line on mid-body, deeply embedded in skin. Lateral line is little angulated just above the pectoral fin, slightly irregular and straight posteriorly without scutes. A single series of (4–8) blotches rounded to vertically oblong is present, and first 6–7 blotches intersecting the lateral line. Caudal peduncle groove is absent (**Figure Y1**).

Otolith shape: lanceolated-triangular. Margins: dorsal sinuate, ventral serrate. Sulcus acusticus: heterosulcoid, ostial, median. Ostium: funnel-like. Cauda: tubular slightly curved posteriorly from the middle region ending very close to the ventral margin. Anterior region: lanceolated; rostrum narrow, long, pointed; antirostrum short, broad, pointed; excisura wide with a medium notch. Posterior region: oblique-irregular (**Figure Y2**).

2.26. Needlescaled Queenfish, *Scomberoides Tol*

The body has fusiform and elongated shape, laterally compressed. Dorsal and ventral profile are identical. Adipose eyelid is developed. Snout length is greater than the eye diameter. Snout blunt, head, and nape are slightly concave. Maxilla extends beyond the posterior border of the pupil. Lower jaw has two rows of teeth. Large conical teeth present in outer row. Small villiform teeth are present in inner row. Dentary in both jaw are sub equal in length. Head length comprises 18% of FL. Gill rakers are large and slender, 5–8 upper, and 16–20 lower.

First dorsal fin with 1 embedded spine and 6–7 free short spines, and second dorsal fin with 1 spine followed by 17–20 soft rays. Pectoral fin is not falcate and equal in length of pelvic fin. Pelvic fin is short in shallow groove. Soft dorsal and anal fin is falcate, equal in length, posteriorly consisting of semidetached finlet. Two detached anal spine are present. Anal fin with 1 spine followed by 18–21 soft rays. Soft dorsal and anal fin comprises 10% of FL and the distal half of soft dorsal fin are pigmented black. Caudal fin is deeply forked and comprises 23% of FL, anal fin origin in a line with second dorsal fin. Scales are needle like, below the lateral line and mid-body, deeply embedded in skin. Lateral line is little angulated just above the pectoral fin, slightly irregular, and straight posteriorly without scutes. A single series of (5–8) blotches rounded to vertically oblong is present, first 4–5 blotches intersecting the lateral line. Caudal peduncle groove is absent (**Figure Z1**).

Otolith shape: lanceolated, slightly curved. Margins: dorsal sinuate, ventral crenate. Sulcus acusticus: heterosulcoid, ostial, median. Ostium: funnel-like. Cauda: tubular slightly curved ending very close to the ventral margin. Anterior region: lanceolated; rostrum narrow, long, pointed; antirostrum short, broad, peaked; excisura wide with a shallow notch. Posterior region: oblique (**Figure Z2**).

2.27. Bigeye Scad, Selar Crumenophthalmus

Body has fusiform and elongated shape, moderately compressed. Ventral profile of body is more convex than dorsal. Eye is large and greater than snout length and covered with adipose eyelid except for vertical slit centered on pupil. Snout is pointed. Lower jaw with a single row of minute is uniserrate tooth. Villiform tooth is also arranged in vomer, palatines, and on central band on tongue. Maxilla extends to below anterior two-third of eye. Head length is about 30% of FL. Gill rakers: 10–11 upper and 30–34 lower in first gill arch. Smooth shoulder girdle margin with deep furrow having 2 papillae, large papilla above, and lower papilla near lower edge. First dorsal fin is membranous, with 8 spines, and second dorsal fin with 1 spine followed by 20–21 rays. Two detached anal spines are present, anal fin with 1 spine and 19–22 soft rays. Pectoral fin is long and falcate about 24–27% of FL and shorter than HL. Pelvic fin is short. Soft dorsal fin is low, anal fin slightly falcate. Dorsal and anal fins without detached terminal finlet. Caudal fin is deeply forked about 24% of FL. The distance from snout to anal fin origin is 57% of FL. Curved lateral line is longer than straight line, and junction is below twelfth–fourteenth soft dorsal fin rays. Curve lateral line with 48 scales and straight lateral line with 11 scales and 30–37 scutes. Scales in body are cycloids and small, entirely covered the body except small portion behind pectoral fin (**Figure AA1**).

Otolith shape: elliptic. Margins: dorsal, entire, ventrally, sinuate. Sulcus acusticus: heterosulcoid, ostial, median. Ostium: oval. Cauda: tubular, markedly curved posteriorly ending close to the ventral margin. Anterior region: peaked, broad with one big protuberance; rostrum broad, small, pointed; antirostrum very short; excisura tiny notch. Posterior region: round (**Figure AA2**).

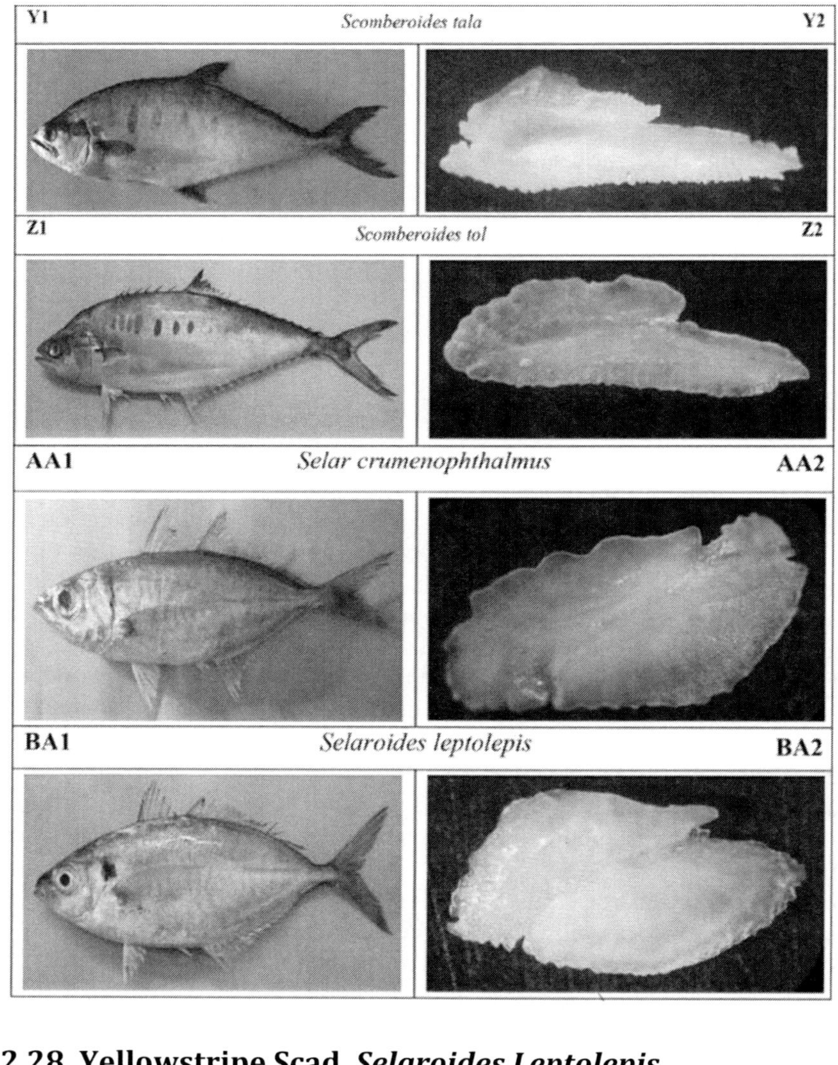

2.28. Yellowstripe Scad, *Selaroides Leptolepis*

Body has elliptic and oblong shape, laterally compressed. Dorsal and ventral profile is identically and equally convex. Yellow stripes on broad eyes equally or slightly greater than snout length. Eye is covered with adipose eyelid posteriorly. Maxilla is protractible and concave above. No tooth in upper jaw and lower jaw with minute tooth. Head is small and comprised of 24–27% of FL. Gill rakers: 10–12 upper, 21–24 in lower limb of first gill arch. Breast is completely covered with First dorsal fin is membranous with 8 spines and second dorsal fin with 1 spine followed by 26 soft rays. Dorsal and anal fin rays are low, not falcate. No ventral grooves are present. Third spine of first dorsal fin is about double than second dorsal fin. Anal fin with two detached spines. Anal fin with 1 spines followed by 19–22 soft anal rays is present. Caudal fin is deeply forked about 28% of FL. Lateral line is deeply concave above pectoral fin, straight part is greater than curved part, and their junction is below the twelfth ray of soft ray (**Figure BA1**).

Otolith shape: elliptic. Margins: dorsal, entirely smooth, ventral, crenate. Sulcus acusticus: heterosulcoid, ostial, median. Ostium: funnel-like. Cauda: tubular, strongly curved posteriorly ending close to the ventral margin. Anterior region: pointed; rostrum elongated, broadly pointed; antirostrum narrow, short, blunt; excisura wide with a shallow notch. Posterior region: oblique (**Figure BA2**).

2.29. Greater Amberjack, *Seriola Dumerilli* (New Record)

Body has fusiform shape, moderately compressed and shallow. Eye is moderate, lower than snout length. Ventral profile is more convex. Maxilla is broad at end extending posterior margin of eye. Super maxilla is broad. Teeth are arranged in a broad band in both jaws. Head length is 25–28% of FL. Gill rakers: upper 6–8, lower 14–16 in first gill arch.

First dorsal spinous fin is short and membranous with 7–8 spines, and second dorsal fin consists of 1 spine followed by 28–34 soft rays. Anterior ray of second dorsal fin is slightly elevated and 13–14% of FL. Pectoral fin is shorter than pelvic fin. Anal fin with 2 detached spines (embedded in larger specimen) followed by 1 spine and 19–21 soft rays. Caudal fin is deeply forked and about 23–26% of FL. The distance from snout to origin of anal fin is 60–64% of FL. Lateral line without scutes, moderately curved above pectoral fin and smoothly straight posteriorly. A cutaneous keel on each side of caudal peduncle is present. Body color is olive dorsally and laterally and silvery belly, snout to the origin of second dorsal fin (**Figure CA1**).

Otolith shape: lanceolated. Margins: dorsal lobed, ventral serrate-dentate. Sulcus acusticus: heterosulcoid, ostial, median. Ostium: funnel-like. Cauda: tubular, strongly curved ending posteriorly far from the posterior margin. Anterior region: lanceolated; rostrum narrow, long, pointed; antirostrum short, narrow, pointed; excisura wide with an acute shallow notch. Posterior region: rounded-irregular (**Figure CA2**).

2.30. Blackbanded Trevally, *Seriolina Nigrofasciata*

Body shape is oblong and elongated, moderately shallow and compressed. Snout is rounded and greater than the eye. Super maxilla is slender and rounded. Maxilla rounded at the end extending beyond the posterior border of the pupil. Head consists of 25–29% of FL. Head profile is steeply raising to interorbital region and point a curve moderately to spinous dorsal fin. Minute villiform teeth arranged in series in both jaws and in central band on tongue. Gill rakers of first gill arch comprised of 2–5 upper, 7–9 lower such as a knob-like masses, including rudiments.

First dorsal spinous fin is short and membranous with 7- to 8-week spines, of which some are embedded in skin, and second dorsal fin consists of 1 spine followed by 33–37 rays. A ray of dorsal and anal fin is slightly elevated but not falcate. Pectoral fin is short, not falcate. Pelvic fin is greater than pectoral fin in length 18% of FL. Anal fin 1 embedded spine followed by 18–20 soft rays. The distance from snout to anal fin origin is 60% of FL. Caudal fin consists of 23–28% of FL. A cutaneous keel on each side of caudal peduncle is developed in adult. Curve lateral line is arched moderately below 24 soft ray of second dorsal fin. In juveniles, 6–7 dark oblong band and blotches is present which disappear with age (**Figure DA1**).

Otolith shape: lanceolated. Margins: dorsal lobed, ventral crenate. Sulcus acusticus: heterosulcoid, ostial, median in position. Ostium: funnel-like. Cauda: tubular, strongly curved posteriorly ending close to the ventral margin. Anterior region: lanceolated; rostrum narrow, long, pointed; antirostrum short, narrow, pointed; excisura wide with a shallow notch. Posterior region: angled (**Figure DA2**).

2.31. Small Spotted Dart, *Trachinotus Baillonii*

Body has elliptical and oblong shape, laterally compressed. Dorsal and ventral profile is identical. Eye is equal to snout length. Snout is pointed. Maxillae extend beyond the anterior border of eye. Palatine and villiform teeth are arranged in bands in jaws and vomer in a triangular patch. Tongue without teeth. Head is small and comprises 25–27% of FL. Gill rakers on upper 6–7, lower 14–18 on first gill arch.

First dorsal fin modified into 6–7 short free spines, and second dorsal fin with 1 spine followed by 22–24 soft rays. Dorsal fin is long, falcate, and comprises of 35% of Fl. Pectoral fin greater than pelvic fin, but relatively shorter than dorsal and anal fin. Two detached anal spine are present. Anal fin consists of 1 spine followed by 18–21 soft rays. Anal fin originated just behind the dorsal fin. Lateral line is slightly irregular, weekly convex above pectoral fin, containing 2–5 black spot (less than eye diameter) along the line. No scutes and caudal pedunclee groove are present. Caudal fin deeply forked about 40% of FL (**Figure EA1**).

Otolith shape: elliptic-fusiform. Margins: Dorso-ventrally crenate-serrate. Sulcus acusticus: heterosulcoid, ostial, median. Ostium: funnel-like shorter than cauda. Cauda: tubular, slightly curved posteriorly ending close to the ventral margin. Anterior region: peaked with a small protuberance; rostrum long, broad, peaked; antirostrum short, broad, round; excisura wide without notch. Posterior region: round (**Figure EA2**).

2.32. Snubnose Pompano, *Trachinotus Blochi*

Body shape is elliptical and oblong, laterally compressed. Dorsal and ventral profile is identical. Eye is smaller than snout length. Snout is rounded. Maxillae extend beyond the anterior border of eye. Small villiform teeth are arranged in band in both jaws. No teeth on tongue. Head is small and comprises 22–28% of FL. Gill rakers on upper 6, lower 8–12 on first gill arch.

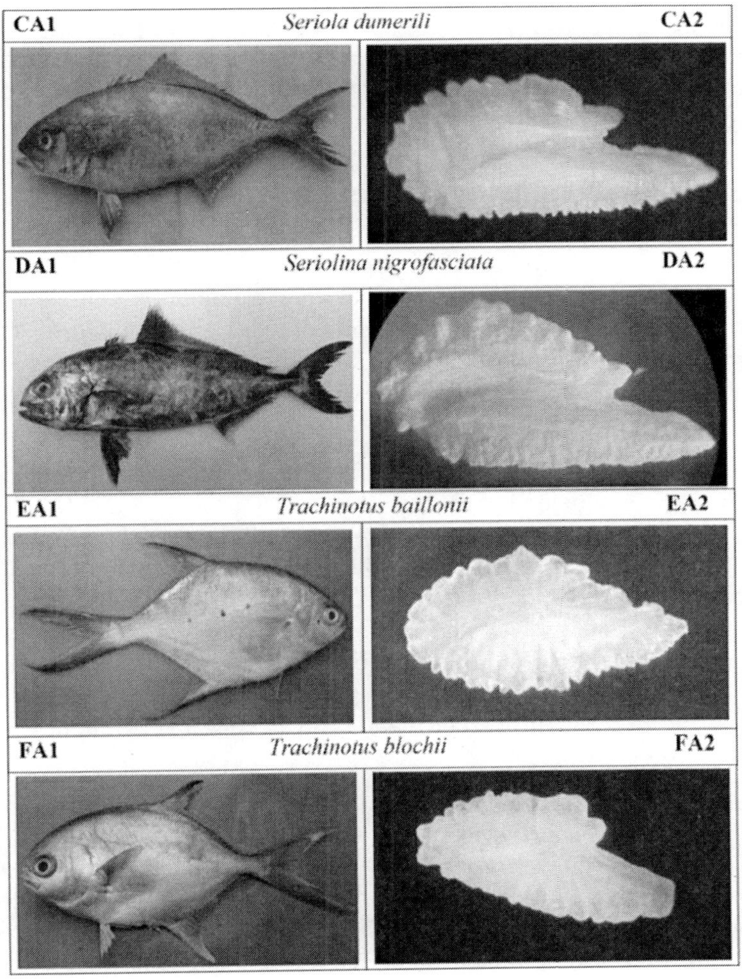

First dorsal fin modified into 6–7 short free spines, often embedded in adult, and second dorsal fin with 1 spine followed by 22–23 soft rays. Pectoral fin is greater than pelvic fin, but relatively shorter than dorsal and anal fin. Two detached anal spine present. Anal fin consists of 1 spine followed by 17–18 soft rays. Anal fin originated just behind the dorsal fin. Snout to anal fin distance is 56–62% of FL. Height of second dorsal fin lobe is 28% of FL. Lateral line is slightly irregular, weekly convex above pectoral fin and straight posteriorly. First predorsal bone inverted teardrop-shaped, and supra-occipital bone is thin and blade like in young. No scutes and caudal peduncle groove are present. Caudal fin deeply forked about 26–34% of FL (**Figure FA1**).

Otolith shape: fusiform. Margins: crenate-lobed. Sulcus acusticus: heterosulcoid, ostial, median. Ostium: funnel-like. Cauda: tubular, slightly curved ending posteriorly close to the ventral margin. Anterior region: peaked; rostrum long, broad, pointed; antirostrum short, broad, blunt; excisura wide with a shallow notch. Posterior region: oblique-irregular (**Figure FA2**).

2.33. Indian Pompano, *Trachinotus Mookalee*

Body has ovate shape, laterally compressed. Dorsal and ventral profile is identical. Eye is smaller than snout length. Snout is rounded. Maxillae extend beyond the anterior border of eye. Small villiform teeth are arranged in band in both jaws. Narrow patches of teeth are present in a tongue. Head is small and comprises 25–26% of FL. Gill rakers on upper 6–8, lower 10–15 on first gill arch.

First dorsal fin modified into 6–7 short free spines, and second dorsal fin with 1 spine followed by 20–22 soft rays. Pectoral fin is greater than pelvic fin, but relatively shorter than dorsal and anal fin. Two detached anal spine are present. Anal fin consists of 1 spine followed by 17–18 soft rays. Anal fin is originated just behind the dorsal fin. Snout to anal fin distance is 55% of FL. Height of second dorsal fin lobe is 28% of FL. Lateral line is slightly irregular, weekly convex above pectoral fin. First predorsal bone shaped as an inverted-L with the anteriorly projecting arm. No scutes and caudal peduncle groove are present. Caudal fin deeply forked about 37% of FL (**Figure GA1**).

Otolith shape: elliptic-fusiform. Margins: Dorso-ventrally crenate. Sulcus acusticus: heterosulcoid, ostial, median. Ostium: funnel-like. Cauda: tubular, slightly curved posteriorly ending close to the ventral margin. Anterior region: irregular with a prominent protuberance joining to antirostrum; rostrum long, wide pointed upward; antirostrum very short, broad, peaked; excisura wide without notch. Posterior region: oblique (**Figure GA2**).

2.34. Large Spotted Dart, *Trachinotus Botla* (*Russelii*) (New Record)

Body has elliptical shape, laterally compressed. Dorsal and ventral profile is identically convex. Snout rounded. Teeth villiform are arranged in bands in both Jaws and absent in adult. Gill rakers: 7–9 on upper limb and 11–15 lower limb of first gill arch.

First dorsal fin with 6 short free spines, second soft dorsal fin with 1 spine followed by 23 soft rays. Pectoral fin is short and shorter than pelvic fin. Anal fin with 1 spine followed by 19 rays. Dorsal fin lobe is greater than anal fin lobe. Anal fin with 1 spine and 22 soft rays. Caudal fin is deeply forked. Lateral line is strongly concave above pectoral fin and deeply arch. Scales are minute and deeply embedded in the skin. A series of 5 plumbeous spots present, first three large spots intersecting the lateral line and other small two spots touching the lateral line. Scutes are absent in lateral line (**Figure HA1**).

Otolith shape: fusiform. Margins: irregular to dentate. Sulcus acusticus: heterosulcoid, ostial, median. Ostium: funnel-like. Cauda: tubular, slightly curved posteriorly ending close to the ventral margin. Anterior region: peaked; rostrum moderately long, broad, rounded; antirostrum short, broad, irregular upward; excisura wide with an acute and deep notch. Posterior region: round-irregular (**Figure HA2**).

2.35. Arabian Scad, *Trachurus Indicus*

Body shape is elongate, laterally compressed; dorsal and ventral profile is identical. Eye is moderate, slightly smaller than snout. Adipose eyelid is covered the entire eye except the vertical slit of the eye. Maxilla extends to the middle of pupil. Small villiform teeth are arranged single row in both jaws. Opercular dark crescent shape bloch is present. Head length comprises of 29% FL. Gill rakers: 12–15 in upper limb, 35–40 in lower limb of first gill arch.

Two separate dorsal fin. First dorsal fin is membranous with 7–8 spines. Soft dorsal fin with 1 spine followed by 32 soft rays. Pectoral fin is long and falcate and greater than HL. Pelvic fin is moderate and hyaline. Anal fin with two detached anal spines and 1 spine followed by 28 soft rays. Finlet absent. Caudal fin is deeply forked about 28% FL. Curve lateral line is longer than straight lateral line. Scutes on curve lateral line are 38, and scutes on straight lateral line are 37–40. Junction of CLL and SLL is below the ninth–tenth ray of second dorsal fin. Dorsal accessory lateral line is terminating at the base of caudal peduncle (**Figure IA1**).

Otolith shape: lanceolated. Margins: sinuate. Sulcus acusticus: heterosulcoid, ostial, median. Ostium: funnel-like. Cauda: tubular, slightly curved ending close to the ventral margin. Anterior region: peaked; rostrum long, broad, peaked; antirostrum short, broad, blunt; excisura wide without notch. Posterior region: oblique (**Figure IA2**).

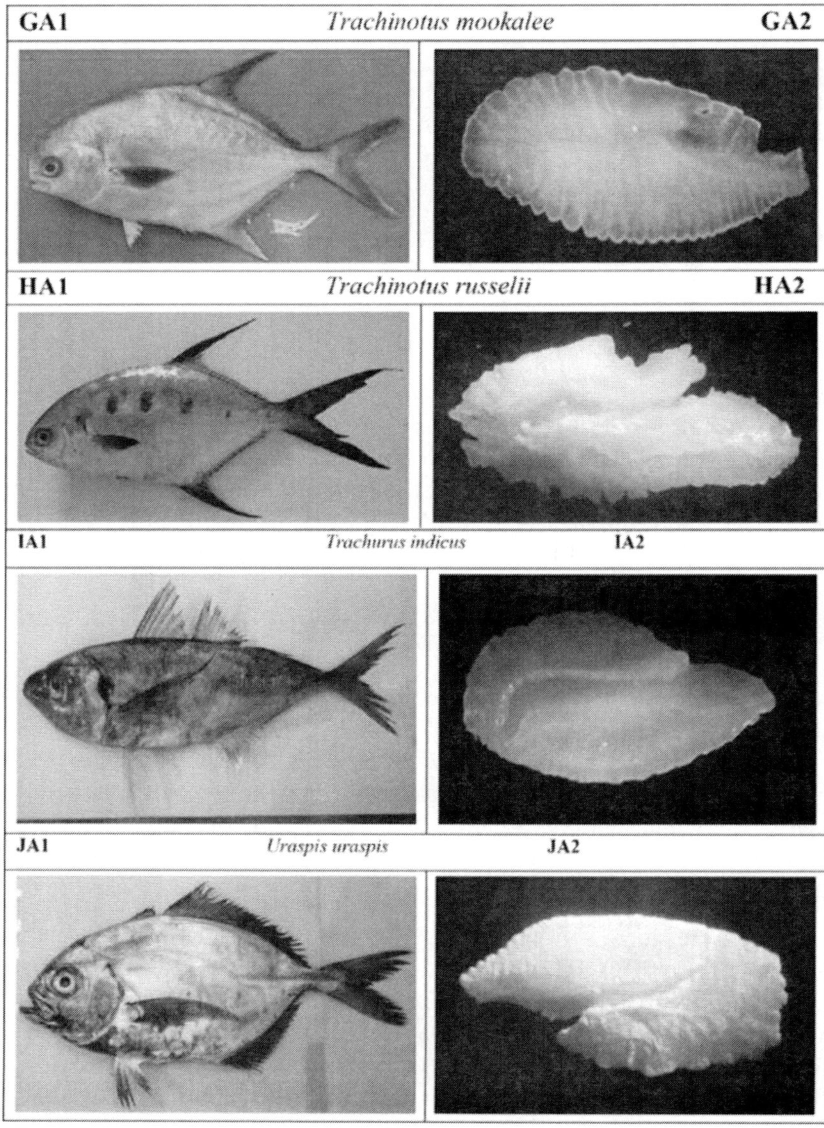

2.36. Whitemouth Jack, *Uraspis Uraspis* (New Record)

Body shape is oval; compressed, ventral profile is slightly convex to isthmus than straight to the origin of second dorsal fin. Maxilla is extended to the anterior margin of the eye. Small pointed teeth are arranged in bands on both jaws. Tongue and floor of mouth is white with dark blue to black border. Naked area of breast extends to half of the pectoral fin base. Head length is 30.6% of FL. Gill rackers upper 3–5 and 13–15, lower limbs of the first gill arch.

Two detached dorsal fin with 3–4 posteriorly embedded spines before first dorsal fin. First dorsal fin section is small, membranous and consists of 7–8 spine and second dorsal fin with 1 spine followed by 27–32 spines. Pectoral fin is long, falcate and reaches to the junction of curved and straight lateral line. Pectoral fin length is 31.8% of FL. Pelvic fin length decreases with age. Anal fin with two detached and embedded spines followed by 19–23 soft rays. Caudal fin is deeply forked 33.6% of FL. Straight lateral is 1.02–1.05% shorter than straight lateral line. Scutes on straight lateral line are strong to moderate consist of 35–38 antrose (projecting anteriorly) (**Figure JA1**).

Otolith shape: elliptic-lanceolated. Margins: dorsal irregular, ventral sinuate. Sulcus acusticus: heterosulcoid, ostial, median. Ostium: funnel-like. Cauda: tubular, strongly curved posteriorly ending close to the ventral margin. Anterior region: peaked; rostrum long, narrow, peaked. Anti-rostrum short, broad, pointed; excisura wide with an acute and deep notch. Posterior region: oblique-flattened (**Figure JA2**).

3. ACKNOWLEDGEMENTS

This chapter is the part of Ph.D dissertation of Dr. Nazia Qamar. We thank Victor Tusset, Spanish Research Council for candid comments and suggestions. The Director, CEMB, University of Karachi is greatly acknowledged for the facilities provided during the study.

REFERENCES

1. Bianchi G.. FAO species identification sheets for fishery purposes-Field guide to the commercial marine and brackish-water species of Pakistan, prepared with the support of PAK/77/033/ and FAO (FIRM) Regular Programme. FAO, Rome. 1985: 200.
2. Nelson, J.S.. Fishes of the world. John Wiley & Sons, New York. 2006.

3. Riede K.. Global register of migratory species – from global to regional scales. Final Report of the R&D Project 808 05 081, Federal Agency for Nature Conservation, Bonn, Germany. 2004; 329.
4. Smith-Vaniz W.F. and Carpenter K.E.. Review of the crevalle jacks, *Caranx hippos* complex (Teleostei: Carangidae), with a description of a new species from West Africa. Fisheries Bulletin. 2007; 105(2): 207-233.
5. Sudekum A. E., Parrish J.D., Radtke R. L. Ralston S.. Life history of large jacks in undistributed, shallow, oceanic communities. Fisheries Bulletin, 1991; 89(3): 493-513.
6. Thompson R. Munro J. L.. The biology, ecology, and bionomics of the jacks, Carangidae. In: Munro JL (ed) Caribbean coral reef fishery resources. Int. Cent. Living Aquat. Res. Manage, Manila, Philippines. 1983; 82–93.
7. Qamar N., Panhwar S. K., Jahangir S.. Seasonal variation in diet composition of Torpedo Trevally, *Megalaspis cordyla* (Linneaus, 1758) depending upon its size and sex. Pakistan Journal of Zoology. 2015; 47(4): 1171-1179.
8. Fischer W. Bianchi G.. FAO species identification sheets for fishery purposes In: Field guide to the commercial marine and brackish-water species of Pakistan prepared with the support of PAK/77/033/ and FAO (FIRM) Regular Program. FAO, Rome. 1983; 200.
9. Panhwar S. K., Qamar N., Jahanghir, S.. Fishery and stock estimates of Talang queenfish, *Scomberoides commersonnianus* (Fam: Carangidae) from the Arabian sea coast, Pakistan, Pakistan journal of agricultural sciences. 2014; 51(4): 1111-1116
10. Qamar N., Panhwar S. K. Brower S.. Population characteristics and biological reference point estimates for two carangid fishes *Megalaspis cordyla*and *Scomberoides tol* in the Northern Arabian Sea, coast of Pakistan. Pakistan Journal of Zoology. 2016;48(3): 869-874.
11. Qamar N., Jahangir S., Waryani B., Panhwar S. K., Bhutto A. H.. Anomalous torpedo trevally, *Megalaspis cordyla* l. (Pisces: Carangidae) found in Pakistan. International journal of biology and biotechnology. 2015;12 (2): 193-196.
12. Hand book of the fisheries statistics of Pakistan, Marine Fisheries Department. 2012; 20: 1-217.
13. FAO.. Fishery and aquaculture country profile, FAO" Fisheries Department, Rome, 2012; 1-18.
14. Abdussamad E. M., Prathibha Rohit K. P., Said Koya O. M. M. J., Habeeb M. Jeyabalan, K.. Carangids (Family Carangidae) in the seas around Indian subcontinent with description of macro-taxonomic characters for the field identification of genera and species. Indian Journal of Fisheries. 2013; 60 (2): 21-3.
15. Matsunuma M., Motomura H, Matsuura K., Shazili N. and Ambak M.. Fishes of Terengganu East coast of Malay Peninsula, Malaysia, National Museum of Nature and Science, Tokyo, University Malaysia Terengganu, Terengganu, and Kagoshima University Museum, Kagoshima. 2011; 25.
16. Quigley D. T., Flannery G. K., Shea J. O.. Fish note: trigger fish *Balistes capriscus Gmelin*. Irish Naturalists Journal. 1993; 24: 223–228.
17. Kuiter R. H.. The complete divers and fishermans guide to coastal fishes of south-eastern Australia. Gary Allen, Sydney. 2000; 437.
18. Gunn J. S.. A revision of selected genera of the family Carangidae (Pisces) from Australian waters. Record of Australian museum, Supplement. .1990; 12: 1-77.
19. Hoda S. M. S.. Fishes from the coast of Pakistan. Biologia (Lahore). 1988; 34: 1-38
20. Ahmad M. F., Niazi, M. S.. Important edible fishes of Pakistan. Zoological Survey Department, Government of Pakistan.1988: 1-31.
21. Laroche W. A. Smith- Vaniz W.F. Richardson, S. L.. Carangidae: Development. In: Moser H. G., Richards W. J., Cohen D. M., Fahay M. P., Kendall. A. W., Richardson S. L. (EDS.), Ontogeny and systematic of fishes. Special publication 1, American Society of Ichthyologists and Herpetologists. 1984; 510-522.
22. Nelson J.S.. Fishes of the world. John Wiley & Sons, New York. 1984.

23. Jalil S. A. Khaliluddin M. A.. checklist of marine fishes of Pakistan. Government of Pakistan. 1972: 1-16.
24. Bannikov, A. F.. On the taxonomy, composition and origin of the family Carangidae. Journal of applied ichthyology.1987; 24: 4156–158. (Originally published in Russian in Voprosy Ikhtiologii,(6): 833–839.
25. Eschmeyer W. N.. Catalog of the genera of recent fishes. California Academy of Science, San Francisco. 1990: 697.
26. Smith-Vaniz, W.F. Opistognathidae. Jaw fishes. p. 2588-2589. In K.E. Carpenter and V.H. Niem (eds.) FAO species identification guide for fishery purposes. The living marine resources of the Western Central Pacific. Volume 4. Bony fishes part 2 (Mugilidae to Carangidae). FAO, Rome. 1999.
27. Smith-Vaniz, W. F. Carangidae. In: Fischer W. and Bianchi,G. (Eds.).. FAO species identification sheets for fishery purpose, Western Indian Ocean (Fishery area 51), Vol. 1,FAO, Rome. 1984.
28. Webb P. W.. Hydrodynamics and energetics of fish propulsion. Bulletin of the fisheries research board of Canada. 1975;190 :1-156.
29. Brown-Peterson N. J., Wyanski D. M., Saborido-Rey F., Macewicz B. J. Lowerre-Barbieri S. K.. A standardized terminology for describing reproductive development in fishes. Marine and Coastal Fisheries: Dynamics, Management, and Ecosystem Science [online serial] 2011; 3: 52–70.

CHAPTER 11

The first record of a frogfish, *Fowlerichthys scriptissimus* (Antennariidae, Lophiiformes), from Korea

Song-Hun Han[1], Joon Sang Kim[2] and Choon Bok Song[3*]

[1]*Jeju Fisheries Research Institute, National Institute of Fisheries Science, Jeju 63068, Korea.*

[2]*Korea Fisheries Resources Agency, Jeju Branch, Jeju 63005, Korea.*

[3]*College of Ocean Sciences, Jeju National University, Jeju 63243, Korea*

ABSTRACT

This is the first report of *Fowlerichthys scriptissimus* (Lophiiformes, Antennariidae) from Korea. A single specimen (291.0 mm SL) was collected off the coast of Jejudo Island by gill net on 28 March 2012 and identified with morphological and molecular approaches. The specimen is characterized by having all five pelvic fin rays bifurcate and possessing 20 vertebrae, 13 pectoral-fin rays, and a basidorsal ocellus on the side of the body. This species is distinguishable from other Korean taxa by the number of pectoral fin rays, the bifurcate form of the pelvic rays, and the vertebral count. We add this species to the Korean fish fauna and suggest new Korean names, "Byeol-ssin-beng-i-sok" and "Byeol-ssin-beng-i" for the genus and species, respectively.

Keywords: *Fowlerichthys scriptissimus,* Antennariidae, New record, Jejudo Island, Korea

BACKGROUND

The frogfishes (Antennariidae), which belong to order Lophiiformes, occur in all tropical and subtropical seas except the Mediterranean Sea (Nelson 2006). Worldwide, the family includes 46 species in 13 genera (Arnold and Pietsch 2012) and four species in two genera in Korea (Kim et al. 2005; Kim et al. 2011). All members of Antenariidae have the first dorsal fin spine modified into a fishing pole (illicium) and gill openings below or behind the base of the pectoral fin (Nelson 2006; Arnold and Pietsch 2012). The genus *Fowlerichthys* was originally suggested by Barbour (1941) when he described the new species, *Fowlerichthys floridanus*. However, as some of his morphological descriptions were not clear, most ichthyologists regarded the genus *Fowlerichthys* as a synonym of *Antennarius* (Pietsch 1984; Senou 2002; Manilo and Bogorodsky 2003). Later, Arnold and Pietsch (2012) reconstructed evolutionary relationships within Antennariidae with molecular phylogenetics and recognized that *Fowlerichthys* as a valid genus. They demonstrated that it can be separated from *Antennarius* by having all five pelvic fin rays bifurcate instead of one in the genus *Antennarius*.

A single specimen of *Fowlerichthys scriptissimus*, which is otherwise unknown from Korea, was collected in the coastal waters of Jejudo Island by gill net on 28 March 2012. Here, we describe the morphological characters of *F. scriptissimus* and report the results of a molecular barcode determination of the specimen's identification using the COI gene.

METHODS

Counts and measurements followed the method of Hubbs and Lagler (1964). When conducting this study, we adhered to the ethical guideline of the International Council for Laboratory Animal Science (ICLAS) for researchers. The specimen was fixed in 10% buffered formalin and then transferred to 70% ethanol. Vertebrae were counted from radiographs (REX-525R, listem). The examined specimen was deposited at the Jeju National University (JNU), Korea, and is available from the corresponding author by reasonable request.

Total DNA was extracted from 25 mg of the muscle tissue with an AccuPrep Genomic DNA Extraction Kit (Bioneer Inc.) according to manufacturer's protocol. The mitochondrial cytochrome *c* oxidase subunit I (COI) gene was PCR-amplified with the primers, Asn-F1 (AAA HWC TTA GTT AAC AGC TAA) and Ser-R1 (GGG GTT CDA YTC CYC CCT TTC T). The polymerase chain reaction (PCR) was performed on a thermal cycler (TP600, Takara Bio Inc.) with a final volume of 40 μL in a 0.2 mL PCR tube containing 10 mM Tris-

HCl (pH 8.3), 50 mM KCl, 2 mM MgCl2, 0.2 mM dNTP mix, 0.5 mM of each primer, 50 ng template DNA solution, and 1 U Ex *Taq*DNA polymerase (Takara Bio Inc.) The PCR cycles consisted of an initial denaturing step of 94 °C for 2 min, followed by 30 cycles of 30 s at 94 °C for denaturation, 1 min at 54 °C for primer annealing, and 1 min and 50 s at 72 °C for extension, and an additional 7 min interval at 72 °C for a final extension. The PCR products were sequenced using the BigDye Terminator v3.1 Cycle Sequencing Kit (Applied Biosystems Inc.) and ABI PrismTM 3730XL DNA Analyzer. Sequencing reactions were conducted with the amplification primers in two directions. The DNA sequence obtained was deposited in the National Center for Biotechnology Information (NCBI) as GenBank accession number, KY195977. For molecular identification, we compared the specimen's COI sequence with the GenBank DNA sequences of the seven anglerfish species: *Antennarius hispidus*(FJ582855), *Antennarius pictus* (FJ582858), *Antennarius striatus* (AB282828), *Histrio histrio* (AB282829), *Fowlerichthys avalonis* (DQ0279840), *F. scriptissimus* (GU188480), and *Lophius litulon* (KJ020931). The DNA sequences were aligned and edited using Clustal W (Thompson et al. 1994) and BioEdit version 7 (Hall 1999). Genetic distances were calculated using MEGA 6 (Tamura et al. 2013) based on the Kimura two-parameter (K2P) model (Kimura 1980). A neighbor-joining (NJ) tree was constructed using the K2P model and 10,000 bootstrap replications in MEGA 6.

RESULTS

Genus *Fowlerichthys* Barbour 1941

(New Korean genus name, Byeol-ssin-beng-i-sok)

Fowlerichthys Barbour 1941; 12 (type species, *Fowlerichthys floridanus* Barbour 1941). Arnold and Pietsch 2012, 128 (Rarotanga, Cook Islands); Stewart 2015, 886 (New Zealand).

The genus *Fowlerichthys* includes five species diagnosed by the following combination of characters: one or three darkly pigmented ocelli on side of body; illicium about as long as second dorsal spine; esca in form of tuft of slender filaments, a simple, oval-shaped appendage, escal pigment spots absent; dorsal rays 12–14 (usually 13); pectoral rays 11–14 (usually 12 or 13); all five pelvic rays bifurcate; anal rays 7–10 (usually 8), all bifurcate; vertebrae 20 (Pietsch and Grobecker 1987; Arnold and Pietsch 2012).

F. scriptissimus (Jordan 1902)

(New Korean name, Byeol-ssin-beng-i) (Table 1; Fig. 1)

Table 1 Morphological characters compared between the present specimen and previous studies on *F. Scriptissimus*

Morphological characters	Present study	Jordon (1902)	Araga (1984)	Pietsch and Grobecker (1987)	Senou (2002)
Total length (mm)	350.0 (n = 1)	–	350.0 (n = 1)	- (n = 5)	–
Standard length (mm)	291.0	–	–	99.0–280.0	280.0
Count					
Dorsal fin rays (bifurcate rays)	13 (7)	III-12	III-13	13 (7–9)	13
Pectoral fin rays	13	–	13	13	13
Pelvic fins rays	I, 5	–	5	5	I, 5
Anal fin rays	8	8	8	8	8
Caudal fin rays	9	–	–	9	9

Figure. 1 *F. scriptissimus* (Jordan 1902), JNU-637, 291.0 mm SL, gill net, Hanlim-eup, Jeju-si, Jejudo Island, Korea. *Scale bar* = 50 mm

Antennarius scriptissimus Jordan, 1902, 373 (type locality, Bōsō Peninsula, Chiba Prefecture, Japan); Pietsch 2000, 597 (South China Sea); Senou 2002, 456 (Japan); Manilo and Bogorodsky 2003, S99 (Oman).

Antennarius sarasa Pietsch 1984, 36 (Japan); Araga 1984, 103 (Japan); Pietsch and Grobecker 1987, 123 (Réunion Island, New Zealand, and Philippines).

F. scriptissimus Arnold and Pietsch 2012, 128 (Rarotanga, Cook Islands)

MATERIAL EXAMINED

JNU-637, one specimen, 291.0 mm in standard length (SL), gill net, Hanlim-eup, Jeju-si, Jejudo Island, Korea, 28 March 2012.

Description

Meristic counts appear in Table 1. Measurements as a percentage of SL are as follows: body depth 67.6; body width 21.3; snout length 24.0; eye diameter 5.1; interorbital length 14.7; illicium length 6.9; first predorsal fin length 18.3; second predorsal fin length 20.1; third predorsal fin length 35.5; prepectoral fin length 36.7; preanal fin length 67.3; prepelvic fin length 9.5; length of second dorsal spine 8.4; length of third dorsal spine 10.8; length of pectoral fin ray 13.0; length of anal fin ray 12.5; caudal peduncle depth 12.1; and caudal peduncle length 9.2.

Body round and compressed; skin of body very rough and covered with close-set dermal spinules; head and eyes small; eyes lateral; mouth large and extremely oblique; first dorsal spine (illicium) shorter than second spine; esca simple and oval-shaped with a tuft; second and third dorsal spine rough and curved posteriorly; gill opening located below base of pectoral fin; depression between second and third dorsal spines; membrane behind second dorsal spine extending posteriorly, dividing area between second and third dorsal spines and nearly reaching to base of third (Fig. 3).

Live coloration

Whole body uniformly greenish brown; entire head, body, and fins with mottled dark brown reticulations except for inner surface of paired fins; a single, darkly pigmented, basidorsal ocellus on each side of body; scattered beige spots on body; esca with dark pigment at base.

Color in preservative

Body uniformly pale greenish brown; mottled dark brown spots and ocellus; beige spots no longer apparent.

Distribution

Widely known from Indo-West Pacific, including Réunion Island, New Zealand, the Philippines (Pietsch and Grobecker 1987), the Cook Islands (Arnold and Pietsch 2012), the South China Sea (Pietsch 2000), Japan

(Jordan 1902; Pietsch 1984; Araga 1984), and Korea (Jejudo Island, present study).

Molecular identification

To confirm and verify the accuracy of the morphological species identification, we also analyzed 634 base pairs of the mitochondrial COI gene. The DNA sequence of the specimen was almost identical to a previously published sequence (GU188480) of *F. scriptissimus* (genetic distance, $d = 0.003$). The NJ tree also clustered the specimen with the GenBank sample of *F. scriptissimus*, with a strong 100% bootstrap value (Fig. 4).

DISCUSSION

The specimen in question has all the previously reported diagnostic morphological characters of *F. scriptissimus*, including five bifurcate pelvic fin rays, 20 vertebrae, 13 pectoral fin rays, and a basidorsal ocellus on the side of the body (Table 1 and Fig. 2). The form of the esca as a simple, oval-shaped appendage with numerous, more or less parallel, vertically aligned folds also matches the morphology of *F. scriptissimus*. Additionally, the form of the membrane behind the second dorsal spine matches *F. scriptissimus*, by lacking division into naked dorsal and ventral portions, extending posteriorly to the area between the second and third dorsal spines, and nearly reaching to the base of the third spine (Fig. 3). Thus, the morphological characteristics of the specimen clearly fit the species descriptions given by previous studies (Table 1). In the molecular analysis of the mitochondrial COI gene, the small genetic distance ($d = 0.003$) between the new specimen and a previously sequenced specimen of *F. scriptissimus*, and the close clustering of those two samples in a neighbor-joining tree (Fig. 4) including the seven anglerfish species confirmed the morphological identification. Thus, both morphological and molecular approaches indicated the specimen in the present study to be *F. scriptissimus.*

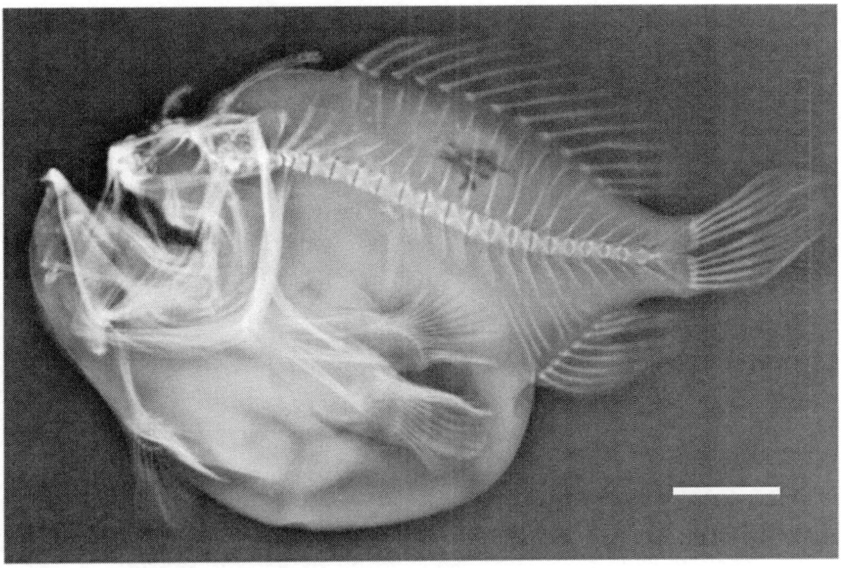

Figure. 2 Radiograph of JNU-637, *F. scriptissimus. Scale bar* = 50 mm

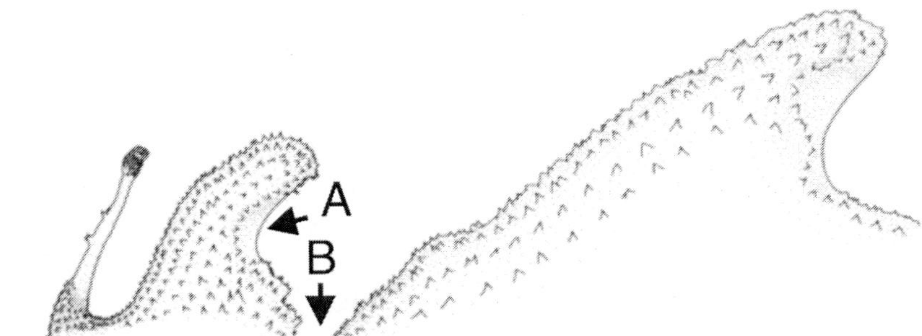

Figure. 3 Morphology of the dorsal fin spines of JNU-637, *F. scriptissimus*, showing the pattern of dermal spinules on the membrane behind the second spine (*A*) and membrane dividing the area between the second and third dorsal spines (*B*)

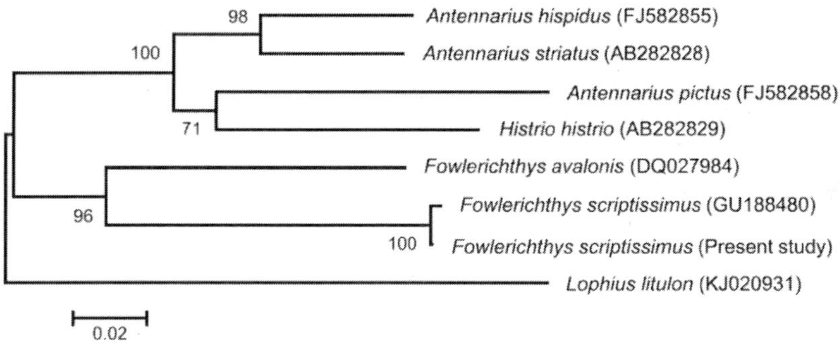

Figure. 4 Neighbor-joining tree showing the relationship of the specimen with the seven species of anglerfishes. *Numbers above nodes* indicate bootstrap probabilities based on 10,000 replications. *Bar* indicates a K2P genetic distance of 0.02

The genus *Fowlerichthys* was firstly suggested when Barbour (1941) described *Fowlerichthys floridanus*, but his description and diagnosis of the genus were unclear. He mentioned that the strong and sharp dorsal spines of the species were unlike the dorsal fin rays in the species of *Antennarius*, but whether these spines were bare in life is difficult now to determine certainly. Thus, most ichthyologists have regarded the genus *Fowlerichthys* as a synonym of *Antennarius*. Recently, Arnold and Pietsch (2012) demonstrated that *Fowlerichthys* is a valid genus that can be separated from *Antennarius* by having all five pelvic fin rays bifurcate (all other antennariids have four simple and one bifurcate rays). It can be further separated from all members of *Antennarius* except *Antennarius commerson* by possessing 20 vertebrae (Table 2). Currently, the genus *Fowlerichthys* is recognized as valid, with five known species worldwide *Fowlerichthys radiosus*, *F. avalonis*, *Fowlerichthys senegalensis*, *Fowlerichthys ocellatus*, and *F. scriptissimus* (Froese and Pauly 2016).

Table 2 Comparison in the number of pectoral fin rays, bifurcate pelvic fin rays, and vertebrae among five frogfish species inhabiting Korea

Species	Pectoral fin rays	Bifurcate pelvic rays	Vertebrae	References
F. scriptissimus	13	5	20	Pietsch and Grobecker (1987)
A. commerson	10–11 (usually 11)	1	19–20 (usually 19)	
A. maculatus	10–11 (usually 10)	1	19	
A. pictus	9–11 (usually 10)	1	19	
H. histrio	9–11 (usually 10)	0	18–19 (usually 19)	

F. scriptissimus is easily distinguished from the other Korean frogfishes (*A. pictus, A. commerson, A. maculatus*, and *H. histrio*) by having 13 pectoral fin rays (vs. 10 in *A. commerson*, 11 in *A. maculatus*, 10 in *A. pictus*, and 10 in *H. histrio*), five bifurcate pelvic rays (vs. one in the genus *Antennarius* and none in *H. histrio*), and 20 vertebrae (vs. 19 in *H. histrio and* all members of *Antennarius* except *A. commerson*) (Table 2).

We propose new Korean names, "Byeol-ssin-beng-i-sok" and "Byeol-ssin-beng-i" for the genus and species, respectively. The Korean name "Byeol-ssin-beng-i" was given for *F. scriptissimus* because of its scattered beige spots on the body that look alike stars. The Korean words "byeol" and "ssin-beng-i" mean star and frogfish, respectively.

NOTES

An erratum to this article is available at http://dx.doi.org/10.1186/s41240-017-0050-3.

ABBREVIATIONS

COI: Cytochrome c oxidase subunit 1

d **:** Genetic distance

ACKNOWLEDGEMENTS

This work was supported by a grant from the National Institute of Fisheries Science (R2016034). We thank Prof. B. Sidlauskas, Oregon State University, for the valuable discussion.

Funding

This study was funded by a grant from the National Institute of Fisheries Science (R2016034).

Availability of data and materials

All datasets generated during and/or analyzed during the current study are available from the corresponding author on reasonable request.

REFERENCES

1. Araga C. Family Antennarioidae. In: Masuda H, Amaoka K, Araga C, Uyeno U, Yoshino T, editors. The fishes of the Japanese Archipelago. Tokyo: Tokai Univ Press; 1984. p. 102–3.
2. Arnold RJ, Pietsch TW. Evolutionary history of frogfishes (Teleostei: Lophiiformes: Antennariidae): a molecular approach. Mol Phylogenet Evol. 2012. doi:10.1016/j.ympev.2011.09.012.
3. Barbour T. Notes on pediculate fishes. Proc New England Zool Club. 1941;19:7–14.
4. Froese R, Pauly D, editors. FishBase. World Wide Web electronic publication. www.fishbase.org, version (10/2016); 2016.
5. Hall TA. BioEdit: a user-friendly biological sequence alignment editor and analysis program for Windows 95/98/NT. Nucleic Acids Symp Ser. 1999;41:95–8.
6. Hubbs CL, Lagler KF. Fishes of the Great Lakes region. Bull Granbrook Inst Sci. 1964;26:19–27.
7. Jordan DS. A review of the pediculate fishes or anglers of Japan. Proc US Nat Mus. 1902. doi:10.5479/si.00963801.24-1261.361.
8. Kim IS, Choi Y, Lee CL, Lee YJ, Kim BJ, Kim JH. Illustrated book of Korean fishes. Seoul: Kyo-Hak Publishing; 2005. p. 615.
9. Kim BY, Kim MJ, Song CB. First record of the frogfishes *Antennarius pictus* (Antennariidae, Lophiiformes). Korean J Ichthyol. 2011;23:168–71.
10. Kimura M. A simple method for estimating evolutionary rates of base substitutions through comparative studies of nucleotide sequences. J Mol Evol. 1980;16:111–20.
11. Manilo LG, Bogorodsky SV. Taxonomic composition, diversity and distribution of coastal fishes of the Arabian Sea. J Ichthyol. 2003;43:S75–S149.
12. Nelson JS. Fishes of the world. 4th ed. New Jersey: Wiley; 2006. p. 601.
13. Pietsch TW. The genera of frogfishes (family Antennariidae). Copeia. 1984. doi:10.2307/1445032.
14. Pietsch TW. Antennariidae. In: Randall JE, Lim KKP, editors. A checklist of the fishes of the South China Sea, Raffles Bull Zool Suppl, vol. 28. 2000. p. 569–667.
15. Pietsch TW, Grobecker DB. Frogfishes of the world: systematics, zoogeography, and behavioral ecology. California: Stanford Univ Press; 1987. p. 420.
16. Senou H. Antennariidae. In: Nakabo T, editor. Fishes of Japan with pictorial keys to the species. Tokyo: Tokai Univ Press; 2002. p. 454–8.
17. Stewart AL. Family Antennariidae. In: Roberts CD, Stewart AL, Struthers CD, editors. The Fishes of New Zealand, vol. 3. 2015. p. 577–1152.
18. Tamura K, Stecher G, Peterson D, Filipski A, Kumar S. MEGA6: molecular evolutionary genetics analysis version 6.0. Mol Bio Evol. 2013;30:2725–9.
19. Thompson JD, Higgins DG, Gibson TJ. CLUSTAL W: improving the sensitivity of progressive multiple sequence alignment through sequence weighting, position specific gap penalties and weight matrix choice. Nucl Acids Res. 1994;22:4673–80.

CHAPTER 12

Novel Method to Depurate B-Lactam Antibiotic Residues by Administration of a Broad-Spectrum B-Lactamase Enzyme in Fish Tissues

Young-Sik Choe, Ji-Hoon Lee, Soo-Geun Jo and Kwan Ha Park*

Department of Aquatic Life Medicine, College of Ocean Sciences, Kunsan National University, San-68 Miryong-Dong, Gunsan City, Jeonbuk, South Korea

ABSTRACT

As a novel strategy to remove β-lactam antibiotic residues from fish tissues, utilization of β-lactamase, enzyme that normally degrades β-lactam structure-containing drugs, was explored. The enzyme (TEM-52) selectively degraded β-lactam antibiotics but was completely inactive against tetracycline-, quinolone-, macrolide-, or aminoglycoside-structured antibacterials. After simultaneous administration of the enzyme with cefazolin (a β-lactam antibiotic) to the carp, significantly lowered tissue cefazolin levels were observed. It was confirmed that the enzyme successfully reached the general circulation after intraperitoneal administration, as the carp serum obtained after enzyme injection could also degrade cefazolin ex vivo. These results suggest that antibiotics-degrading enzymes can be good candidates for antibiotic residue depuration.

Keywords: β-Lactamase, Carp, Antibiotics, Depuration

BACKGROUND

Concerns have risen over the presence of antimicrobial residues in food fishes because the contaminants can cause human pathogenic bacteria to develop resistance to therapeutically valuable antimicrobials (Cabello, 2006). Development of antibacterial resistance will then require higher doses of the same agent, or more potent drugs for efficacious treatment of bacterial pathogens. Of several known mechanisms for resistance development, degradation of antimicrobial agents is regarded as the most frequent one (De Pascale and Wright 2010).

Soon after the introduction of β-lactam antibiotics such as penicillins and cephalaosporins into clinical use, bacterial resistance to these drugs were reported (Abraham and Chain, 1988). Those bacteria have been found to harbor plasmid-borne genes producing enzymes which can catalyze a few β-lactam antibiotics nullifying the efficacy of such drugs (Matagne et al. 1999). However, the efficacy has become increasingly limited by the advent of β-lactamase that catalyzes a broader range of β-lactam antibiotics compared with previous enzymes (Livermore 1995). Those enzymes are called extended-spectrum β-lactamase (ESBL) and they are usually represented by TEM-52.

If β-lactamase can catalyze β-lactam antibiotics in fish tissues, it could be useful to remove residual antibiotics remaining after use to food fish species. To test this possibility, we assessed the catalytic activity in vitro of purified TEM-52 enzymes obtained by recombinant DNA techniques, followed by an in vivo activity test in carp after administration of cefazolin, a β-lactam antibiotic.

METHODS

β-Lactamase enzyme preparations

The gene for TEM-52 was obtained from a Korean clinical isolate after necessary cloning processes (Pai et al., 1999). Its sequence was confirmed to match 100% by query coverage tests with those of extended-spectrum enzymes reported from *Salmonella enterica* subspp. enterica serovar Typhimurium (gb|AY883411.1|) and *Klebsiella pneumoniae* (emb|Y13612.1). This gene was used to build a recombinant plasmid and inserted into *E. coli* and then cultured at 30 °C for about 48 h in 2-L complex media. When the culture reached the density of ca. 0.7 optical density at 650 nm, cells were harvested by centrifugation (×13,000*g*, 10 min, 3 °C).

Cells were disrupted by sonication and the lysate was processed for purification purposes with Q-Sepharose FF column chromatography after refolding steps. The refolding process was composed of decreasingly stepwise dilutions at 8, 2, 1, and 0.4 M in denaturant urea. The TEM-52 enzyme was eluted in 20 mM Tris buffer (pH 7.0) and adjusted to contain 500 mU/ml β-lactamase activity when assayed for 5 min using nitrocefin (50 μg/ml, Oxoid, Basingstoke, UK) as the substrate (O'Callaghan et al., 1972). The final purified TEM-52 enzyme is known to have a molecular mass of 28 KDa (Perilli et al., 2002).

In vitro degradation activities against various antimicrobials

Degradation activity of TEM-52 against various classes of antimicrobials was assessed incubating the β-lactamase enzyme solution (25 mU/ml) with various antimicrobials. Ten different antimicrobials (cefotaxime-Na, cefazolin-Na, penicillin-G-K salt, amoxicillin, pivampicillin, oxytetracycline-HCl, ciprofloxacin, erythromycin, josamycin, and streptomycin) were incubated with TEM-52 in 50 mM sodium phosphate buffer (pH 7.0) for 30 min at 25 °C, and then followed by bioassay for semi-quantitative analysis of remaining drug levels. These antimicrobials were obtained from Sigma (St. Louis, MO, USA) except pivampicillin (Dong-Hwa Pharmaceuticals, Seoul, Korea) and josamycin (Wako Chemicals, Japan). Paper disks (8 mm diameter, Advantec Toyo, Dublin, CA, USA) were soaked with 50 μl of incubated enzyme-antimicrobial mixtures and placed on top of BHI agar in Petri dishes streaked with *Staphylococcus epidermidis* ATCC10145 (precultured in BHI to 1.8×10^7 CFU/ml). The dishes were incubated at 36 °C for 12 h, and inhibitory zones beyond the disk areas were measured.

In vivo and ex vivo degradation activities against cefazolin

Of several β-lactam antibiotics, cefazolin was selected as the model antibiotic because of the simplicity in chemical analysis for both in vivo and ex vivo studies. To assess degradation activity of TEM-52 more quantitatively than the above in vitro study, cefazolin was incubated with TEM-52 (25 mU/ml in 50 mM phosphate buffer, pH 7.0) for 30 min and remaining concentration of cefazolin was chemically analyzed with a high-performance liquid chromatography (HPLC) method.

The HPLC-UV method for cefazolin analysis (Nadai et al., 1993) was composed of a Waters 2690 separation module and Waters 2487 dual wavelength UV-visible detector at 274 nm (Waters, Milford, MA, USA). Samples were injected into a C_{18} reverse-phase HPLC column

(250 × 4.6 mm, 5 μm particle size, Shiseido, Japan) at 10–50 μl range and eluted with a mobile phase composed of 30 mM sodium phosphate buffer (pH 5.0) and methanol (88:12 ratio). Flow rate of the mobile phase was 1.2 ml/min.

For in vivo cefazolin degradation experiment, common carp *Cyprinus carpio* weighing 40–100 g maintained at 23 °C were used. Cefazolin (sodium salt, Sigma) was dissolved in sterile saline and administered intramuscularly (im) around the lateral line to render doses of 10 and 30 mg/kg of drug base. TEM-52 enzyme was immediately injected into the peritoneal cavity at 75 mU/100 g body weight. This intraperitoneal (ip) injection volume was 150 μl/100 g and an equivalent volume of saline was also administered to control fish. One hour after TEM-52 injections, fish were anesthetized with MS-222 (Sigma) and blood was sampled for cefazolin analysis through the caudal vessels. Serum was subsequently obtained following centrifugation at ×3600*g* for 20 min under refrigeration (3 °C). Liver and muscle were also isolated for cefazolin analyses. Serum and fish tissues were mixed with HPLC mobile phase in tenfold volumes and vigorously vortex-shaken before filtering through membrane filters, and finally injected into the HPLC column. This method led to an almost complete cefazolin recovery (>90%, $n = 3$). With cefazolin-spiked liver and muscle tissues (10, 20, 50, 100, and 200 ng/g), linearity of the standard curve was confirmed ($r^2 = 0.967$). The limits of detection (LOD) and quantification (LOQ) were 10 and 30 ng/g, respectively.

In some fish, TEM-52 was administered into the peritoneal cavity (ip) at 75 mU/100 g to naive carp, and serum was obtained 1 h later to assess ex vivo degradation activity against cefazolin. All experiments using carp were performed in accordance with the guidelines approved by the Institutional Animal Care and Use Committee, Kunsan National University, Korea.

Statistics

Data were expressed as mean ± S.D. Statistical significance was examined with unpaired *t* tests at the significant level of $p < 0.05$.

RESULTS AND DISCUSSION

Table 1 shows in vitro enzymatic activity of TEM-52 to degrade different classes of antibacterial agents. TEM-52 is a wide spectrum β-lactamase obtained from *E. coli* through recombinant DNA technique. The enzyme preparation reduced antibacterial activity of β-lactam antibiotics without effects on drugs of other classes, indicating highly selective degradation activity toward β-lactams. Other classes of antibacterials, i.e., tetracycline, quinolone, macrolide, and aminoglycoside, were not influenced at all. These data reflect that the enzyme preparation contains exclusively β-lactamase efficacy and hydrolyzes β-lactams, cephalosporins, and penicillins (Bush et al., 1995; Pai et al., 1999; Poyart et al., 1998; Shahada et al., 2010), without any activity on other antibacterials.

Table 1 Bioassay of degradation activity of TEM-52 enzyme against various classes of antimicrobials

Class	**Antimicrobials**	**Inhibition zone (mm, mean ± SD)**	
		Control drug disk	**TEM-52-treated drug disk**
β-Lactam	Cefotaxime	9.3 ± 0.6	0.0 ± 0.0
	Cefazolin	6.0 ± 1.7	2.3 ± 1.2
	Penicillin G	8.0 ± 0.6	0.0 ± 0.0
	Amoxicillin	8.7 ± 0.6	0.0 ± 0.0
	Pivampicillin	7.3 ± 0.6	0.0 ± 0.0
Tetracycline	Oxytetracycline	9.8 ± 0.9	9.7 ± 0.6
Quinolone	Ciprofloxacin	8.2 ± 0.0	8.0 ± 0.4
Macrolide	Erythromycin	10.0 ± 1.0	10.0 ± 2.0
	Josamycin	8.6 ± 0.6	8.7 ± 0.5
Aminoglycoside	Streptomycin	8.5 ± 0.4	8.7 ± 0.6

Concentrations of antimicrobials in incubation were adjusted to produce 6–11 mm clear zones beyond the disk areas (0.5–500 μg/disk) Triplicate determinations

Of the five β-lactam antibiotics examined with the in vitro bioassay, cefazolin was chosen for in vivo studies mostly because the procedures for HPLC analysis are quite simple. The results obtained after simultaneous administration of TEM-52 and cefazolin to carp are demonstrated in Fig. 1. In all three tissue samples examined, the levels of cefazolin were significantly lower in TEM-52 co-administered fish compared with the cefazolin-alone group. With cefazolin doses 10 and 30 mg/kg, and regardless of tissues, the degree of reduction was ~50% (Fig. 1a–c). In this in vivo experiment, TEM-52 was administered into the peritoneal cavity whereas cefazolin into the muscle. These results therefore indicate that β-lactamase activity included in TEM-52 and the β-lactam antibiotic cefazolin could come into direct contact in the fish body leading to cefazolin degradation. These results thus clearly demonstrate that TEM-52 is active not only in vitro system but also in the fish body, i.e., in vivo following injections.

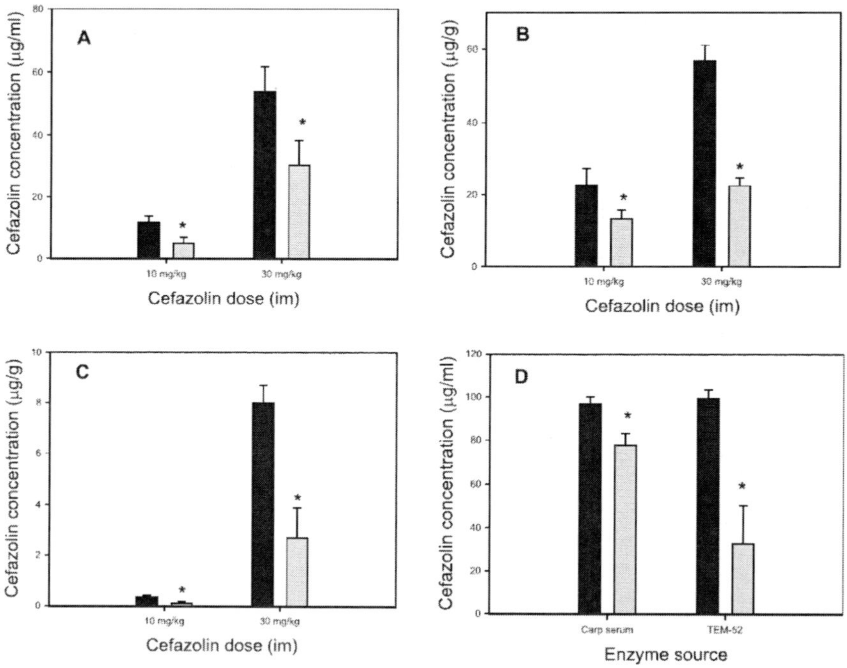

Figure. 1 In vivo and ex vivo cefazolin degradation activity of TEM-52 in carp. **a–c** In vivo cefazolin changes in serum (**a**), liver, (**b**) and muscle (**c**) following intraperitoneal injection of TEM-52, $N = 5–10$; *filled bar,* control carp; *shaded bar,* TEM-52-injected carp (**d**) ex vivo cefazolin changes when TEM-52-injected carp serum were incubated with cefazolin, $N = 5$; *filled bar,* without enzyme; *shaded bar,* incubated with TEM-52 itself or TEM-52-injected carp serum. $^*p < 0.05$ with unpaired *t* tests when compared to control groups

It was additionally checked whether TEM-52 was actually present in carp serum after intraperitoneal injection to the fish. For this, TEM-52 was injected and serum was sampled 1 h later for ex vivo degradation tests. After incubation of TEM-52-pretreated carp serum with cefazolin ex vivo, a significant reduction ($p < 0.05$) in cefazolin residue was observed (Fig. 1d, Carp serum columns). In a parallel test, it was observed that if a higher reduction of cefazolin concentration occurred, cefazolin was incubated directly with TEM-52 in the presence of naive serum (Fig. 1d, TEN-52 columns). Although it is difficult to compare directly because these two test conditions were not identical, a significant portion of intact TEM-52 enzymes seemed to reach in serum when blood was taken (1 h post-injection).

Because TEM-52 was active for cefazolin degradation in carp, it can be deduced that the administered enzyme must have been sufficiently absorbed from the peritoneal cavity. It is not well described in fishes whether there is an inhibiting barrier to proteins injected into the peritoneal cavity. However, the fact that peptides are systemically active following intraperitoneal injection in fishes (Hong and Secombe, 2009; Murashita et al., 2010) suggests a loose barrier within the peritoneum in fishes. In rats too, absorption of proteins from the peritoneum is comparable to that of isotonic solution (Flessner, 2005).

CONCLUSIONS

In summary, this study is the first attempt to seek a utilization of sorts of antibiotics degradation enzymes to depurate drug residues in animals. We observed that TEM-52, an enzyme active against a broad range of β-lactam antibiotics, can be a novel tool to remove residues from drug-treated fish bodies. It may be needed to expand tests against other β-lactam antibiotics and also in different fish species. There has not been any attempt in idea similar to this up to now. Various aspects of studies need to be carried out in order to establish practical usefulness. Safety of the test enzyme to humans is not known at all, for example.

Conclusively, however, a similar strategy will also be applicable to various classes of antimicrobials to which antibiotic-degrading enzymes have been reported.

ACKNOWLEDGEMENTS

This work was supported by the fund of Fisheries Research Institute of Kunsan National University in the year 2016. The authors are grateful for the production and supply of enzyme preparations to Dr. CS Shin at Advanced Protein Technologies Corp. (Gyeonggi Bio-center, Suwon, Korea) and Dr. SK Moon at MTC Korea (Gyeonggi Techno-Park, Ansan City, Gyeonggi-Do, Korea).

REFERENCES

1. Abraham EP, Chain E. An enzyme from bacteria is able to destroy penicillin. 1940. Rev Infect Dis. 1988;10:677–8.
2. Bush K, Jacoby GA, Medeiros AA. A functional classification scheme for β-lactamases and its correlation with molecular structure. Antimicrob Agents Chemother. 1995;39:1211–33.
3. Cabello FC. Heavy use of prophylactic antibiotics in aquaculture: a growing problem for human and animal health and for the environment. Environ Microbiol. 2006;8:1137–44.
4. De Pascale G, Wright GD. Antibiotic resistance by enzyme inactivation: from mechanisms to solutions. Chem Biol Chem. 2010;11:1326–34.
5. Flessner MF. The transport barrier in intraperitoneal therapy. Am J Ren Physiol. 2005;288:F433–42.
6. Hong S, Secombe CJ. Two peptides derived from trout IL-1β have different stimulatory effects on immune gene expression after intraperitoneal administration. Comp Biochem Physiol Part B. 2009;153:275–80.
7. Livermore DM. β-lactamases in laboratory and clinical resistance. Clin Microbiol Rev. 1995;8:557–84.
8. Matagne A, Dubus A, Galleni M, Frere JM. The beta-lactamase cycle: a tale of selective pressure and bacterial ingenuity. Nat Prod Rep. 1999;16:1–19.
9. Murashita K, Jordal AO, Nilsen TO, Stefansson SO, Kurokawa T, Björnsson BT, Moen AG, Rønnestad I. Leptin reduces Atlantic salmon growth through the entral pro-opiomelanocortin pathway. Comp Biochem Physiol Part A. 2010;158:79–86.
10. Nadai M, Hasegawa T, Kato K, Wang L, Nabeshima T, Kato N. Antimicrob Agents Chemother. 1993;37:1781–5.
11. O'Callaghan CH, Morris A, Kirby SM, Shingler AH. Novel method for detection of beta-lactamases by using a chromogenic cephalosporin substrate. Antimicrob Agents Chemother. 1972;4:283–8.
12. Pai H, Lyu S, Lee JH, Kim J, Kwon Y, Kim JW, Choe KW. Survey of extended-spectrum β-lactamases in clinical isolates of *Escheria coli* and *Klebsiella pneumoniae*: prevalence of TEM-52 in Korea. J Clin Microbiol. 1999;37:1758–63.
13. Perilli M, Segatore B, De Massis MR, Pagani L, Luzzaro F, Rossolini GM, Amicosante G. Biochemical characterization of TEM-92 extended-spectrum b-lactamase, a protein differing from TEM-52 in the signal peptide. Antimicrob Agent Chemother. 2002;46:3981–3.
14. Poyart C, Mugnier P, Quesne G, Berche P, Trieu-Cuot P. A novel extended-spectrum TEM-type β-lactamse (TEM-52) associated with decreased susceptibility to moxalactam in *Klebsiella pneumoniae*. Antimicrob Agents Chemother. 1998;42:108–13.
15. Shahada F, Chuma T, Dahshan H, Akiba M, Sueyoshi M, Okamoto K. Detection and characterization of extend-spectrum β-latamase (TEM-52)-producing *Salmonella* serotype infantis from broilers in Japan. Foodborne Pathogens Dis. 2010;7:515–21.

CHAPTER 13

Sub-Regionalisation of Fisheries Governance: The Case of The Western and Central Pacific Ocean Tuna Fisheries

Alice MM Miller[1*], Simon R Bush[1] and Paul AM van Zwieten[2]

[1]Environmental Policy Group Wageningen University, Hollandseweg 1, 6706 KN Wageningen, Netherlands.

[2]Aquaculture and Fisheries Group Wageningen University, De Elst 1, 6708WD Wageningen, Netherlands.

ABSTRACT

Shifting political alliances and new environmental challenges are prompting debate over processes of sub-regionalisation and whether the interplay between multiple scales of governance leads to positive synergistic outcomes or negative institutional disruption. Regional management of tuna fisheries in the Western and Central Pacific Ocean is an example where a web of treaties, conventions and institutional frameworks underlie international cooperation. Through examining the interplay between the regional Western and Central Pacific Fisheries Commission and sub-regional Parties to the Nauru Agreement, this paper explores the extent to which the PNA and WCPFC interact in the management of regional tuna fisheries. The results demonstrate that for contested marine resources such as fisheries, international sub-regions can go beyond functional units to also present wider opportunities to shift power relations in favour of small island states. Additionally, the presence of sub-regional groups like the PNA has served to challenge the performance of the WCPFC, stimulating greater debate and progress within the regional body. The paper concludes that the combined work of the PNA and the WCPFC puts them ahead on many issues and may represent a testing ground for a functional multilateralism utilising both regional and sub-regional governance platforms for the management of shared resources.

Keywords: Regionalism Governance Tuna West and Central Pacific Parties to the Nauru Agreement Multilateralism Transboundary

INTRODUCTION

The perceived crisis in the effectiveness of multilateral institutions has led to a new round of debate over the form and function of environmental international regimes, especially around complex environmental problems (Young [2011]; Biermann et al. [2012]; Conca [2012]). While some scholars have gone so far as to suggest a complete disbandment of international regimes, others have focused on processes of sub-regionalisation in response to shifting political alliances and new environmental challenges (Balsiger and VanDeveer [2010]; Balsiger and VanDeveer [2012]). Questions are being raised as to what the most appropriate region or 'scale' of governance is, defined not only by the functional spatial extent of a resource or an ecosystem, but also by the social relations which produce and reproduce cooperation and ultimately governance (Paasi [2004]). The process of sub-regionalisation is demanding new questions of how multiple scales of governance interplay with each other leading to positive synergistic outcomes or negative institutional disruption (Young [2006]; Oberthür and Stokke [2011]; Van Leeuwen and Kern [2013]). It also questions the extent to which conformity or conflict of interests can be resolved and lead to governance innovation (Young [2006]).

Regional management of tuna fisheries in the Western and Central Pacific Ocean (WCPO) is made up of a web of treaties, conventions and institutional frameworks that underlie regional cooperation (see Tsamenyi [1999]; Tutangata and Power [2002]; Wright et al. [2006]). The main regional fisheries management organisation (RFMO) is the Western and Central Pacific Fisheries Commission (WCPFC), formed under the United Nations Convention on the Law of the Sea (UNCLOS). Additionally, two treaty-defined sub-regional bodies that pre-date the WCPFC, the Parties to the Nauru Agreement (PNA) and Pacific Islands Forum Fisheries Agency (FFA), are also involved in tuna fisheries management (see Figure 1). However, unlike regional architectures in other environmental governance regimes, such as the climate governance regime (e.g. Biermann et al. [2009]), these two sub-regional bodies supported the formation of the regional WCPFC. The complex set of governance and institutional arrangements that manage tuna fisheries in the WCPO have been referred to as some of the most sophisticated sets of cooperative tools in the world

(Hanich et al. [2010]), providing a variety of normative structures from which a range of international and domestic conservation and development policies emerge.

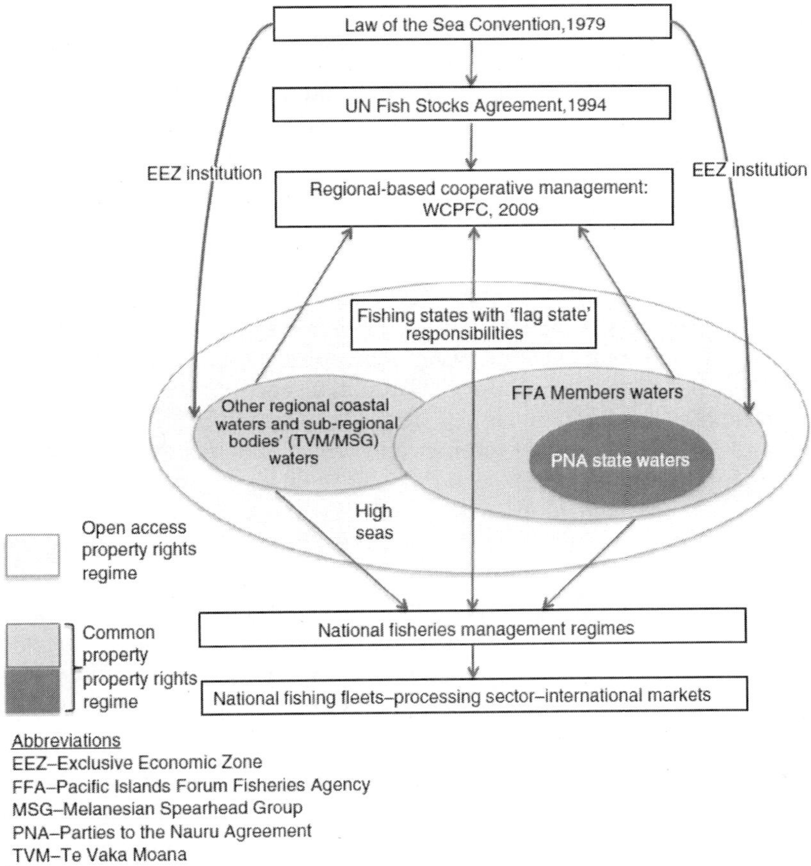

Figure 1 Institutional map of the regional and sub-regional governance regimes in the West and Central Pacific Ocean (Adapted from Parris[2010a]).

Despite being the youngest of the seven RFMOs, the WCPFC has been described, with various qualifications, as one of the most successful in terms of regulating oceanic tuna and billfish species (Cullis-Suzuki and Pauly [2010]; Aranda et al. [2012]). However, its critics maintain that many of the conservation and management measures (CMMs) the RFMO has generated are insufficient, and that the institutional practices of the WCPFC lack transparency (Parris [2010a]; Gilman and Kingma [2013]). Amidst such debate over the effectiveness of the WCPFC, the PNA (made up of

Federated States of Micronesia, Kiribati, Marshall Islands, Nauru, Palau, Papua New Guinea, Solomon Islands and Tuvalu), has received conspicuous support from academics and NGOs alike for their recent successes in agreeing on and implementing CMMs, new access arrangements around a vessel day scheme (VDS), and the successful Marine Stewardship Council (MSC) certification of free-school purse seine skipjack tuna fisheries in 2012. However, some WCPFC members, like the EU, have criticised the emergence of the PNA as a competing sub-regional fisheries management body that limits wider regional coherence.

Most analyses of governance strategies for tuna fisheries in the region have focused on the trade-offs and complementarities between multilateral cooperation vs. domestic strategies in terms of management and economic development (Parris and Grafton [2006]; Barclay and Cartwright [2007]; Parris [2010a],[b]). Other questions have focused on models of Pacific regionalism that enable cooperation and capacity for effective engagement in the "global ocean community" (Tutangata and Power [2002], p. 883). Less attention has been given to a better understanding of the roles of the PNA and WCPFC, how they interact and what influence each has over the governance of tuna in the region. If the goal is to create management systems that are sensitive to the spatio-temporal complexity of fishery resources (Wilson [2006]), then can the interplay between regional and sub-regional management platforms lead to innovative management outcomes? What is the relationship between regional and sub-regional bodies? Do sub-regional groupings like the PNA present a means of stimulating innovation and change towards greater effectiveness at the RFMO level, do they make them less effective, or is there combination of both?

We explore how the PNA, as a sub-regional body, has been able to be more experimental in developing tuna fisheries conservation and management measures, how durable the outcomes of this experimentation are in terms of providing long-term change, and ultimately the extent to which the PNA and WCPFC interact in the management of regional transboundary tuna fisheries. The interplay between these different scales of environmental governance in Western and Central Pacific tuna fisheries offers a relevant and timely lens to understand how these two scales of governance interact and what influence they have over each other in terms of creating innovative and effective management outcomes. More specifically, we ask what implications a shift toward sub-regional governance holds for the distribution of decision-making power between the PNA and WCPFC over ongoing contested management issues.

The research is based on attendance at the 9th WCPFC Meeting 2012, where one of the authors participated as a volunteer with the Secretariat, the contribution of all three authors in a European Parliamentary hearing in part focusing on the effectiveness of PNA conservation and management measures, and an in depth a review of WCPFC documentation. In addition we interviewed four WCPO regional experts to follow up on specific points of clarification. The paper begins with an introduction of key questions asked around regional environmental governance. We then turn to a discussion of the emergence of sub-regional governance in the context of transboundary tuna fisheries before going into a detailed description of the multi-level architecture of fisheries governance in the WCPO. Our analysis then focuses on the tensions that exist between the WCPFC and PNA in setting fishing reference points, administering CMMs and balancing the interests of their respective members. Finally, we return to the implications of sub-regionalism in the WCPO and what implication it holds for governance regimes such as RFMOs aimed at innovative governance for the long-term sustainability of transboundary and high seas tuna fisheries.

FISHERIES REGIONALISM

Emergence of regionalism in fisheries governance

In resource governance terms, regionalisation is often referred to as a politico-administrative process of establishing spatially defined scales of management that devolve decision-making and create more responsive and adaptive management decisions (Symes [2005]; Balsiger and VanDeveer [2012]). Although there is often an aspiration to base the definition of a region on the geography of a resource, it can also be based on a process of regionalisation; referring to the socio-political exercise of creating new political territories (Symes [2012]). This can entail either moving 'down and out' through the decentralisation of administrative functions and the devolution of decision making (e.g. co-management), or moving 'up and in' through centralisation and concentration to an aggregate scale (e.g. the EU or RFMO) (Oberthür and Stokke [2011]). Examining fisheries regionalism therefore opens up questions about the political and economic rationale for creating functional and geographic scales of management on the basis, for example, of resource use sustainability. In the case of fisheries, the strength of regionalisation and regionalism varies according to how the geography of fishery resources and relevant political scales are brought together.

The realisation of regional cooperation over marine resources began with the lead up to the formulation and ratification of UNCLOS, under which 'global' transboundary fisheries, defined by their ecological extent spanning exclusive economic zones (EEZs) and the high seas, were divided into functionally defined RFMOs (Valencia [1978]; Morgan [1989]). Since then functional approaches to marine regionalisation and governance have based themselves on a mix of treaty-based agreements, such as the EU Common Fisheries Policy (CFP) and the Palau Agreement in the Pacific (Tsamenyi [1999]; Symes [2005]), and alternative ecosystem-based regional configurations, such as large marine ecosystems (LMEs). LMEs, for instance, are an attempt to combine geographical scales of the marine environment with functional administrative scales of management (Vallega [2001]; Fanning et al. [2007]).

The 2012 reforms to the EU-CFP raised a heated round of debate about the role of regionalisation within European seas. Paralleling discussions of multi-level governance, regionalisation in the EU is seen as a process of 'moving down', thereby enabling lower level authorities to take control of tailor-made management for particular spatial areas, and 'moving out', referring to the increased involvement of private actors in fisheries management (Raakjaer and Hegland [2012]). The exact institutional design of regional fisheries management in the EU is still under debate, with proposals ranging from sub-EU RFMOs (as distinct from UNCLOS defined RFMOs) to nationalisation (Hegland et al. [2012]). Ultimately this represents a wider double-movement; the creation of common pool management by centralising management at the EU level through the CFP and a political process that decentralises control to formal (geographical) and functional (administrative) sub-regions (Symes [2012]).

In the Western and Central Pacific, the potential for political sub-regionalisation has been proposed by Hanich et al. ([2010]). Under their model the administrative burden imposed on Pacific Island countries by the various supra-national treaties, such as the PNA and WCPFC, would be mitigated through joint management by sub-regional groupings of three or four Pacific Island countries. This is nowhere more relevant than in the WCPO where 40 (WCPFC) member and non-member states make up a mix of states with sovereign interests over the resources in their EEZs and states with distant water licences to fish in these EEZs. There are also joint interests of all states over the four WCPO high seas pockets, international waters enclosed by EEZs (see Figure 2) (Tsamenyi [1999]; Hanich [2009]). These interactions have led to a complex multi-level interaction, with scientific input from the FFA and Secretariat of the Pacific Community (SPC), of treaties and measures such as the Harmonised Minimum Terms and Conditions of Access for Foreign Fishing Vessels, the VDS, the FFA Vessel Monitoring Scheme and the Niue Treaty to manage transboundary stocks (see Hanich et al. [2010] for detail). Hanich and colleagues argue that a collective sub-regional strategy would require states to "pursue their own national interests within their vision of a collective strategy" (p. 89) under a sub-regional collective management authority that would act on behalf of sovereign states to administer licencing or access arrangements in their EEZs.

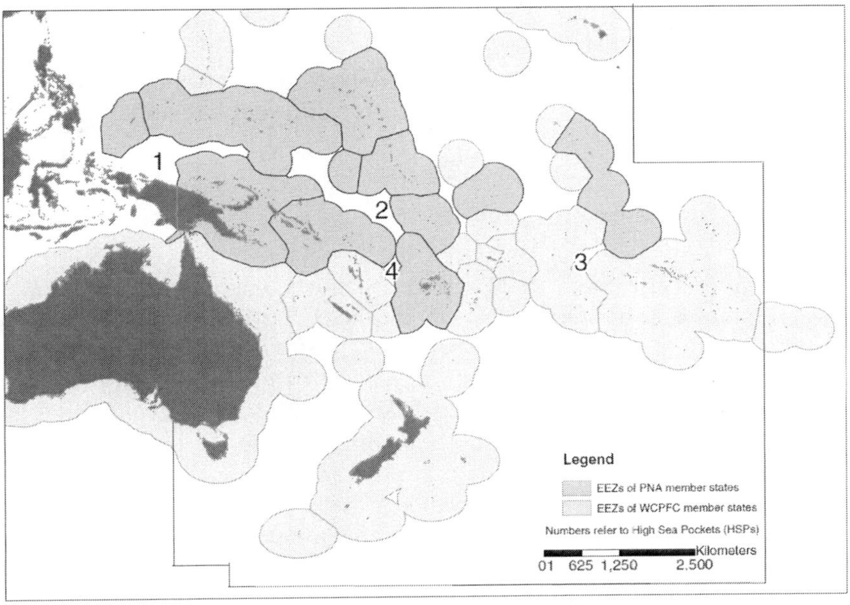

Figure 2 WCPFC Area and corresponding EEZs. The areas numbered 1 to 4 correspond to the high seas pockets. Pockets 1 and 2 are closed to purse seine fishing since 2010 under the WCPFC. Pockets 1 to 4 are closed to any purse seine vessels licensed to fish within the EEZs which are part of the PNA".

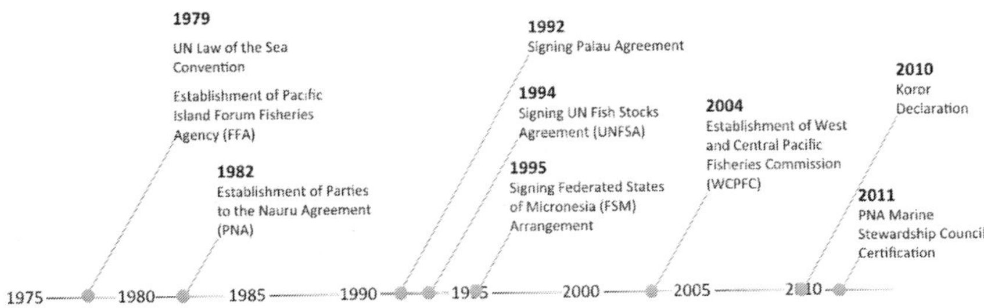

Figure 3 Timeline illustrating the developments in WCPO fisheries governance from 1979-2011.

These different models emphasise the importance being given to regionalisation and regionalism in the marine environment, which as noted by Symes ([2005]) is made all the more complex by how the "shifting distribution of fish populations and the dynamic nature of ecosystems ... [make] natural boundaries ... both permeable and unstable" (p. 87). As he goes on to argue, under such changing conditions, "regions are bound to be socially constructed rather than naturally occurring and their boundaries inevitably reflect a compromise between overlapping sets of distributions and ecosystems." (p. 87). This very contestation opens up questions on the performance and interplay between different regional scales of management.

Evaluating regionalisation

The proposals of both the EU and the governing bodies in the WCPO provide a set of (explicit and implicit) normative goals for sub-regionalisation: subsidiarity provides more adaptive management; decentralisation and participation leads to more effective decisions; improved implementation can meet sustainability and development goals of nation states. But while sub-regionalisation has the potential to create geographically, ecologically and administratively effective platforms, there is nothing to say that smaller regional groupings would always improve control of fishery resources or create adequate incentives for meaningful cooperation that would lead to more sustainable outcomes. We have identified four interrelated factors that can be used to evaluate the specific outcomes of the process of sub-regionalisation and the interplay between regional levels: scale, the creation of normative structures, conformity vs. divergence of interests, and incentives for cooperation (see Table 1).

Table 1 Factors for analysing sub-regionalisation in the WCPO

Factors	Potential outcomes
1. Scale	• Ideological motivation for 'region' formation (regionalism)
	• Mobilisation of biophysical, political-administrative, socio-economic or cultural-symbolic dimensions to include/exclude actors
	• Upward or downward transfer of governing innovation between levels
2. Creation of normative structures	• Equity, justice and (perceived) legitimacy
	• Criterion of inclusive, cooperative arrangements that reduce complexity of multiple interests
	• Interplay of normative structures between jurisdictional levels
3. Conformity vs. divergence of interests	• Diversity of functional areas
	• Degree of institutional-ecological concordance
	• Power relations and discourse
4. Distribution of the costs and benefits of conservation	• Incentives for cooperation
	• Degree of cooperation around conservation
	• Creation of mechanisms for (re)distribution of benefits

Scale refers broadly to the spatially and temporally bounded extents at which a natural or social phenomenon is observable (Cash and Moser [2000]). In resource management terms scale also refers to the spatial and temporal extent at which institutions are set and socially or politically organised (e.g. local, national, global) – as such they are social constructs open to politicisation (e.g. Meadowcroft [2002]; Swyngedouw [2004]). For example, scale helps to identify ideological motivation for the definition of a 'region': a social and political construct created to include and exclude actors from a resource or decision making process through the mobilisation of biophysical, political-administrative, socio-economic, or cultural-symbolic dimensions (Balsiger and VanDeveer [2010]). Scale therefore presents an institutional design challenge – ensuring that management institutions and the organisations tasked with their oversight are set at levels that are concordant with the social, political and ecological extents they govern (Cash et al. [2006]). Following Young ([2006]), one key strategy for ensuring concordance, especially when there are multiple scales of governance, is institutional interplay; where governing functions are distributed among regimes located at higher and lower levels on the jurisdictional scale (see also Cash et al. [2006]). This institutional interplay also covers the potential of upward or downward transfer of governing innovations between levels.

Underscoring the creation of normative structures that govern control over transboundary fishery resources by states and fishers alike are tensions between resource access, sovereignty and economic development. Although functionally specific regional bodies may have a higher degree of legitimacy than non-specific bodies, participation and control remain centrally political issues – especially when dealing with perceptions of equity and justice (Lebel et al. [2005]). The challenge for these regional bodies is therefore to maintain a degree of inclusiveness and cooperation while effectively reducing the complexity of addressing multiple interests.

The complexity of regional-level governance requires a balance of multiple state and non-state parties' interests. In models of sub-regionalism that move down in scale and out to include a wider range of actors, subsidiarity and democratisation of decision-making are core principles (Raakjaer and Hegland [2012]). But in many other contexts sub(-national) regions may not be feasible or politically desirable, and increased participation may exacerbate existing (political) complexity in rule making and implementation. Instead, sub-regionalism may also involve a shift of power to a smaller group of actors that reduce participation of other parties in order to capture control over fishery resources. Returning to Young ([2006]), it is again the interplay between scales of governance, represented by functional regions, that determines the extent to which the conformity or divergence in the interests of different actors can be resolved.

Finally, the distribution of costs and benefits of conservation influence the degree of and incentives for cooperation within and between particular regions. The common assumption is that fisheries sub-regionalism is underscored by adequate incentives for cooperation. However, as Bailey et al. ([2012]) note, the theoretical evidence for cooperative governance arrangements of fisheries, including tuna fisheries, is in stark contrast to their successful implementation. The voluminous literature on cooperation around tuna fisheries in the WCPO focuses on how 'resource rent rivalry' has been driven by a combination of competition for access by distant water fishing nations (DWFNs), the dependency of many Pacific Island countries on tuna for national income, and aspirations for domestic social and economic development (Campbell [1989]; Barclay and Cartwright [2007]; Havice [2010]; Parris [2010b]; Havice and Reed [2012]; Gagern and van den Bergh [In Press]). Incentives for cooperation around tuna, like many other marine fishery regions around the world, are therefore underlined by the need for generating domestic wealth from shared resources – which under pressure from divisive treaties and aid relations represents an archetypal prisoner's dilemma. The degree to which regional management bodies can establish (re)distribution mechanisms of this wealth, and the strength of these mechanisms in the face of strong external pressure provides an indicator for on-going cooperation.

The balance between regional and sub-regional governance in WCPO tuna fisheries opens up a discussion on the architecture of effective arrangements through the interplay between these different scales of governance. The rest of the paper explores this interplay between the WCPFC and the PNA as two regional fisheries governance arrangements, their effectiveness in responding to complex fishing practices that operate at different ecological and political scales, and the potential of regionalism to act as a driver for innovation in regional and sub-regional sustainable fisheries governance.

Regionalism in the WCPO

The WCPFC was established in 2004 after the Convention for the Conservation and Management of Highly Migratory Fish Stocks in the WCPO was ratified by 17 member states. The WCPFC is based on an international fisheries agreement that aims to foster the long-term conservation and sustainable use of highly migratory fish stocks in the WCPO. Reflecting the complexity of multilateral platforms, the Commission holds the potential to represent the interests of 65 states (as listed in Annex A of the Convention), but until now has been limited to 25 members, seven participating territories and nine cooperating non-members and observers. In addition the WCPFC has three technical subsidiary bodies that meet annually, before the full session of the Commission: the Scientific Committee, the Technical and Compliance Committee, and the Northern Committee. Unlike all other oceans and tuna RFMOs that predominantly cover fisheries in the high seas, WCPFC covers both coastal states' EEZs and the high seas. This places those states whose EEZs fall in the Convention area in a position of power regarding decision-making at the WCPFC but adds further to the complexity of this governing body.

Three sub-regional groups, that preceded and were responsible for creating the WCPFC, are the SPC, the FFA, and the PNA. All three bodies closely coordinate their activities through joint participation in regional meetings involving their members and more formal annual consultations between the Secretariats. While the SPC plays a more neutral role by providing independent scientific stock appraisal information and advice to Scientific Committee members, the FFA and PNA are multilateral sub-regional governance bodies that over time have developed different levels of advisory and regulatory authority.

The FFA was established in 1979 in response to changes resulting from the third Law of the Sea Conference (see Tsamenyi [1999]). The main role of the FFA is as an advisory and support body to 17 Pacific Island countries – who are also members of the of the Forum Fisheries Committee (FFC) in the WCPFC – targeting sustainable management of fish stocks in their states' EEZs while also maximizing their social and economic benefits. The FFA does not have any authority to enforce the decisions of its governing council: all member states maintain sovereign control over fisheries in their EEZs and archipelagic waters. Instead, the FFA provides support to Pacific Island countries by facilitating capacity building and regional cooperation through providing technical and policy advice to its members participating in international forums like the PNA and WCPFC (Langley et al. [2009]).

The PNA came into force in 1982 with the objective of member countries controlling the terms and conditions of allowing foreign fishing vessels in their EEZs (Nauru Agreement [1982]). Their role strengthened in 1992 under the Palau agreement, which set arrangements for regular management meetings for tuna stocks and established the role and responsibilities of the PNA office. Since its establishment, the PNA has produced a series of measures that have reinforced their control over tuna fisheries throughout their EEZs. As summarised by Havice ([2010]), this was first seen in 1994 when the PNA countries reduced foreign purse seine licences by 10 per cent and reallocated them to domestic/locally based vessels. In 1995, they then developed the FSM (Federated States of Micronesia) Arrangement for Regional Fisheries Access, which gave vessels from PNA countries discounted fishing licences and reciprocal access to all PNA waters. After the WCPFC was established, the PNA has continued to create its own access arrangements and management systems. In 2010,

PNA members signed the Koror declaration underlined by the PNA Implementation Arrangements, which confirmed support for a range of far reaching fisheries management measures. These were: the purse seine Vessel Day Scheme (VDS), which allocates days to member countries who then distribute them to distant water fishing nations (DWFNs) based on their licencing agreements; high seas pocket closures as part of EEZ licensing arrangements (see Figure 2); establishment of Minimum Terms and Conditions for foreign vessels; seasonal closures for the use of fish aggregating devices (FADs) (Palau Arrangement [1992]); catch retention measures; and a regional observer programme for purse seine vessels (Shanks [2010]; Havice [2013]). The PNA also created an independent secretariat to administer control over the Implementation Arrangements, most notably the purse seine VDS, and the MSC certification of the free-school (non-FAD) skipjack fishery that was obtained in 2011. This strengthening of the PNA's regulatory capacity responded in part to the perceived increased power of the DWFNs under the WCPFC Convention.

The PNA is therefore a functionally important sub-region in the WCPO, with a clear mandate for management aimed at both domestic economic development and at the sustainability of their tuna resources. Demonstrating a case of upward transfer of governing innovations, the PNA is promoting the uptake of innovative management measures at the WCPFC level with the aim of increasing the functional scale of governance across all tuna in the Western and Central Pacific. Given the unique experience of RFMO formation in the WCPO, and the persistence of sub-regional groupings like the PNA, we now turn to a comparison of these two scales of governance, exploring the relative success each has had in tuna management and their influence on each other's overall performance (Figure 3).

Comparison of WCPFC and PNA performance

The interaction of the WCPFC and the PNA on specific points of fisheries governance is illustrative of the tensions, both creative and constraining, between the two regional bodies. The following section examines three examples where they have influenced each other's performance in terms of establishing reference points; designing and implementing CMMs and the distribution of the conservation burden; and defining spatial jurisdiction through changing access arrangements to high seas pockets.

Reference points: the precautionary approach

In fisheries, taking a precautionary approach involves management policies and strategies that account for the risk of overexploitation given uncertainties in assessing the status of fish stocks and fishing pressures. Guidance on the application of the precautionary management of highly migratory and straddling fish stocks was introduced in Annex II of the United Nations Fish Stocks Agreement (UNFSA) in 1995. Fundamental to this approach is setting and employing two reference points for fisheries management: a limit reference point aimed at constraining catch within safe biological limits; and a target reference point aimed at meeting management objectives, such as desired biological, social, and economic outcomes.

The WCPFC provides for the application of the precautionary approach. The Commission is currently developing reference points to inform the development of operational objectives and performance measures for longer-term management strategies. At the 9th Regular Meeting of the WCPFC, members agreed upon limit reference points, according to recommendations from the SPC. Special mention was also given to the WCPFC prior to the meeting on their application of the precautionary approach in the WCPFC Performance Review, which stated "The Commission and Scientific Committee are to be commended on progress made in developing limit reference points, particularly for bigeye, skipjack and yellowfin" (WCPFC [2012c], p.157). The WCPFC is ahead of most other tuna RFMOs on this: the Inter-American Tropical Tuna Commission (IATTC) is currently the only other RFMO that makes explicit reference to the application of the precautionary approach and the use of reference points (de Bruyn et al. [2013]). Therefore, compared to most other RFMOs the WCPFC is among the leaders in advancing the development of a formalised precautionary approach framework.

In spite of the WCPFC's comparative progress, the PNA are also looking to take the lead toward setting both limit *and* target reference points, should the WCPFC fail to make progress with target reference points. The PNA's concern about the WCPFC's potential for implementing these reference points was highlighted during the MSC certification of the PNA free-school skipjack tuna fishery in 2011. One of the conditions of the certification pertained specifically to setting both limit and target reference points for skipjack stocks in PNA waters within five years of the certification. Initially, there was pressure from non-PNA actors engaged with the certification procedure for WCPFC to be responsible for setting these limits. However, one PNA official described how they fought this because, "if it became a Commission initiative and had to be decided under consensus, all it would take would be for one party to block it and we would lose the certification". Instead, the PNA negotiated that either "PNA *and/or* WCPFC" could take the initiative on setting the reference points. The official argued that the PNA would go ahead with setting both reference points, while maintaining an option for deferring overall responsibility to the Commission, should they fail to reach an agreement.

The position the PNA takes on this issue is telling for a number of reasons. Firstly, it provides evidence of the difficulties that have been faced in negotiating regional measures through the WCPFC, particularly in terms the conflicts that arise from consensus voting. Secondly, it demonstrates a degree of PNA dependency on the WCPFC, should they fail to agree amongst themselves on applying limits to their shared fishing resources. Finally, the condition for setting both target and limit reference points came from the MSC certification procedure and illustrates the extent to which external market-defined sustainability concepts can steer state-based decision-making.

Conservation burden

Under the WCPFC, the primary CMM concerning bigeye and yellowfin tuna was agreed on in 2008 (CMM-2008-01). Broadly speaking, it covers purse seine effort limits both within the EEZs of coastal states and in high seas pockets, longline effort limits and seasonal FAD closures. However, under Article 30 of the Convention, it underlines "the need to ensure that such measures do not result in transferring, directly or indirectly, a disproportionate burden of conservation action onto developing States Parties, and territories and possessions" (WCPFC [2000] p.19). This is reiterated in Paragraph 6 of the CMM where, "Unless otherwise stated, nothing in this measure shall prejudice the legitimate rights and obligations of those small island developing state members and participating territories in the Convention Area seeking to develop their own domestic fisheries" (WCPFC [2008] p.3). These clauses attempt to accommodate what Hanich ([2012]) has labelled the 'conservation burden' associated with small island developing states' implementation of CMMs.

Acknowledging the different status of Pacific Island countries, the implementation of CMMs has been accomplished through the incorporation of a number of exemptions, including: 1) no limits on bigeye catch by domestic longline fleets; and 2) exclusion of archipelagic waters from CMMs (Parris [2010a]). These exemption clauses could be problematic from a sustainability perspective because they could mean a substantial portion of the tuna industry may remain unregulated under the WCPFC and therefore undermine the effectiveness of the CMMs. Thus far, the so-called exemptions-based management approach, has not led to measurable improvements in the fishery. In fact, the PNA have rapidly increased their fishing mortality above their 2004 levels for the years 2005–2007 (Parris [2010a]). Concern has specifically been raised around the access of foreign (Philippine) vessels to Papua New Guinea's EEZ, as part of licence agreements that are aiming to develop processing capacity, and the poor regulation of domestic fishing activities within their archipelagic waters (Hamilton et al. [2011]).

In spite of indications that the PNA fails to act in a way consistent with the sustainable management of fish stocks when it comes to the implementation of WCPFC CMMs, outside the Commission it is providing a platform for developing and implementing innovative management measures for its purse seine fisheries. These include FAD closures and high seas pocket closures. As evidence of PNA's leadership in this regard, the Commission has since adopted both these PNA measures.

The result is a trade-off between Pacific Island countries calling for special consideration under Article 30 to reflect their conservation burden, while at the same time developing innovative CMMs, some of which have been taken up by the WCPFC. At the 9th Regular Meeting of the WCPFC in 2012, continued reference was made to Article 30 when negotiating the terms of the new CMMs. An example of this was during discussions about expanding the length of seasonal FAD closures for the purse seine fishery. There was firm resistance from some of the smallest Pacific Island countries on the basis that this discriminated against the purse seine fisheries, which provide their primary source of income. They argued that longer closures than those already in place would be economically devastating, owing to national dependence on revenue from the sale of purse seine vessel days. Citing Article 30 and the conservation burden, they explained that they were already experiencing revenue loss from the 800mt of tuna 'saved' (not caught) because of the existing closures. And therefore, they would need compensation for additional closures and/or evidence that longliners, that were not included in the CMM, were also having effort restrictions placed on them.

The economic advantage the PNA states currently hold through revenue from the purse seine fishery operating in their EEZs provides an incentive to support measures that distribute much of the conservation burden onto longline fleets (Hanich [2012]). As longline fleets fish primarily on the high seas, PNA states do not stand to benefit significantly from their activities and therefore want emphasis to also be placed on managing their effort. This also encourages them to oppose conservation measures that do not distinguish between high seas and EEZs, or apply new measures over their EEZs. Therefore, on the one hand the PNA is 'feeding' measures they have developed up to the commission level but on the other hand, under Article 30 there is also the option to retain control over the implementation of these measures to ensure the conservation burden is not transferred back to them. This reinforces the strong negotiating position of the PNA.

However, there has also been resistance among some WCPFC Members over the PNA's role as leader in the sustainable exploitation of its tuna stocks within the WCPFC. This is most evident when looking at the VDS. Some DWFNs like the EU have been reticent to fully endorse the scheme, on the basis that the PNA is insufficiently transparent about how they are allocating vessel days and of evidence that PNA members were overspending their allocation of days (Havice [2013]). In spite of this, the WCPFC Science Committee has indicated that the VDS has had positive impacts on data quality by improving log-sheet data and requiring 100 per cent observer coverage. This has contributed to providing the Committee with better information for their stock assessments. It also indicates that sub-regional management measures can have indirect benefits to regional-level governance.

Contested spatial jurisdictions

To date, the legal role of WCPFC is undefined given it has "no legal authority to allocate rights to fish in any manner that undermines the sovereign rights or sovereignty of coastal states" (Hanich [2009], p. 226). This has created a major tension in connection with wider issues of allocation of tuna between Pacific Island countries wishing to develop their domestic fishing capacity within their EEZs and the WCPFC, which is also responsible for management of fish stocks in the high seas (Langley et al. [2009]). DWFNs want the measures agreed and set out by the WCPFC to apply equally across both the high seas and EEZs. They also want to see effort reduction measures in place that would counter any expansion of domestic fleets. The PNA have indicated they are amenable to these reductions but as one official stated, "on our terms".

Tensions between the PNA and the WCPFC over control of the high seas areas are illustrative of an underlying conflict between *de facto* and *de jure* definitions of spatial jurisdiction. The Convention provides the WCPFC with the remit to manage transboundary stocks across their full extent, including both high seas and EEZs in the Convention Area. In their 2012 CMM for bigeye, yellowfin and skipjack tuna, the WCPFC made it clear that "This Measure applies to all areas of high seas and EEZs in the Convention Area except where otherwise stated in the Measure" (WCPFC [2012b] p.5). The PNA maintains ownership over the resources in their EEZs, but not in the high seas. However, the PNA is extending control over the practices of distant water fleets, from states that are members of the WCPFC, in two distinct ways. First, they control these fleets by setting the adherence to CMMs for fishing in their EEZs. Second, the PNA countries have closed high seas fishing to DWFN vessels that have licences to fish in their EEZs. This represents a *de facto* extension of the PNA's control beyond their spatial jurisdiction, and a spatio-legal challenge to the *de jure* control exercised by the WCPFC under UNCLOS.

Under the 3rd Implementing Arrangement of the Nauru Agreement, the PNA agreed to close two high seas pockets to purse seining. The WCPFC subsequently adopted the closures through the CMM 2008-01, which came into force from January 1, 2010. Explaining "the serious impact on the bigeye stock from fishing by distant water longliners and purse seiners in the high seas and that the high seas continues to provide a safe haven for IUU fishing" (PNA [2010] p.2), PNA leaders agreed to the closure of two additional high seas pockets to all purse seine vessels licensed to fish in the EEZs of the PNA in 2011 (Figure 1). Any licence granted to a DWFN to fish in the EEZs of a PNA member means the vessel cannot fish in these high seas pockets. On the basis that high seas pocket closures have shown no demonstrable decrease in fishing effort, and that in fact effort has been transferred to other areas (SPC-OFP [2012]), Japan, the Philippines, Indonesia and the EU have opposed this CMM. Thus, when at the 7thRegular Session of the WCPFC in 2010, the PNA sought support from WCPFC to mandate the additional closures, it was rejected (Ride [2010]).

Further, in the 8th regular session of the WCPFC in March 2012, the Philippines negotiated access for 36 boats to High Seas Pocket 1 (see Figure 2). The position of the Philippines has been that the closures have put undue stress on their domestic fisheries and that the decision has led to increased fishing effort in its national waters. In exchange for the Philippine fishing access, the Philippine government is supposed to report its domestic and international tuna catch. Philippine vessels must also apply for international fishing permits before entering High Seas Pocket 1 and must allow 100 per cent regional observer coverage on board their vessels. It is notable, however, that civil society surveillance has indicated that not all Philippine vessels are complying with this (Greenpeace [2013]).

As the body responsible for introducing the closures, the PNA vehemently opposed this re-opening, insisting that the high seas pockets should remain closed permanently or at least until the Commission decides otherwise. However, the permission granted to the Philippine vessels still remains largely under the control of the PNA who have affirmed they will only licence DWFN vessels to fish in PNA waters if they voluntarily forego fishing on the high seas. Therefore, fishing will only take place by Philippine vessels and maybe a few other vessels that do not need access to PNA waters. Thus, in spite of the WCPFC ruling, fishing access remains primarily under the control of the PNA countries. This illustrates the contrast between PNA and WCPFC levels of decision-making and demonstrates the resilience of measures implemented and controlled at the sub-regional (PNA) level, especially when leverage over fishing access is involved.

Emerging regionalism in the WCPO

Whether and how innovation is occurring in the WCPO needs to be understood in the context of the on-going interaction between the WCPFC with the PNA and, although not our particular focus, the FFA. The WCPFC is different to other RFMOs because of the existence of these sub-regional groups, which provide an additional, meso-level of governance. We now turn to a discussion of the main factors that affect the emergence of (sub-)regional fisheries management in the WCPO and the effect regionalism has on the innovation of conservation and management measures over transboundary fishery resources.

Not by scale alone

The powerful position the PNA has secured has meant that, unlike many global governance regimes, sub-regionalisation enables a process of feeding tested policy strategies upwards for regional implementation. The uptake of PNA policies at the WCPFC level indicates that, instead of representing a crisis point for multilateralism (Conca [2012]), strong sub-regional governance has acted in part as a catalyst for regional policy convergence. It also appears to demonstrate a degree of scalar concordance; indicating that decisions over management measures are being distributed over levels most effective at dealing with the diverse interests of the actors involved (Cash et al. [2006]). Additionally, through this convergence, the perceived legitimacy of PNA policies is further scrutinised by the wider global community associated with the WCPFC. The interplay between the PNA and WCPFC has therefore led to positive creative tension, as well as a greater degree of transparency and openness than would otherwise have been the case had governance remained at the more exclusive sub-regional PNA level.

The results also indicate that the PNA's high degree of functional control over the fishery has been translated into a political process of fishery regionalisation. Reflecting findings in other resource sectors (e.g. Balsiger and VanDeveer [2010]), regionalisation has enabled the PNA to exclude some groups from decision making over resource access by mobilising biophysical, socio-economic and cultural symbolic dimensions of sustainable Pacific Island tuna fisheries. This is evident by their success in reinforcing the functional extent of the fishery with a perceived pro-active concern in the purse seine fishery (Hanich [2012]), strategically using the institutional backing of WCPFC Article 30, and reinforcing their legitimacy through the MSC certification of the PNA free-school skipjack fishery. The result has been the convergence of both normative and functional leverage (Symes [2012]) of the PNA as a negotiating body within the WCPFC.

However, it also raises questions about whether these arrangements are replicable at similar sub-regional scales within the WCPO and in other RFMOs, or if they are specific to the context of the PNA.

Indeed, the power of cooperation through the PNA's current position to act as a coherent and notably powerful unit when interacting regionally through the WCPFC has not been lost on other Pacific Island countries. Two additional sub-regional groupings in the Pacific, the Melanesian Spearhead Group Fisheries Technical Advisory Group and the Te Vaka Moana arrangement between certain Polynesian states and a group of countries in Indian Ocean, have shown interest in replicating the PNA model. The emergence of these new groupings points to the value that can be gained from fishery sub-regionalisation, bringing together a single functional area and natural phenomena (Symes [2012]). However, such groups within the WCPO are unlikely to achieve the same kind of leverage with the WCPFC because of the smaller proportion of fish under their control – the MSC certification accounts for 70 per cent of the tuna stock in its combined EEZ (PNA [2014]). In the Indian Ocean they are less likely due to the more open geography of national EEZs. As we now go to argue, creating concordance between functional, jurisdictional and natural scales can facilitate a process of sub-regionalisation but will not automatically lead to successful political and functional integration into larger multilateral institutions.

Mutually reinforcing normative structures

Although fundamentally different in aim and legal set-up, the normative structures of the WCPFC and the PNA have produced a creative interplay for tuna management. The RFMO structure is legislated through UNCLOS, which mandates the inclusion of DWFNs. The explicit governance objective at this level is therefore not about furthering the interests of member states but fulfilling the broader conservation and resource management requirements. The RFMO system is therefore designed to create an up and out 'double movement' which manages the interests of historical fishing rights of port and flag states, rather than a down and out movement designed to stimulate functional subsidiarity (cf. Hegland et al. [2012]). At this level, no project exists to decentralise formal and functional control to sub-regions because there is no political will of the member countries at the WCPFC level.

In contrast the PNA has created an opposite 'double movement' akin to that seen in the last round of the EU CFP reforms (Raakjaer and Hegland [2012]); a reduction in the number of states involved in the decision making and the promotion of conservation and management measures up to the WCPFC. As such, the PNA is structured as an exclusive bottom up coalition excluding external actors from governance decisions. Again reflecting a process of sub-regionalisation, and in contrast to the WCPFC, the PNA is explicitly working to further the interests of its members in the face of pressure from DWFNs seeking access to tuna resources. Through this, they are aiming to capture development benefits, equity and justice for members by promoting sustainable tuna fishing. This is illustrated by the exclusion of Philippine vessels to the high seas pockets now controlled by the PNA, despite the Philippines being a WCPFC member with mandated access.

Moreover, the MSC certification allows the member countries of the PNA to pursue national interests within a collective strategy of market-driven sustainability (cf. Hanich et al. [2010]). The strategy of the PNA to have their free school skipjack fishery certified therefore plays an important role in supporting its internal normative structures. While the recognition of the PNA's capacity for producing innovative conservation and management strategies is evidenced by the uptake of these measures at the WCPFC level, gaining MSC certification also adds further recognition to the powerful position the PNA holds globally and increases the legitimacy of its governance (See for example Gulbrandsen [2013]). Strategically, the certification also adds market value to the PNA skipjack tuna, connecting the PNA to powerful external markets like the EU and the US, further ensuring member countries' economic stronghold over this part of the tuna resource. This illustrates that as both a market actor and sub-regional governance body, the PNA now plays a strong functional role in the WCPO governance architecture.

In spite of the different structures and objectives, both governance bodies focus on multilateral engagement to produce cooperative arrangements that reduce the complexity of negotiations over and between multiple interests in tuna fisheries. The interplay of these normative structures between levels greatly influences both the sub-regionalisation of fisheries management and the political regionalism of the PNA and WCPFC. On the one hand, the PNA are feeding innovative management measures from the bottom up, where shared socioeconomic interests allow the body to 'speak in one voice'. On the other hand, scrutiny of these measures within the WCPFC provides top down oversight and legitimacy. Therefore, while these bodies are fulfilling different roles one could not replace the other. They reinforce a creative tension through their interplay that would not be possible should only one level exist and are as such mutually dependent on each other for management innovation.

Balancing a diversity of interests and the conservation burden

As a region characterised by small island developing states, the process of sub-regionalisation allows for collective authority and strategic capacity to be generated among countries whose individual capacity in strategic analysis and strategy development is otherwise limited both nationally and when negotiating under a wider regional regime (Hanich [2010]). Our results show that balancing the diversity of interests through an active process of fisheries regionalism by the PNA is strategic and closely related to negotiating a more equitable distribution of the conservation burden in international fora like the WPCFC. We therefore discuss the two evaluative criteria of tension between converging/diverging interests and benefit distribution together.

In spite of playing an instrumental role in the creation of the WCPFC, its presence as the highest governing body within the region has seen the Pacific Island countries consolidate their position as a sub-regional group, which is less open to outside, country by country, negotiation over access. As opposed to the partnership-oriented approach intended by RFMO level governance, the PNA have placed more of a focus on internally controlled implementation, treating DWFNs increasingly as licensees. The proactive role of the PNA in searching for innovative ways to control the WCPO tuna fishery also provides an indication of underlying regional political tensions. As outlined by Aqorau ([2009]), the PNA's response is a direct challenge to DWFNs that are seen as "intent on vitiating the gains the PNA have made through the VDS" (Aqorau [2009], p.599). This can be seen as an on-going exercise in capturing control over tuna fisheries that was lost in previous attempts to reduce licence numbers under the Palau agreement and increased power of the DWFNs under the WCPFC Convention. In turn this is illustrated by the PNA's shift from imposing a cap on vessel numbers to fishing days just after the WCPFC emerged, thereby strengthening their management and control over access agreements (Havice [2010]). Although criticised for its lack of transparency in allocating vessel days, and for consistently exceeding the number of days apportioned, the control the PNA nations exercise over fishing effort within limits consistent with resource sustainability both put in place a stronger identity of the PNA as a management region, while also putting the associated countries in a powerful position to secure control over DWFNs represented at the WCPFC.

The PNA's success in shaping these broader governance processes is still however, a relatively new aspect of the overall WCPO governance architecture. While the PNA itself has been around for a long time, it has only really presented a demonstrable challenge to the status quo of regional-level governance since 2008. Therefore, the longevity of this kind of governance interplay is unknown. While strategic interactions between the different levels illustrate the value of sub-regionalisation in WCPO tuna fisheries governance, it is dependent on the PNA either retaining their position as a unified group or face what has been referred to in other trans-boundary governance regimes as fragmentation (e.g. Biermann et al. [2009]). For instance, this is being challenged by Kiribati's negotiations with the EU over their fisheries partnership agreement. In these negotiations, the EU has insisted that the three year protocol granting access to 10 EU fishing vessels (four purse seiners and six longliners), does not incorporate the VDS (EU Committee on Fisheries [2013]). This agreement would see Kiribati essentially defecting from the conditions set out by the VDS. Additionally, by insisting on remaining outside the VDS, the EU is maintaining a relatively low access fee that distorts the regional market for access to the tuna fishery. This case demonstrates both the strength and weakness of fishery regionalism in the PNA. It also shows the change in strategy required by the DWFN members of the WCPFC who are marginalised by the powers of exclusion now exercised by the PNA through their conservation and management measures.

Therefore, given the additional importance of DWFNs' fishing interests, the WCPFC provides a critical venue for enforced cooperation amongst its member. This makes governance at the regional level far from redundant. The complex mosaic of jurisdictions in the region means that the challenges that the FFA, PNA and the WCPFC have faced so far will be compounded over the coming years as they come under increasing pressure to respond to over-fishing and over-capacity concerns (Hanich [2010]), the rise of other emerging regional groupings staking their own claims and changes in negotiation strategies by DWFNs.

CONCLUSION

The performance of tuna fisheries governance in the WCPO is increasingly determined by the interplay between regional and sub-regional bodies responsible for designing and implementing conservation and management measures. Sub-regionalisation appears to be a means for overcoming the limitations of large international environmental regimes. However, distributing management functions over geographical scales is only likely to be effective if there is a mix of political will and multilateralism that can create cooperation for sustainable management and reduce competition. In the case of the WCPO, the sub-regional PNA supported higher levels of governance through the creation of the WCPFC, but have since started to reassert themselves and their governance capacity. The case illustrates that for contested marine resources such as fisheries, international sub-regions can go beyond functional units to also present wider opportunities to shift power relations in the favour of small island states.

The PNA has emerged in a strong position to innovate management measures that would not otherwise be possible at the more inclusive WCPFC level. This strength has led some to describe the PNA as a "tuna cartel" with the WCPO becoming the "Saudi Arabia of Tuna" (Brian Jeffries quoted in Aqorau [2009] p.581). The skipjack fishery most clearly demonstrates this 'cartel': it illustrates how functional regionalisation can lead to a wider regionalism of fisheries management, which in turn can lead to positive interplay between levels of governance. MSC certification of the PNA free school tuna fishery provides an added layer of legitimacy to the PNA management measures and indirectly the PNA as a fishery region. But although the MSC appears to play a role in supporting fisheries regionalism, it remains unclear as to whether it offers a strong long term strategy for consolidating cooperation between the member states. Should economic benefits not accrue to these states, defection rather cooperation may prove to be the outcome.

This case contributes a more nuanced understanding to the perceived crisis of multilateralism by identifying how new forms of regionalism can contribute to solving complex environmental problems. Far from complicating the process of governance and decision-making, the presence of sub-regional groups like the PNA have the potential to constructively challenge higher scale regional groupings like the WCPFC. The combined work of the PNA and the WCPFC may represent a testing ground for a functional multilateralism based on shared resources and utilising both regional and sub-regional governance platforms. However, at least in fisheries, but also in other resource sectors, the question remains whether and how other regional/sub-regional arrangements can emerge that that can take into account the complex social and ecological interactions to promote sustainable production, waste or conservation practices.

ABBREVIATIONS

CFP: Common fisheries policy; **CMM:** Conservation and management measure; **DWFN:** Distant water fishing nation; **EEZ:** Exclusive economic zone; **EU:** European unio; **FAD:** Fish aggregating device; **FFA:** Forum fisheries agency; **FFC:** Forum fisheries committee; **LME:** Large marine ecosystem **MSC:** Marine stewardship council **NGO:** Non-governmental organization; **PNA:** Parties to the Nauru Agreement; **RFMO:** Regional fisheries management organization; **SC:** Scientific committee; **SPC:** Secretariat to the pacific community; **UNCLOS:** United Nations Commission for the Law of the Sea; **UNFSA:** United Nations Fish Stocks Agreemen; **US:**

United States of America; **VDS:** Vessel day scheme; **WCPFC:** West and Central Pacific Fisheries Commission; **WCPO:** West and Central Pacific Ocean

REFERENCES

1. Aqorau T: Recent developments in pacific tuna fisheries: The Palau Arrangement and the Vessel Day Scheme. *The International Journal of Marine and Coastal Law* 2009, 24(3):557–581. 10.1163/157180809X455647
2. Aranda M, Murua H, de Bruyn P: Managing fishing capacity in tuna regional fisheries management organisations (RFMOs): Development and state of the art. *Marine Policy* 2012, 36(5):985–992. 10.1016/j.marpol.2012.01.006
3. Bailey M, Flores J, Pokajam S, Sumaila UR: Towards better management of Coral Triangle tuna. *Ocean and Coastal Management* 2012, 63: 30–42. 10.1016/j.ocecoaman.2012.03.010
4. Balsiger J, VanDeveer SD: Regional governance and environmental problems. *The International Studies Encyclopedia* 2010, 9: 6179–6200.
5. Balsiger J, VanDeveer SD: Navigating regional environmental governance. *Global Environmental Politics* 2012, 12(3):1–17. 10.1162/GLEP_e_00120
6. Barclay K, Cartwright I: Governance of tuna industries: the key to economic viability and sustainability in the Western and Central Pacific Ocean. *Marine Policy* 2007, 31(3):348–358. 10.1016/j.marpol.2006.09.007
7. Biermann F, Pattberg P, van Asselt H, Zelli F: The fragmentation of global governance architectures: a framework for analysis. *Global Environmental Politics* 2009, 9(4):14–40. 10.1162/glep.2009.9.4.14
8. Biermann F, Abbott K, Andresen S, Bäckstrand K, Bernstein S, Betsill MM, Bulkeley H, Cashore B, Clapp J, Folke C, Gupta A, Gupta J, Haas PM, Jordan A, Kanie N, Kluvánková-Oravská T, Lebel L, Liverman D, Meadowcroft J, Mitchell RB, Newell P, Oberthür S, Olsson L, Pattberg P, Sánchez-Rodríguez R, Schroeder H, Underdal A, Vieira SC, Vogel C, Young OR, Brock A, Zondervan R: Navigating the anthropocene: improving earth system governance. *Science* 2012, 335(6074):1306–1307. 10.1126/science.1217255
9. Campbell HF: Resource rent rivalry in the western pacific tuna fishery: cooperative and non-cooperative solutions. *Ocean and Shoreline Management* 1989, 12(2):179–187. 10.1016/0951-8312(89)90052-0
10. Cash DW, Moser SC: Linking global and local scales: designing dynamic assessment and management processes. *Global Environmental Change* 2000, 10(2):109–120. 10.1016/S0959-3780(00)00017-0
11. Cash DW, Adger WN, Berkes F, Garden P, Lebel L, Olsson P, Pritchard L, Young O: Scale and cross-scale dynamics: governance and information in a multilevel world. *Ecology and Society* 2006, 11(2):8.
12. Conca K: The rise of the region in global environmental politics. *Global Environmental Politics* 2012, 12(3):127–133. 10.1162/GLEP_a_00132
13. Cullis-Suzuki S, Pauly D: Failing the high seas: a global evaluation of regional fisheries management organizations. *Marine Policy* 2010, 34(5):1036–1042. 10.1016/j.marpol.2010.03.002
14. de Bruyn P, Murua H, Aranda M: The Precautionary approach to fisheries management: How this is taken into account by Tuna regional fisheries management organisations (RFMOs). *Marine Policy* 2013, 38: 397–406. 10.1016/j.marpol.2012.06.019
15. EU Committee on Fisheries. 2013. Recommendation on the draft Council decision on the conclusion of the Protocol setting out the fishing opportunities and financial

contribution provided for in the Fisheries Partnership Agreement between the European Community, on the one hand, and the Republic of Kiribati, on the other (13331/2012 – C7 0036/2013 – 2012/0229(NLE)). European Parliament.

16. Fanning L, Mahon R, McConney P, Angulo J, Burrows F, Chakalall B, Gil D, Haughton M, Heileman S, Martínez S, Ostine L, Oviedo A, Parsons S, Phillips T, Santizo Arroya C, Simmons B, Toro C: A large marine ecosystem governance framework. *Marine Policy* 2007, 31(4):434–443. 10.1016/j.marpol.2007.01.003
17. Gagern A, van den Bergh J: In Press. A critical review of fishing agreements with tropical developing countries. *Marine Policy* , : . http://dx.doi.org /10.1016/j.marpol.2012.06.016 http://dx.doi.org/10.1016/j.marpol.2012.06.016
18. Gilman E, Kingma E: Standard for assessing transparency in information on compliance with obligations of regional fisheries management organizations: Validation through assessment of the Western and Central Pacific Fisheries Commission. *Ocean and Coastal Management* 2013, 84: 31–39. 10.1016/j.ocecoaman.2013.07.006
19. Greenpeace Briefing to the 10th Reguar Meeting of hte WCPFC. Cairns, Australia, Greenpeace; 2013.
20. Gulbrandsen LH: Dynamic governance interactions: Evolutionary effects of state responses to non-state certification programs. *Regulation & Governance* 2013, 8(1):74–92. 10.1111/rego.12005
21. Hamilton A, Lewis A, Campling L: *Report on the Implementation of the derogation to the standard rules of origin granted to the Pacific ACP States in the framework of the Interim Economic Partnership Agreement.* European Union. FWC COM 2011 RFS 2011/266449, Brussels; 2011.
22. Hanich Q: Control, Cooperation and 'Participatory Rights' in the Western and Central Pacific Ocean Tuna Fisheries. In *Navigating Pacific Fisheries: Legal and Policy Trends in the Implementation of International Fisheries Instruments in the Western and Central Pacific Region.* Edited by: Hanich Q, Tsamenyi M. ANCORS, Wollongong; 2009:221–249.
23. Hanich Q: Regional fisheries management in ocean areas surrounding Pacific Islands States. In *Proceedings of the International Seminar on Islands and Oceans.* Edited by: Terashima H. Ocean Policy Research Foundation, Tokyo, Japan; 2010:195–212.
24. Hanich Q: Distributing the bigeye conservation burden in the western and central pacific fisheries. *Marine Policy* 2012, 36(2):327–332. 10.1016/j.marpol.2011.07.008
25. Hanich Q, Teo F, Tsamenyi M: A collective approach to Pacific islands fisheries management: Moving beyond regional agreements. *Marine Policy* 2010, 34(1):85–91. 10.1016/j.marpol.2009.04.018
26. Havice E: The structure of tuna access agreements in the Western and Central Pacific Ocean: Lessons for Vessel Day Scheme planning. *Marine Policy* 2010, 34(5):979–987. 10.1016/j.marpol.2010.02.004
27. Havice E: Rights-based management in the Western and Central Pacific Ocean tuna fishery: Economic and environmental change under the Vessel Day Scheme. *Marine Policy* 2013, 42: 259–267. 10.1016/j.marpol.2013.03.003
28. Havice E, Reed K: Fishing for Development? Tuna Resource Access and Industrial Change in Papua New Guinea. *Journal of Agrarian Change* 2012, 12(2–3):413–435. 10.1111/j.1471-0366.2011.00351.x
29. Hegland T, Ounanian K, Raakjaer J: Why and how to regionalise the Common Fisheries Policy. *Maritime Studies* 2012, 11(1):7. 10.1186/2212-9790-11-7
30. Langley A, Wright A, Hurry G, Hampton J, Aqorua T, Rodwell L: Slow steps towards management of the world's largest tuna fishery. *Marine Policy* 2009, 33(2):271–279. 10.1016/j.marpol.2008.07.009
31. Lebel L, Garden P, Imamura M: The politics of scale, position, and place in the governance of water resources in the Mekong region. *Ecology and Society* 2005, 10(2):18.

32. Meadowcroft J: Politics and scale: some implications for environmental governance. *Landscape and Urban Planning* 2002, 61(2):169–179. 10.1016/S0169-2046(02)00111-1
33. Morgan JR: Marine regions: Myth or reality. *Marine Policy* 1989, 12: 1–21.
34. Nauru Agreement Concerning Cooperation in the Management of Fisheries of Common Interest, Parties to the Nauru Agreement. 1982.
35. Oberthür S, Stokke OS: *Managing institutional complexity: regime interplay and global environmental change*. MIT Press, Michigan; 2011.
36. Paasi A: Place and region: looking through the prism of scale. *Progress in Human Geography* 2004, 28(4):536–546. 10.1191/0309132504ph502pr
37. Palau Arrangement for the Management Of The Western Pacific Fishery as Amended, 1992 (Amended 11th September 2010). 1992.
38. Parris H: Is the Western and Central Pacific Fisheries Commission meeting its conservation and management objectives? *Ocean and Coastal Management* 2010, 53(1):10–26. 10.1016/j.ocecoaman.2009.10.002
39. Parris H: Tuna dreams and tuna realities: Defining the term "maximising economic returns from the tuna fisheries" in six Pacific Island states. *Marine Policy* 2010, 34(1):105–113. 10.1016/j.marpol.2009.04.022
40. Parris H, Grafton RQ: Can tuna promote sustainable development in the pacific? *The Journal of Environment & Development*2006, 15(3):269–296. 10.1177/1070496506289364
41. Additional Closures by the PNA - Republic of the Marshall Islands on Behalf of the Parties to the Nauru Agreement. Federated States of Micronesia, WCPFC, Pohnpei; 2010.
42. PNA Presents to Industry at InfoFish Bangkok. 2014.
43. Raakjaer J, Hegland T: Introduction: regionalising the Common Fisheries Policy. *Maritime Studies* 2012, 11(1):5. 10.1186/2212-9790-11-5
44. Ride A: *Unfinished business remains as WCPFC ends*. Forum Fisheries Agency, Honiara; 2010.
45. Shanks S: Introducing a transferable fishing day management regime for Pacific Island countries. *Marine Policy* 2010, 34(5):988–994. 10.1016/j.marpol.2010.02.005
46. Review of the Implementation and Effectiveness of CMM 2008–01. Republic of Korea, Busan; 2012.
47. Swyngedouw E: "Scaled geographies: Nature, place, and the politics of scale". Scale and geographic inquiry: Nature, society, and method. In *Scale and geographic inquiry: Scale and geographic inquiry: Nature, society, and method*. Edited by: McMaster R, Sheppard E. Blackwell Publishers, Oxford; 2004:129–153. 10.1002/9780470999141.ch7
48. Symes D: Regionalisation of Fisheries Governance: An Empty Vessel or a Cornucopia of Opportunity. In *Participation in Fisheries Governance*. Edited by: Gray T. Springer, Netherlands; 2005:85–102. 10.1007/1-4020-3778-3_5
49. Symes D: Regionalising the Common Fisheries Policy: context, content and controversy. *Maritime Studies* 2012, 11(1):6. 10.1186/2212-9790-11-6
50. Tsamenyi M: The institutional framework for regional cooperation in ocean and coastal management in the South Pacific. *Ocean and Coastal Management* 1999, 42(6–7):465–481. 10.1016/S0964-5691(99)00033-2
51. Tutangata T, Power M: The regional scale of ocean governance regional cooperation in the Pacific Islands. *Ocean and Coastal Management* 2002, 45(11–12):873–884. 10.1016/S0964-5691(02)00111-4
52. Valencia MJ: The South China Sea Prospects for marine regionalism. *Marine Policy* 1978, 2(2):87–104. 10.1016/0308-597X(78)90002-7
53. Vallega A: Ocean governance in post-modern society—a geographical perspective. *Marine Policy* 2001, 25(6):399–414. 10.1016/S0308-597X(01)00024-0
54. Van Leeuwen J, Kern K: The External Dimension of European Union Marine Governance: Institutional Interplay between the EU and the International Maritime

Organization. *Global Environmental Politics* 2013, 13(1):69–87. 10.1162/GLEP_a_00154

55. Convention on the conservation and management of highly migratory fish stocks in the Western and Central Pacific Ocean. WCPFC, Pohnpei; 2000.
56. Conservation and management measure for Bigeye and Yellowfin Tuna in the Western and Central Pacific Ocean. 2008.
57. Commission for the Conservation and Management of Highly Migratory Fish Stocks in the Western and Central Pacific Ocean. Scientific Committee Report. W. S. Committee. WCPFC, Kolonia, Pohnpei; 2012.
58. Conservation and management measure for Bigeye, Yellowfin and Skipjack tuna in the Western and Central Pacific ocean, Commission 9th Regular Session Manila, Philippines, 2–6 December 2012. 2012.
59. Recommendations from the Review of the WCPFC, WCPFC-SC8–2012/GN-WP-08. 2012c.
60. Wilson JA: Matching social and ecological systems in complex ocean fisheries. *Ecology and Society* 2006, 11(9):9. [online] URL:http://www. Ecology andsociety.org/vol11/iss1/art9/ [online] URL:http://www. Ecologyand society. org/vol11/iss1/art9/
61. Wright A, Stacey N, Holland P: The cooperative framework for ocean and coastal management in the Pacific Islands: Effectiveness, constraints and future direction. *Ocean and Coastal Management* 2006, 49(9–10):739–763. 10.1016/j. ocecoaman.2006.06.013
62. Young O: Vertical interplay among scale-dependent environmental and resource regimes. *Ecology and Society* 2006, 11(1): . [online] URL:http://www. Ecologyand society.org/vol11/iss1/art27/ [online] URL:http://www.ecologyandsociety.org/vol11/iss1/art27/
63. Young OR: Effectiveness of international environmental regimes: Existing knowledge, cutting-edge themes, and research strategies. *Proceedings of the National Academy of Sciences* 2011, 108(50):19853–19860. 10.1073/pnas. 1111690108

CHAPTER 14

Oil and Gas Platforms in the Gulf of Mexico: Their Relationship to Fish and Fisheries

James H. Cowan* and Kenneth A. Rose

[1]*Department of Oceanography and Coastal Sciences, Louisiana State University, Baton Rouge, USA*

ABSTRACT

There are over 2300 standing oil and gas platforms in the northern Gulf of Mexico (GOM). It has been argued that platforms provide reef-like habitat that increases the growth and survival rates of fishes by increasing prey availability and affording shelter for protection from predators, provide additional spawning substrate, and by acting as a visual attractant for organisms not otherwise dependent upon hard bottom. Platforms differ from most natural habitats, and from traditional artificial reefs, in that their vertical profile extends upward through the water column into the photic zone and the sea surface. Increased habitat quality on, or immediately around, oil and gas platforms are thought to be derived from increased *in situ* food production associated with encrustation by fouling organisms. In this chapter, we address the issue of how to evaluate the role of artificial reefs by first establishing levels of evaluation for individual fish species found on oil and gas platforms in the GOM. The levels of evaluation relate to the amount and adequacy of the available information, which was populated with an extensive literature and data search. Three levels of assessment are established, analogous to the levels of analysis established National Oceanographic and Atmospheric Administration (NOAA) Fisheries for identification of Essential Fish Habitat. More than 1300 documents, including reports, stock assessments, other gray literature, and papers published in the primary literature, were used to complete this chapter. When available, published literature was the preferred source of information.

Keywords: Gulf of Mexico, oil and gas platforms, fish, fisheries, biomass production

1. INTRODUCTION

In the waters of the northern Gulf of Mexico (GOM) there are over 2300 standing oil and gas platforms that together constitute the largest *de facto* artificial reef program in the world (**Figure 1**). It has been argued that platforms provide reef-like habitat that increases the growth and survival rates of fishes by increasing prey availability, and affording shelter for protection from predators, provide additional spawning substrate, and by acting as a visual attractant for organisms not otherwise dependent upon hard bottom. Platforms differ from most natural habitats, and from traditional artificial reefs, in that their vertical profile extends upward through the water column into the photic zone and the sea surface. Increased habitat quality on, or immediately around, oil and gas platforms are thought to be derived from increased *in situ* food production associated with encrustation by fouling organisms.

Artificial reefs, such as oil and gas platforms, may be useful tools for fishery managers if they increase reef fish biomass production, but many researchers question whether or not they are a positive influence on fish stock dynamics. If artificial reefs constitute habitat that is otherwise limiting for reef-associated fishes, then they may be viable management tools. If they are simply attracting fish, then they may be promoting overfishing. Unfortunately, the extent to which structures have influenced the status of exploited fish stocks, either directly via population production rates or indirectly through changes in fishing mortality rates, is still not well understood. The structures may alter fish populations and communities as a result of altering ecosystem structure and function. The effects on exploited fish stocks could also be detrimental if the high levels of fishing mortality rates do not result in compensatory processes that lead to increases in stock production.

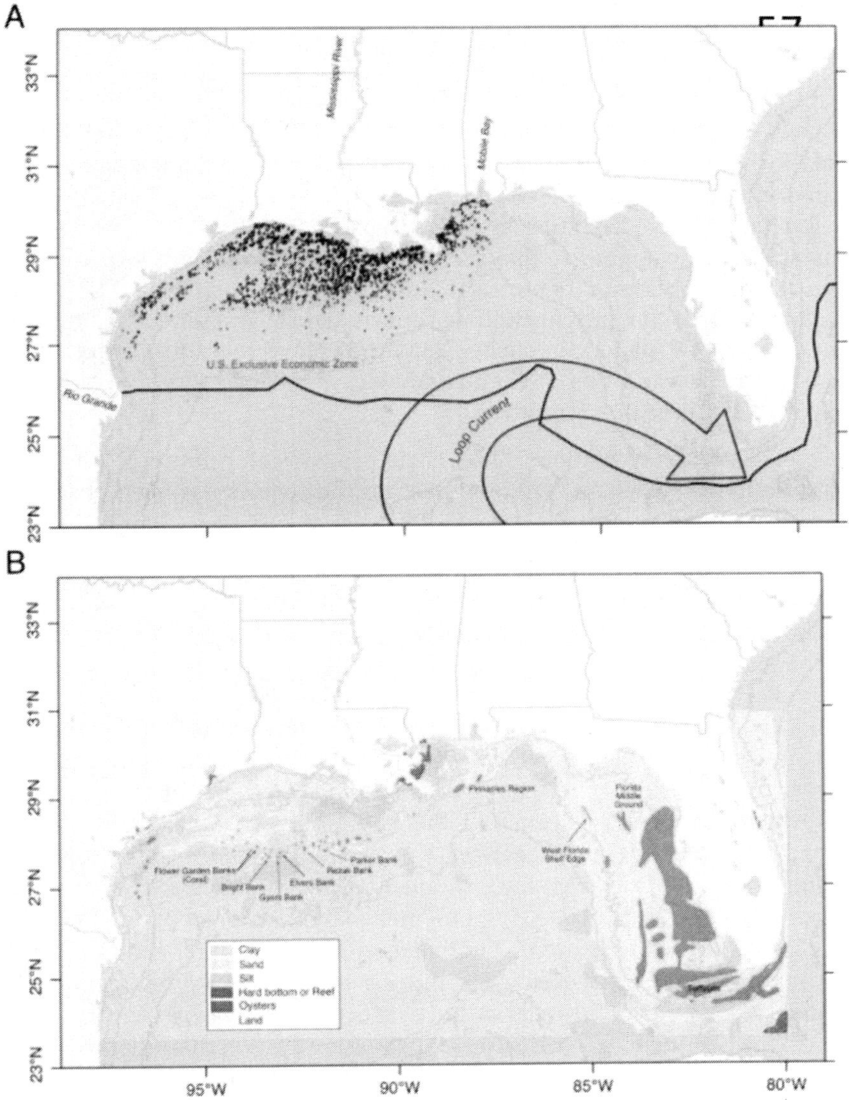

FIGURE 1. Map of the northern Gulf of Mexico. Panel A depicts oil and gas structures currently in place (black dots), the U.S. exclusive economic zone, and depth contours for 20, 200, and 2000 m. The continental shelf follows the 200-m contour approximately. Panel B (adapted from Gulf of Mexico Fishery Management Council 2005) depicts the primary bottom substrate in the area. Labeled banks were sampled during fishery independent reef fish surveys.

2. METHODS

In this chapter, we evaluate the role of artificial reefs by first establishing levels of evaluation for individual fish species found on oil and gas platforms in the GOM. The levels of evaluation are dependent upon the amount and adequacy of the available information, which was populated with an extensive literature and data search. Three levels of assessment are established, analogous to the levels of analysis established National Marine Fisheries Service for identification of Essential Fish Habitat (EFH). More than 1300 documents, including reports, stock assessments, other gray literature, and papers published in the primary literature, were used to complete this chapter. When available, published literature was the preferred source of information.

Level 1–For species about which little process information is known, the evaluation is based simply upon whether the species has been observed in association with platforms or artificial reefs.

Level 2–Here we used the conceptual model of Bohnsack [1] (qualitative, **Figure 2**). This conceptual model centers on the attraction vs. production issue, which encompasses much of the debate about the ecological role of artificial reefs (including oil and gas platforms) in a complex and dynamic coastal geography. The difference between Level 1 and Level 2 assessments is the degree of inference at the process level about the species in question, even if the process information (e.g., fishing mortality, site fidelity) is poorly documented. As such, relative knowledge of where a species falls along the continuum of data availability and confidence for several process-related variables provides significant insight into how that species may be affected by platforms. The evaluation presented here is based upon information reported In FishBase® (http://www.fishbase.org/), combined with expert opinion, and is used to provide relative species-specific assessments of: (1) Site fidelity (High, Moderate, Low); (2) Whether or not a directed fishery exists for this species (Yes or No, includes recreational fishing); (3) Whether diet is derived directly from reef habitat (Reef, Benthic, Pelagic); (4) Whether population size is believed to be limited by recruitment or habitat limitation (Habitat limited, Recruitment limited); (5) Type of behavior of adults (Reef, Demersal, Pelagic, Highly Migratory); and, (6) A summary judgment about whether the species is reef or habitat dependent, and the type of habitat on which some dependent species are most often found (e.g., Sargassum, Sea Grass, Hard Bottom).

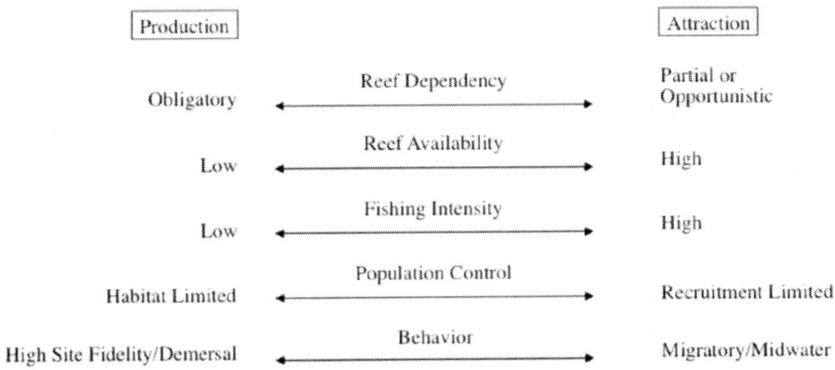

FIGURE 2. Bohnsack's [1] conceptual model for addressing the role of reefs in fisheries (redrawn from Bull. Mar. Sci. 44: 631–645).

Level 3–Based upon our extensive review of the literature, data for only five species of fishes were deemed sufficient for more complex analysis of estimating production; these are red snapper *Lutjanus campechanus*, blue runner, *Caranx crysos*, sheepshead *Archosargus probatocephalus*, Atlantic spadefish *Chaetodipterus faber*, and bluefish *Pomatomus saltatrix*. As might be expected, these species are abundant around standing platforms in the Gulf of Mexico and have been studied in some detail. Here evaluations are based upon various semi-quantitative or quantitative models found in the published literature.

The first model used in Level 3 evaluations is the semi-quantitative model described in Powers and colleagues [2], which uses the species-specific fish biomass production of a population on a reef (here a platform) weighted by the degree to which growth (biomass production) is attributable to prey resources produced on the reef. The production estimate for each species is multiplied by an index of reef exclusivity (IRE) derived from quantitative diet data. Applying the IRE, annual production (P) of a species attributed to a platform (AP; kg platform^{-1} yr^{-1}) is calculated by:

$$AP_i = IRE_i * P_i \qquad (1)$$

where AP is a measure of relative species-specific production attributable to a platform. In the original equation, there was a term for the difference between pre-structure and post-structure biomass, but we are unable to provide pre-structure data because almost all of the platforms on the continental shelf were employed more than 25 years ago. We also believe that an estimate of biomass on a seafloor lacking structure, when compared to biomass after a structure has been constructed, may result in an overestimate of new biomass production because of the likelihood that individuals of many species are simply attracted to standing platforms.

The second method of estimating annual production for Level 3 is an empirical model [3]:

$$P = 0.00051 * B^{0.69} * T^{1.04} \tag{2}$$

where P is production (g dw d^{-1}), B is biomass (kg), and T is temperature (°C). We assumed that g dw = g wet weight * 0.20. The data used to estimate B (g) is based upon data obtained from published literature on biomass, estimated daily somatic production, and ambient water temperature for 62 species of fishes collected from numerous locations in Australia and elsewhere. Temperature data were summarized from several Gulf of Mexico studies on the continental shelf.

The previous and following methods of estimating production require an estimate of biomass on a platform. To make this estimate, we first calculated the simple arithmetic mean number of fish by species on a platform by summing all of the available estimates of numbers observed, based mostly upon visual surveys using scuba diving. In addition to numbers of individuals, length ranges (cm) are also reported for each species.

The third estimate of production for Level 3 is based upon methods described in by Ricker [4] where annual production is estimated by:

$$\hat{P} = \overline{B} * (G) \tag{3}$$

where $\hat{P}$ is biomass production, $\overline{B}$ is annual mean biomass, and G is specific growth rate yr^{-1}. Annual mean biomass is estimated by using:

$$\overline{B} = \frac{B^{(1-e^{(-Z-G)})}}{Z-G} \text{ when } G > Z \tag{4}$$

$$\overline{B} = \frac{B(e^{(G-Z)} - 1)}{G-Z} \text{ when } Z > G \tag{5}$$

Where $\overline{B}$ is annual mean biomass, B is biomass per platform in kg, G is specific growth rate yr^{-1}, and Z is specific mortality rate yr^{-1}. In actuality, Z is the sum of F (annual fishing mortality rate) plus M (annual natural mortality rate). In our Level 3 assessment, we ignore F in calculations, but we discuss the implications of this omission in the discussion. Biomass estimates were calculated based upon age and growth relationships reported in the literature from samples collected in the GOM for all species except Atlantic spadefish. For Atlantic spadefish, we used data collected in South Carolina. To estimate length at age, we used the Von Bertalanffy growth model (**Tables 1** and **2**):

$$TL_t = L_\infty(1 - e^{-k(t-t_0)}) \tag{6}$$

TABLE 1. Literature values for mean number per platform (MN), size range at platform (SR, cm FL), mean length of individuals observed (ML, cm FL), age range (AR yrs), and biomass per platform (B kg wet wt) for five abundant species of fishes collected from Gulf of Mexico oil and gas platforms.

Species	MN	SR	ML	AR	B (kg)
Red snapper	1884 (range 905–4632)	25.5–79.1	295.3	2–10	886
Bluefish	1438 (range 282–4000)	45–50	475	1–6	1489
Atlantic spadefish	4177 (range 10 – 5323)	10–50	30.0	1–8	2618
Sheepshead	2250 (range 150–17,000)	22–50	360	2–5	1774
Blue runner	6260 (range 427–25,188)	30–36	33.5	2–6	4152

TABLE 2. Literature values for maximum age, estimated growth and mortality rates, and Von Bertalanffy length-at-age parameters used in production calculations.

Species	T_{max}	G	M	AD	L_∞	K	t_0	a	b	Source
Red snapper	57	0.31 (at T_{max} G=0.05)	0.10 (at T_{max} M=0.07)	2–10	94.1	0.18	−0.55	0.0165	3.03	[5–7]
Blue runner	11	0.39	0.38	2–4	41.2	0.35	−1.17	0.0524	2.690	[8]
Sheepshead	20	0.23	0.22	2–5	41.9 (M)	0.417	−0.901	0.000448	2.88	[9]
					44.7 (F)	0.367 to 1.025		0.000530	2.85	
Atlantic spadefish	8	0.41	0.58	1–8	49.0	0.340	−0.18	0.0927	2.64	[10]
Bluefish	8	0.41	0.58	1–6	94.4	0.18	1.033	−10.02	2.80	[11]

T_{max} = maximum age; M = natural mortality rate; G = specific growth rate (yr); L_∞ (cm TL) is the asymptotic maximum length, K is a constant (the Brody growth coefficient), and t_0 is a constant representing the age (yr) at 0 length. Letters in parentheses following L_∞ indicates sex if males and females were dimorphic. Fish length is converted to wet weight using a length-weight equation with constants a and b. Age distributions of fish (AD) were derived from empirical data for size ranges at age of fishes observed at platforms, except for red snapper that are most abundant from ages 2 to 6 (Wilson and Nieland 2001).

where *TLt* is total length (TL) at age t, *L* is the asymptotic TL, *k* is the Brody growth coefficient, *t* is age in years, and *t0* is a hypothetical age when TL is zero. Using species-specific versions of this equation, we determined the age range of each species observed on platforms that correspond to the observed length ranges. Length-length relationships available for each species allowed for conversion among total length (TL), fork length and standard length as necessary for consistency among units of length. Length at age was converted to wet weight at age using:

$$TW_i = a * TL^b \tag{7}$$

where *TWi* is wet weight in g of species i, *TL* (cm) and a and b are constants derived for each species. The values for a and b for each species are reported in **Table 3**. Growth (G) is estimated using the equation:

$$G = (\ln W_{t=j} - \ln W_{t-i}) + (t_j - t_i) \tag{8}$$

where G is the specific growth rate in kg yr^{-1}, W is weight in kg and t is time.

TABLE 3. Estimated relative production attributable to oil and gas platforms (AP).

Species	Trophic Level (SE)	Diet Composition (source)	IRE	T°C	P_r	P_e	AP_r	AP_e
Red snapper	4.01 ± 0.59	Benthic inverts, demersal fishes, squid, pelagic zooplankton [12-15]	0.05	25.9	306	115	15	6
Blue runner	4.40 ± 0.77	Fish, decapods, hyperid amphipods, chaetognaths, other [16, FishBase]	0.10	25.9	1627	333	163	33
Sheepshead	3.53 ± 0.53	Portunid crabs, shrimp, barnacles, fish, copepods, bryozoans, amphipods, sargassum [17, FishBase]	0.90	25.9	410	185	369	167
Atlantic spadefish	3.50 ± 0.47	Sponges and tunicates, cnidarians, worms, ascidicans, plants, benthic inverts, echinoderms, zooplankton (FishBase)	0.95	25.9	987	243	938	231
Bluefish	4.50 ± 0.55	Pelagic and demersal fish and macrocrustaceans (from soft bottoms) [18]	0.01	25.9	561	164	6	2

The index of reef exclusivity (IRE) is an estimate of species utilization of resources associated with platform habitat compared to resources from nearby natural habitat (Powers et al. 2003). The IRE is based upon diet information from the sources provided. Trophic level for each species was obtained from FishBase (http://www.fishbase.org/), T°C is the annual averaged sea surface temperature in degrees centigrade obtained from the NOAA Data Buoy Center for years 2013-2015 and 2005 at SPLL1 (28.87 N, 90.48 W) and MRSL1 (29.44 N, 92.06 W) on the Louisiana Shelf. P_r = annual biomass production in kg $platform^{-1}$ yr^{-1} based upon calculations using Ricker (1975). P_e = annual biomass production in kg $platform^{-1}$ yr^{-1} estimated by using the empirical relationship in Edgar and Shaw (1990). AP_r = IRE*P_r, AP_e = IRE*P_e. Numbers proceeded by [+] refer to literature resources listed on the table.

3. RESULTS

Levels 1 and 2: In total, 246 species of fishes have been reported from oil and gas platforms in the GOM (**Appendix Table 1**, hereafter **ATable 1**). Of these, 33 species are Caribbean expatriates (23 of which are reported to be reef dependent) that occur sporadically in low numbers, and are not believed to contribute to overall stock productivity because their larvae are nearly absent in waters of the northern GOM [17]. One hundred-two species have life history strategies that conclusively exclude them from being reef associated or dependent (N in **ATable 1**), even though these species have been reported in collections of fishes from platforms. Thirty-six species are conclusively considered to be reef dependent (Y in **ATable 1**; note that Y= reef dependent, Y plus habitat descriptor HB, SG, S are habitat dependent on a specified habitat), which here indicates that reef habitat is required for these species to complete their life cycles, or that their diet is almost exclusively derived from the reef [18]. Reef dependent species that are not expatriates are: *Balistes capricus; Cantherhines pullus; Cephalopholis cruentatus; Chaetodipterus faber; Clepticus parrae; Parablennius marmoreus*; and *Xyrichthys novacula*. Thirteen species have life history strategies that appear to preclude reef dependency, or are documented to occur on structured, non-reef habitats, but are listed in FishBase as reef associated. In **ATable 1**, these 13 species are denoted with an N under the Dep category, followed by the habitat type that is reported to be of greatest importance.

The species for which we assigned an N have life history and behavioral characteristics that are qualitatively similar to the attraction end of the Bohnsack's continuum [1]. These species are directly fished or overfished, exhibit low site fidelity, are less or not dependent upon the reef for food, are not dependent upon the reef for completion of their life cycles, and are pelagic and/or migratory, and thus less likely to habitat limited. In contrast, the species for which we assigned a Y in **ATable 1** have life history and behavioral characteristics that are qualitatively more similar to production end of the continuum [1]. These species have relatively high site fidelity, the need for reef or structured habitat to complete the life cycle, and a significant fraction of their diet is comprised primarily of reef-associated prey.

There are numerous species for which expectations about reef dependence are more difficult to describe, even qualitatively. To provide some interpretation, we use both a qualitative assessment relative to the Bohnsack [1] conceptual model, and insight derived from the Level-3 assessments to make comparisons among the reef-associated species in **ATable 1**. Where possible, we identify species that are comparable with respect to ecology, life history, and behavioral characteristics to the Level-3 species. It is fortunate that the latter group is comprised of species that appear to differ significantly in their relative ecological dependence on reefs and, by extension, to platforms. We also consulted additional reference materials to make our determinations [18–24].

In all, 46 species are listed as reef associated (RA in **ATable 1**). Of these, many are known to be pelagic and/or highly migratory. Among this group are several species of jacks (fm. Carangidae, genus *Caranx*(6 species), genus *Seriola* (3 species), *Elagatis bipinnulata, Selar crumenophthalmus, Oligoplites saurus*, and *Trachurus lathami*), mackerels (fm. Scombribdae, *Scomberomorus* (2 species)), clupeids (*Harengula jaguana, Opisthonema oglinum, and Sardinella aurita*), barracudas (genus *Sphyraena* (3 species)), cobia (*Rachycentron canadum)*, and ocean triggerfish (*Canthidermis sufflamen*). Although listed as reef associated in FishBase, they exhibit life history and behavioral characteristics that are more typical of fishes at the attraction end of Bohnsacks continuum, and appear to be most comparable in their use of reefs and platforms to bluefish and/or blue runner. As such, these species may not be significantly dependent upon platforms.

Thirty of the 46 species listed as reef associated (RA) have other habitats listed as primary (fm. Carangidae, mackerels fm. Scombribdae, and clupeiods herrings and anchovies). Many of these species are reported to primarily associate with hard-bottom (HB) habitats (25 species including most of the groupers). This makes sense given the nature of most of the natural reef habitat in the GOM. Other species are reported to associate with sea grass (SG) meadows. Where possible, we have included additional detail in **ATable 1** about primary habitat associations reported for many of the reef-associated species. These primary habitats are consistent with natural habitats reported to occur in the GOM and include reef flats, rocky reefs, coral reefs, oyster reefs, floatsam, shelf-edge banks, offshore rock bottoms, offshore banks, *Oculina* reefs, and structure of all types. This group clearly is the most difficult to assess. We note that the lack of habitat-specific, process-level data makes the following interpretations speculative.

Most of the grouper species reported in **ATable 1** in the genus *Epinephelus*, with the exception of *E. itajara*, and in the genus *Mycteroperca*, with the exception of *M. microlepis*, are managed as a complex in the GOM referred to as the "deep-water groupers." Relatively little is known about the ecology and behavioral characteristics of these species, although they are believed to be long-lived and exhibit relatively low stock productivity [25]. Assessment of the role that platforms play in their life histories would be speculation on our part, but it is likely that association with platforms does not significantly increase their vulnerability to fishing, especially to recreational anglers, given their preferred depth distribution. It also is unlikely that a significant number of platforms are available as habitat for these groupers for the same reason; these fishes are likely to occur only on those structures near or on the shelf-edge banks shown in **Figure 1**. In contrast, *E. itajara*, the goliath grouper, and *M. microlepis*, the gag grouper, are found inshore on a variety of habitats ranging from platforms to artificial reefs, to bridge pilings, piers, docks, seawalls, and other hard structures [26]. Juvenile goliath groupers are most often found in mangroves, which appear to be its primary nursery ground. Goliath groupers are overfished in the US GOM and currently, harvest is limited [27]. They are confined mostly to Florida Bay and the southern portion of the Florida peninsula and the Bay of Campeche where they are harvested in great numbers as juveniles in the GOM, but have been observed occasionally by scuba divers around platforms. As the stock rebuilds and expands northward in the GOM, however, it is plausible that platforms will contain increasing numbers of goliath groupers. The relative role of platforms as sources of stock productivity and fishing mortality should be closely monitored as the goliath grouper stock increases.

Gag groupers are more widely distributed in the GOM, but also are overfished [28]. They are extremely vulnerable to overexploitation because they are haremic as adults, and aggregate to spawn at just a few locations in the northeastern GOM. Juvenile gag groupers are mostly associated with sea grass meadows as nursery areas. To our knowledge, the ecology of gag grouper on platforms has received little study. However, the work of Lindberg and coworkers on the west Florida shelf has demonstrated that the value of artificial reefs as habitat is affected both by size and spatial arrangement of reef modules. The net effect on stock production of reefs is negative when fishing mortality is considered [29, 30]. Despite these results, we caution against drawing inference about the role of platforms as habitat for gag groupers because the aforementioned work was done on relatively small, low-relief, reef modules. Two other species reported as reef-associated in **ATable 1** (*Neoconger mucronatus* and *Ophidion selenops*) also are found in deep waters on or near the shelf edge.

There are several species in **ATable 1** that are reported to be reef-associated, but also occur on a wide variety of habitats including inshore waters, bays, estuaries, and sea grass meadows. Qualitatively, these species have life history and behavioral characteristics that are more similar to those found at the production end of Bohnsack's continuum [1]. These include *Chilomycterus schoepfii, Lagodon rhomboides*, two members of the genus *Opsanus, Sphoeroides spengleri*, two members of the genus *Trachinotus*, and *Stephanolepis hispidus*. This group appears to be most like Atlantic spadefish and sheepshead, for which Level 3 assessments were made.

Similarly, there is another group that qualitatively appears to have life history and behavioral characteristics that are more similar to those found at the production end of the continuum, but also appear to be more restricted in their distribution than the reef-associated group that use many habitats (discussed in the previous paragraph). These more habitat-restricted species are reported to occur in coastal waters, and on shelf-edge banks, but are explicitly identified as not being found on coral reefs. This group includes two members of the genus *Equetus, Gymnothorax nigromarginatus*, two members of the genus *Hypleurochilus*, two members of the genus *Hypsoblennius, Myrophis punctatus, Pseudupeneus maculatus, Saurita normani*, two members of the genus *Syacium, Syngnathus louisianae*, three members of the genus *Synodus*, and the blue phase of *Thalassoma bifasciatum*. Furthermore, blennies are nest builders that depend upon hard substrate. Many members of this group also appear to be most like Atlantic spadefish and sheepshead, for which Level 3 assessments were made.

There are several small, cryptic species listed in **ATable 1** as reef-associated that we believe to be more strongly associated with reefs than the many-habitat and restricted habitat reef-associated species (preceding two paragraphs), and whose life history and behavioral characteristics appear to place them solidly at the production end of Bohnsack's continuum. This group includes *Callionymus bairdi, Coryphopterus punctipectophorus, Ophioblennius atlanticus, Prognathodes aya, Rypticus maculatus*, the yellow phase of *Thalassoma bifasciatum*, and *Trachinocephalus myops*. Although information about the ecology and life history of this group on platforms is lacking, there are no analogues for these among the group for which Level 3 assessments were possible.

In addition, there are several species listed in **ATable 1** as being reef-associated in FishBase, but whose life history and behavioral characteristics (at least qualitatively) do not strongly support placement near either endpoint of Bohnsack's continuum. This group has been reported to occur on a wide variety of natural hard-bottom habitats in the GOM, including platforms, but appear to have only moderate to low site fidelity. Many of these species support directed commercial fisheries, and all appear among the list of species harvested by recreational anglers. This group includes six members of the genus *Lutjanus*, including the red snapper *L. camphechanus* (site fidelity < 1% per year on standing platforms [31]), which is overfished but rebuilding, *Rhomboplites aurorubens, Brotula barbata*, three members of the genus *Centropristis, Haemulon plumierii*, and *Paranthias furcifer*.

Of this group only the red snapper, and to some degree *R. aurorubens*, have been reasonably well studied, but almost all of the work on adults of these two species that has focused on recruits to platforms or has been done on small, low-relief, artificial reefs in the northeastern GOM. Similar to the results reported by Linberg and coworkers for gag grouper [29, 30], studies of red snapper indicate that the value of artificial reefs as habitat is affected both by size and spatial arrangement of reef modules, and that the net effect on stock production of reefs is negative when fishing mortality is considered [32, 33]. In addition, diet studies in the northeastern GOM indicate that adult red snapper rely very little on prey derived explicitly from reef habitats, whether collected on artificial [source 2 in **Table 3**] or natural reefs [34]. Despite these results, we caution against drawing inference about the role of platforms as habitat for red snapper given that much of the work has been done on relatively small, low-relief, reef modules. More recently, Simonsen and colleagues [15] compared diets of red snapper on standing platforms, toppled platforms, and on the natural reefs in the western GOM. They found that diets on both of the platform types were less diverse and provided less nutrition than diets on natural reefs. In addition, members of my laboratory and I [35] recently found that natural reefs offer a wider diversity of prey items, and reef-dependent prey species were found only in the diets at the natural reefs. Red Snapper at the natural reefs were on and above the reef, while feeding at the artificial reefs was predominantly on the surrounding seafloor and provided less caloric intake. Natural reefs found in the northwestern GOM appear to offer better habitat quality with regard to prey resources for red snapper compared to artificial reefs (standing and toppled platforms), a difference that should be taken into account as part of management decisions.

The other lutjanids in this group are much less abundant than red and vermilion snapper. Vermilion snapper, *Rhomboplites aurorubens* was overfished as recently as 2010, but the stock has recovered sufficiently to be considered rebuilt [36]. Of the remaining lutjanids reported in **ATable 1**, the dog snapper *L. jocu* is likely to be the most strongly reef associated, as is the creole fish, *Paranthias furcifer*. We believe the remainder of the species in this group to be reef-associated rather than dependent. This group as a whole is most similar to red snapper among the Level 3 species.

Level 3 Results: Sufficient life history information was available for five species to conduct a Level 3 analysis: red snapper, blue runner, sheepshead, Atlantic spadefish, and bluefish. **Tables 1** and **2** provide data and their sources used in the calculations.

Calculated estimates of biomass production per year per platform using Rickers's method ranged from 306 kg platform^{-1} by sheepshead to 1627 kg platform^{-1} for blue runner (**Table 3**). Estimates using an empirical approach [3] were consistent in pattern, but averaged less than half of the values derived from Ricker's [4] method (**Table 3**). This difference was largely because we estimated specific growth rates (G) for each species over a period that was shorter than their reported life span. For red snapper, blue runner, and sheepshead, the age classes we used were those in which high growth occurred during that period of their life cycle. Biomass production in the Ricker's method is sensitive to the ratio of G/Z i.e., specific growth rate and natural mortality rate, and the ratio is close to, or less than, one for all, but red snapper (**Table 3**). When the ratio is less than one, there is a net loss in population biomass. It is also important to note that the often highly productive pre-recruit period was not included in our calculations, which could have a large effect on the overall production estimates.

The results from the Powers [2] model showed that annual production for each species was dependent upon the index of reef exclusivity (**Table 3**). Species for which platforms provide only a small fraction of prey resources (e.g., red snapper, blue fish, and blue runner) are less dependent upon reefs compared to species such as Atlantic spadefish and sheepshead that depend heavily upon the fouling community on platform legs for food. Low annual production values for red snapper, bluefish, and blue runner also imply that platforms are more likely attracting individuals from surrounding natural habitats rather than producing new population biomass. These finding are consistent with recent diet studies of red snapper, and point out the sensitivity of production estimates to subtle changes in the G/Z ratio that are not considered in the less complex methods [2, 3].

4. FINAL THOUGHTS AND RED SNAPPER

The most controversial fishery in U.S. waters of the GOM is for northern red snapper *Lutjanus campechanus*, which collapsed in the late 1980s when stock biomass became too low to be fished commercially in almost half of the stock's former range (east of the Mississippi River). Red snapper management began in earnest in 1989, and the stock now is showing strong signs of recovery. More information about the history of red snapper management is available elsewhere [37–39] but the conflict among competing stakeholders has made stock recovery and sustainable fishing especially difficult to achieve.

Like the examples described by Hilborn [40], a 'faith-based fisheries' argument has been used to defer effective management of red snapper, and consequently has greatly strained the relationships among science, management, and stakeholders in the GOM. It has been argued that mass deployment of artificial reefs has substantially increased productivity of the red snapper stock. The premise is that artificial reefs have transformed less desirable fish biomass into red snapper biomass at locations on the shelf where the latter was not previously abundant [41]. The specifics of this argument were elucidated in a management perspective [42] which postulated that oil and gas platforms that began appearing in the western GOM in the late 1940s function secondarily as large artificial reefs, as well as a myriad of other artificial reef structures in the northcentral and eastern GOM since the 1970s, has enhanced biomass production of red snapper. True, results of the last benchmark stock assessment for red snapper [7] indicate that recruitment and stock productivity may have increased since the late 1980s. However, several other possible causes for this putative increase have been identified beyond artificial reef deployment [39]. The perspective [42] speaks to none of the other possible (more likely?) causes for change and further claims that red snapper were not present in the northwestern GOM until oil and gas platforms began being deployed offshore in the early 1940s. They disregard considerable information showing that a well-established red snapper fishery in the northwestern GOM began as early as 1892 [43].

The arrows in **Figure 3** indicate when artificial reefs began to be deployed in large numbers relative to the estimated spawning stock biomass of red snapper. These deployments took the form of oil and gas platforms in the western GOM and all manner of materials in the east. In both cases, there is no obvious indication that artificial habitats have increased spawning stock biomass because overfishing was occurring until only recently, and changed in response to strong year classes. The artificial reef argument put forth in the perspective is simply not supported by the available information.

We recognize that the stock assessment process for red snapper and other reef-associated species is controversial and sometimes difficult to understand, so in this chapter we have used simple models using a few parameters, to make our case. In **Table 3** our results indicate that oil and gas platforms produce new red snapper biomass. However, even if only small increase in Z in the Ricker based method of estimation [4] is added (i.e., as fishing mortality ($Z=F + M$)), the production outcome for red snapper turns negative. We provided tables that include procedural details for readers that wish to see how we estimated production of red snapper on oil and gas platforms. It is entirely plausible that oil and gas platforms make some reef-associated species, including red snapper, more vulnerable to fishing. Our analysis does not address spatial variation in demographic rates due to the intrinsic habitat quality of essential fish habitat or artificial reefs at scales smaller than regional e.g., between Alabama, Louisiana, and Texas.

This is not a blanket recrimination of artificial reefs and artificial reef programs. There are clear examples in the literature where artificial reefs benefit fishes and ecosystems in which they have been employed. Good examples are where artificial reefs are used to mitigate for loss or injury to natural reefs, or used to reduce destructive diving and fishing pressure on natural reefs [44–47], to name a few.

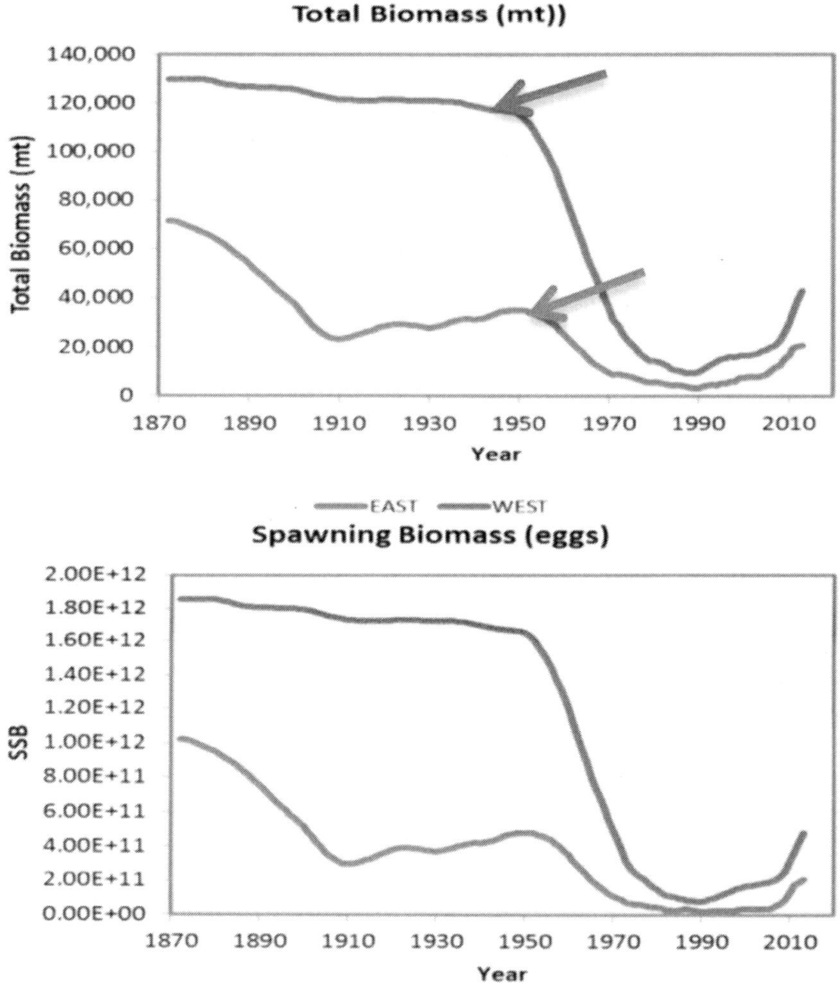

FIGURE 3. Model estimated spawning stock biomass in the eastern (blue line) and western (red line) US GOM of Mexico. Spawning stock biomass is the estimated weight of all mature females in each region of the GOM. The arrows indicate the year when oil and platforms (red arrow, 1942) and artificial reefs (blue arrow, 1950) began to be deployed in each area (The estimate is taken from the 2015 update of and is the most recent benchmark assessment for red snapper). The recent upturn in biomass is attributable to two relatively strong years classes produced in 2004 and 2006; this is the first time this has occurred over the ~30 year history of red snapper management.

Still, the debate about whether artificial habitats attract red snapper from nearby natural habitats or actually enhance production of new biomass (i.e., the attraction vs. production debate) has been called meaningless and unresolvable [40]. This subject often is debated in broad form for all reef-associated species and, as such, may be un-resolvable in the broader context—this issue has mostly been tried in the court of public opinion. However, a more quantitative approach is tractable for a well studies species like red snapper, although difficult due to the scale and complexity of needed studies [48].

Besides red snapper, many reef-associated species that are found on oil and gas platforms and highlighted in red in **ATable 1** are overfished. The role that these structures play in the population dynamics of these species is unknown. Careful consideration and enumeration of potential positive and negative impacts of man-made habitats on dynamic coastal geographies on continental shelves should be made before such habitats are constructed.

APPENDICES

APPENDIX TABLE 1. Fish taxa reported to occur on Gulf of Mexico oil and gas platforms: SF=Site Fidelity (High, Moderate, Low); DF = Directed Fishery (Yes or No, includes recreational fishing); Diet (Reef, Benthic, Pelagic); PC = Population Control (Habitat limited, Recruitment limited; Behavior (Reef, Demersal, Pelagic, Highly Migratory); Dep = Reef or Habitat Dependent (Yes, No, Reef Associated, Sargassum, Sea Grass, Hard Bottom). Red highlight = overfished; Yellow highlight = Caribbean expatriate.

Taxa	SF	DF	Diet	PC	Behavior	Dep.
Abudefduf saxatilis	H	N	R	R?	R, D	Y
Abudefduf taurus	H	N	R	R?	R, D	Y
Acanthocybium solandri	L	Y	P	R	P, HM	N
Acanthurus chirurgus	H	N	R	R	R, D	Y
Acanthurus coeruleus	H	N	R	R?	R, D	Y
Achirus	L	N	B	R	D	N

lineatus						
Albula vulpes	L	Y	B	R	D	N
Aluterus schoepfii	H	N	R, B	R?	R, D	Y
Aluterus scriptus	H	N	R, B	R?	R, D	Y
Amblycirrhitus pinos	H	N	R, B	R?	R, D	Y
Anchoa cubana	L	N	P	R	P	N
Anchoa hepsetus	L	N	P	R	P	N
Anchoa mitchilli	L	N	P	R	P	N
Anchoa nasuta	L	N	P	R	P	N
Anchoviella perfasciata	L	N	P	R	P	N
Ancylopsetta dilecta	L	N	B	R	D	N
Ancylopsetta ommata	L	Y	B	R	D	N
Antennarius ocellatus	H	N	R, B	H	R, D	Y
Apogon maculatus	H	N	R, B	R?	R, D	Y
Apogon pseudomaculatus	H	N	R, B	R?	R,	Y
Archosargus probatocephalus	M	Y	R, B	R	R, D	RA
Ariomma regulus	L	N	B	R	P	S
Arius felis	L	N	B	R	D	N
Bagre marinus	L	Y	B, P	R	D	N

Bairdiella chrysoura	L	N	B, P	R	D	N
Balistes capriscus	H	Y	R, B	H	R, D	Y
Bodianus rufus	H	Y	R, P	R?	R, P	Y
Bollmannia communis	L	N	B	R	D	N
Bregmaceros cantori	L	N	B	R	D	N
Brevoortia patronus	L	Y	P	R	P	N
Brotula barbata	L	N	B	R	D	RA (juv)
Callionymus bairdi	M ?	N	B?	R?	D	RA, SG
Cantherhines pullus	H	N	B	H?	R, D (adults, juv=P)	Y (adults)
Canthidermis sufflamen	L	N	P	R	D, P	N, SG, RA
Canthigaster rostrata	H ?	N	B	H?	R, D	Y?, RA & SG
Caranx bartholomaei	L	N	P	R	HM	RA
Caranx crysos	L	N	P	R	P	RA
Caranx hippos	L	N	P	R	P	RA
Caranx latus	L	N	P	R	P	RA
Caranx lugubris	L	N	P	R	P	RA
Caranx ruber	L	N	P	R	P	RA
Carcharhinus plumbeus	L	Y	P	R	P, HM	N
Caulolatilus intermedius	L	Y	P, B	R	D	N

Centropristis melana	M ?	Y	P, B	R	D	RA
Centropristis ocyurus	M ?	N	B	R	R, D	HB
Centropristis philadelphica	M ?	N	B	R	P, D	HB
Cephalopholis cruentatus	H	N	P	H?	R, D	Y
Chaetodipteru s faber	H	N	B, P	R	P, P	Y?
Chaetodon ocellatus	H	N	R	H?	R, D	Y
Chaetodon sedentarius	H	N	R	H?	R, D	Y
Cheilopogon cyanopterus	L	N	P	R	P	N
Cheilopogon furcatus	L	N	P	R	P	N
Chilomycterus schoepfii	M	N	B	R	D	RA, SG
Chlorophthal mus agassizi	L	N	B	R	D	N
Chloroscombr us chrysurus	L	N	P	R	P	N
Chromis enchrysura	H	N	P	H	D	Y
Chromis mutilineata	H	N	R	H	D	Y
Chromis scotti	H	N	R	H	D	Y
Citharichthys spilopterus	L	Y	B	R	D	N
Clepticus parrae	H	N	P	H	D, P	Y
Coryphaena equiselis	L	Y	P	R	P, HM	N

Coryphaena hippurus	L	Y	P	R	P, HM	N
Coryphopterus punctipectophorus	M	N	?	R	D	RA
Cubiceps pauciradiatus	L	N	P	R	P	N
Cyclopsetta chittendeni	L	Y	B	R	D	N
Cyclopsetta fimbriata	L	Y	?	R	D	N
Cyclothone braueri	L	N	P	R	P	N
Cynoscion arenarius	L	Y	P	R	P	N
Cynoscion nebulosus	L	Y	P	R	P	N, RA
Cynoscion nothus	L	Y	P	R	P	N, RA
Decapterus punctatus	L	Y	P	R	P, D	N
Decodon puellaris	H ?	N	R?	H	D	Y, HB
Dermatolepis inermis	H	Y?*	R, P	H?	D	RA, HB
Diplectrum bivittatum	L	N	B	R	D	N
Diplectrum formosum	L	N	P	R	D	SG, HB
Diplogrammus pauciradiatus	H	N	B	R	D	SG
Diplophos taenia	L	N	P	R	P	N
Diplodus holbrooki	L	N	B	R	D	N, SG

Dormitator maculatus	L	N	P (omnivore)	R	D	N
Echeneis naucrates	L	N	P	R	P	N, RA
Echeneis neucratoides	L	N	P	R	P	N
Echiophis intertinctus	L	N	B	R	D	N
Elagatis bipinnulata	M	Y	P	R	P, HM	RA
Elops saurus	L	N	P	R	P	N
Engraulis eurystole	L	N	P	R	P	N
Engyophrys senta	L?	N?	?	R	D	N?
Epinephelus adscensionis	H	Y?*	R, P, B	H	D	Y, HB
Epinephelus itajara	M	Y	R, P, B	R	D	RA
Epinephelus morio	M	Y	R, B	R	D	RA, HB
Epinephelus nigritus	M	N	R, P, B	R	D	RA, HP
		(closed)				
Epinephelus niveatus	M	Y*	P, B	R	D	RA, HB
Equetus iwamotoi	M?	N	B	R?	D	RA?
Equetus lanceolatus	M	N	B	R?	D	RA?
Etropus crossotus	L	N	B	R	D	N
Etrumeus	L	N	P	R	P	N

teres						
Euthynnus alletteratus	L	N	P	R	P, HM	N, RA
Foetorepus agassizi	L	N	B	R	D	N
Ginglymostoma cirratum	M?	N	B	R	D	N, SG
					(sometimes found on coral reefs)	
Gobiesox strumosus	M?	N	B	R	D	N, SG, HB
						(common on oyster reefs)
Gobionellus oceanicus	L	N	B?	R	D	N
Gymnothorax nigromarginatus	H	N	P, B	R?	D	
Haemulon aurolineatum	H	N	P, B	R	D	Y?, SG, HB
Haemulon plumierii	M	Y	P, B	R	D	
Halichoeres bivittatus	H	N	P, B	H?	D	Y, HB
Halieutichthys aculeatus	L	N	B?	R	D	N
Harengula jaguana	M	N	P, B	R	P	RA
Holacanthus bermudensis	H	N	B	R?	D	Y
Holocanthus ciliaris	H	N	B	R?	D	Y
Holocanthus tricolor	H	N	B	R?	D	Y, RA

Holocentrus ascensionis	H	N	B	R?	D	Y
Hoplunnis macrura	L ?	N	B?	R?	D	N?, RA
Hyperoglyphe perciformis	L	N	P	R	P	N
Hypleurochilus geminatus	H	N	B?	H (nest-builder)	D	RA, HB
Hypleurochilus multifilis	H	N	B?	H (nest-builder)	D	RA, HB
Hypsoblennius hentz	H	N	B?	H (nest-builder)	D	RA, HB (common on oyster reefs)
Hypsoblennius invemar	H	N	B	H (nest-builder)	D	RA, HB
Hypsoblennius ionthas	H	N	B?	H (nest-builder)	D	RA, HB
Ioglossus calliurus	L	N	P	R	D	N
Katsuwonus pelamis	L	Y	P	R	P (associated with	N?

					objects drifting at surface)	
Kyphosus incisor	M ?	N	P?	R	P	Y?, HB. S
		(plants, including Sargassum)				
Kyphosus sectatrix	M ?	N	P, B	R	P	Y?, HB, S
		(plants, including Sargassum, some benthic crustaceans)				
Lachnolaimus maximus	M ?	Y	B	R	D	Y?, HB
Lactophrys quadricornis	L	N	B	R	D	SG
Lagodon rhomboides	M	N	B	R	D	RA, SG, HB
Larimus fasciatus	L	N	B	R	D	N
Leiostomus xanthurus	L	Y	B	R	D	N
Lepophidium profundorum	L	N	B	R	D	N
Lepophidium staurophor	L	N	?	R	D	N
					(deep water)	
Lestrolepis intermedia	L	N	?	R	P	N
					(bathypelagi c)	
Lobotes surinamensis	M	Y	P, B	R	P	RA
					(often found associated with flotsam)	

Lutjanus apodus	M	Y	P, B	R	D	RA
Lutjanus campechanus	M	Y	P, B	R	D, P	RA, HB
Lutjanus griseus	M	Y	P	R	P	RA
Lutjanus jocu	M	Y	R, P, B	R	P	RA, HB
Lutjanus synagris	M	Y	R, P, B	R	D	RA, HB
Lutjanus vivanus	M	Y	R, P, B	R	D	RA, HB
					(common on shelf-edge banks)	
Magnisudis atlantica	L	N	P	R	P	N
					(bathypelagic)	
Makaira nigricans	L	Y	P	R	P, HM	N
Megalops atlanticus	L	Y	P	R	P	N, RA
Membras martinica	L	N	P	R	P	N
Microdesmus lanceolatus	L	N	B?	R	D	N?
Microdesmus longipinnis	L	N	B?	R	D	N
Micropogonias undulatus	L	Y	B	R	D	N
Monolene sessilicauda	L	N	B	R	D	N
					(bathydemersal)	
Mugil cephalus	L	Y	B	R	P	N

			(plants)			
Mugil curema	L	Y	B	R	P	N
Mullus auratus	L	N	B	R	D	N
Mycteroperca microlepis	M	Y	P	R	P, D	RA, SG, HB
					adults offshore on rocky bottoms)	
Mycteroperca phenax	M	Y*	R, P	R	D	RA, HB
				(high-relief rocky bottoms, often found on Oculina reefs)		
Mycteroperca rubra	M	Y*	R, P, B	R	D	RA, HB
						(rocky and sandy bottoms)
Mycteroperca venenosa	M	Y*	R, P	R	P, D	RA
					(rocky and coral reefs, shelf-edge banks in GOM)	
Myrophis punctatus	L	N	B?	R	D	RA, SG
Neoconger mucronatus	M	N	B	R	D	RA (offshore banks)
Ocyurus chrysurus	M	Y	R, P, B	R	P	RA (mostly coral reefs)
Ogcocephalus declivirostris	L	N	B?	R	D	N, HB
Ogcocephalus	L	N	B?	R	D	N, HB

radiatus						
Oligoplites saurus	L	N	P, B	R	P	RA
Ophichthus gomesii	L	N	B	R	D	N
					(common on shrimp grounds)	
Ophidion nocomis	L	N	B?	R	D	N
					(uncommon, shallow sandy bays)	
Ophidion robinsi	L ?	N	B?	R?	D	N
(rare)						
Ophidion selenops	M ?	N	B?	R?	D	RA
(uncommon)						
Ophioblennius atlanticus	H	N	B	H	D	RA
			(plants)			(rocky reefs and corals)
Opisthognathus aurifrons	H	N	R, P	R?	D	Y?
Opisthognathus lonchurus	H	N	R?	R?	D	Y?
Opsanus beta	M	N	B	H?	D	RA, SG
						(common on oyster reefs)
Opsanus pardus	M	N	B	H?	D	RA, HB
Orthopristis	L	N	P, B	R	D	N

chrysoptera						
Parablennius marmoreus	H	N	B	H	D	Y
			(mostly algae)			
Paralichthys albigutta	L	Y	P, B	R	D	N, HB
Paralichthys lethostigma	L	Y	P, B	R	D	N
Paranthias furcifer	H	N	P	R	P	RA, HB
						(coral reefs, hard bottoms)
Pareques umbrosus	L	N	B	R	D	N
Parexocoetus brachypterus	L	N	P	R	P	N
Peprilus alepidotus	L	N	P?	R	P	N
					(bathypelagic)	
Peprilus burti	L	Y	P	R	P, D	N
					(benthopelagic)	
Pogonias chromis	L	Y	B	R	D	N
Polydactylus octonemus	L	N	B	R	D	N
Pomacanthus paru	H	N	R, B	R?	D	Y
Pomatomus saltatrix	L	Y	P	R	P	N

Pontinus longispinis	L	N	B	R	D	N
Priacanthus arenatus	H	N	R, P, B	R?	D	Y
Prionotus roseus	L	N	B	R	D	N
Pristipomoides aquilonaris	L	N	P	R	D	N
			(small fishes)			
Prognathodes aya	H	N	R?	H	D	HB
						(offshore banks)
Pseudupeneus maculatus	M?	N	B	R?	D	RA, HB
Rachycentron canadum	M	Y	P, B	R	P	RA
					(found associated with structure of all types)	
Raja eglanteria	L	N	B	R	D	N
Remora remora	H?	N	P	R	P	RA?
						(usually attached to sharks, turtles)
Rhomboplites aurorubens	M	Y	P, B	R	P	RA, HB
						(HB on shelf-

						edge)
Robia legula	L	N	P	R	P (bathypelagic)	N
Ruvettus pretiosus	L	N	P, B	R	P	N
					(bathypelagic)	
Rypticus maculatus	M?	N	B	R	D	RA?
Sardinella aurita	L	Y	P	R	P	RA
Saurida brasiliensis	L	N	P	R	D	N
			(nekton)			
Saurida normani	M	N	P	R	D	RA
Saurida suspicio		L		N	P	R
Scartella cristata	H	N	R, B	H	D	Y
Schedophilus medusophagus (GOM record is doubtful)	L	N	P	R	P	N
Sciaenops ocellatus	L	Y	P, B	R	D, P	N
Scomber japonicus	L	Y	P	R	P, HM	N
Scomberomorus cavalla	L	Y	P	R	P, HM	RA
Scomberomorus maculatusL	Y	P	R	P, HM		RA

Scorpaena brasiliensis	L	N	B	R	D	HB
Selar crumenophthalmus	L	N	P	R	P	RA
Selene setapinnis	L	N	P, B	R	P	N
					(benthopelagic)	
Selene vomer	L	N	P, B	R	P	N
Seriola dumerilii	M	Y	P	R	P, HM	RA
Seriola fasciata	M	Y	P	R	P, HM	RA
Seriola rivoliana	M	Y	P	R	P, HM	RA
Seriola zonata	M	Y	P	R	P, HM	N
Serranus subligarius	L	N	B	R	D	N
Sphoeroides parvus	L	N	B	R	D	N
Sphoeroides spengleri	L	N	B	R	D	RA
						(SG, reef flats)
Sphyraena barracuda	M	Y	P	R	P	RA, SG
Sphyraena borealis	M	N	P	R	P	RA
Sphyraena guachancho	L	N	P	R	P	N
Stegastes partitus	H	N	R, B	R?	D	Y
Stegastes	H	N	R, B	R?	D	Y

variabilis						
Stellifer lanceolatus	L	N	B	R	D	N
Stenotomus caprinus	L	N	B	R	D	N
Stephanolepis hispidus	H ?	N	B	R?	P, D	RA
Syacium gunteri	M	N	B, P	R	D	RA
Syacium papillosum	M	N	B	R	D	RA
Symphurus civitatium	L	N	B?	R	D	N
Syngnathus louisianae	H	N	P	R	D	RA
Synodus foetens	L	N	B	R	D	HB
Synodus poeyi	L	N	B	R	D	RA, HB
Synodus synodus	M	N	P	R	D	RA, HB
			(nekton)			
Tetragonurus atlanticus	L	N	P	R	P	N
			(jellyfis h)			
Thalassoma bifasciatum	H	N	R, B	R	D	RA, SG
Thunnus albacares	L	Y	P	R	P, HM	N, RA
Thunnus atlanticus	L	Y	P	R	P, HM	N, RA
Thunnus thynnus	L	Y	P	R	P, HM	N
Trachinoceph	M	N	B, P	R	D	RA, HB

alus myops						
Trachinotus carolinus	L	Y	B, P	R	D	N
					(benthopela gic)	
Trachinotus falcatus	M	Y	B	R	D	RA
Trachinotus goodei	M	Y	B, P	R	P	RA
					(benthopela gic)	
Trachurus lathami	M	Y	P	R	P	RA
					(minor, bait)	
Trichiurus lepturus	L	Y	P, B	R	P, D	N
Trichopsetta ventralis	L	Y	B	R	D	N
Trinectes maculatus	L	N	B	R	D	N
Upeneus parvus	L	N	P, B	R	D	N
		(zoobenthos, small fishes)				
Urophycis floridana	L	N	B, P	R	D	N
Vinciquerria nimbaria	L	N	P	R	P	N
					(bathypelagi c)	
Xyrichtys novacula	H	N	B	H?	D	Y?

REFERENCES

1. Bohnsack, J.A. (1989) Are high densities of fishes at artificial reefs the result of habitat limitation or behavioral preference? Bull. Mar. Sci. 44:631–645.
2. Powers, S.P., Grabowski, J.H., Peterson, C.H., Lindberg, W.J. (2003) Estimating enhancement of fish production by offshore artificial reefs: uncertainty exhibited by divergent scenarios. Mar. Ecol. Prog. Ser. 264: 265–277.
3. Edgar, G.J., Shaw, C. (1995) The production and trophic ecology of shallow-water fish assemblages in southern Australia III. General relationships between sediments, seagrasses, invertebrates and fishes. J. Exp. Mar. Biol. Ecol. 194: 107–131.
4. Ricker, W.E. (1975) Computation and interpretation of biological statistics of fish populations. Bull. Fish. Res. Bd. Can. 191: 1–382.
5. Wilson, C.A., Nieland, D.A. (2001) Age and growth of red snapper, *Lutjanus campechanus*, from the northern Gulf of Mexico off Louisiana. Fish. Bull. U.S. 99: 653–664.
6. Fischer, A., Baker, M.S., Jr., Wilson, C.A. (2004) Red snapper (*Lutjanus campechanus*) demographic structure in the northern Gulf of Mexico based on spatial patterns in growth rates and morphometrics. Fish. Bull., US 102: 593–603.
7. SEDAR 31 (2012) Gulf of Mexico Red Snapper Stock Benchmark Assessment Report, Southeast Data and Assessment Review. North Charleston SC., p. 1103 Available online at:http//www.sefsc.noaa.gov/sedar/Sedar_Workshops.jsp?WorkshopNum=31
8. Goodwin, J.M. IV, Johnson, A.G. (1986) Age, growth, and mortality of blue runner, *Caranx crysos* from the northern Gulf of Mexico. Northeast Gulf Sci, 8: 107–114.
9. Beckman, D.W., Stanley, A.L., Render, J.H., Wilson, C.A. (1991) Age and growth-rate estimation of sheepshead *Archosargus probatocephalus* in Louisiana waters using otoliths. Fish. Bull., US 89: 1–8.
10. Hayse, J.W. (1990). Feeding habits, age, growth, and reproduction of Atlantic spadefish *Chaetodipterus faber* (Pisces: Ephippidae) in South Carolina. Fish. Bull. US 88:67–83.
11. Barger, L.E. (1990). Age and growth of bluefish *Pomatomus saltatrix* from the northern Gulf of Mexico and the U.S. South Atlantic Coast. Fish. Bull. US 88: 805–809.
12. McCawley, J.R., Cowan, J.H. Jr., Shipp, R.L., (2006) Diel feeding periodicity and prey habitat preference of red snapper, *Lutjanis campechanus*, on Alabama artificial reefs. Gulf. Mex. Sci. 24: 14–27
13. McCawley, J.R., Cowan, J.R. Jr. (2007). Seasonal and size specific diet and prey demand of red snapper on Alabama artificial reefs: Implications for management. pp. 77–104 *In* Patterson, W.F., Cowan, J.H. Jr., Fitzhugh, G.R., Nieland, D.L. (eds.) Red Snapper Ecology and Fisheries in the US Gulf of Mexico. Am. Fish. Soc., Symposium 60, Bethesda, Maryland.
14. Randall, J.E., (1967). Food habits of reef fishes of the West Indies. Stud. Trop. Oceanogr. Miami 5: 665–847.
15. Simonsen, K.A., Cowan, J.H. Jr., Boswell, K.M., (2015). Habitat differences in the feeding ecology of red snapper (*Lutjanus campechanus*, Poey 1860): a comparison between artificial and natural reefs in the northern Gulf of Mexico. Environ. Biol. Fish. 98: 811–824.
16. Scharf, F.S., Manderson, J.P., Fabrizio, M.C., Pessutti, J.P., Rosendale, J.E., Chant R.J., Bejda, A.J., (2004). Seasonal and interannual patterns of distribution and diet of bluefish within a middle Atlantic bight estuary in relation to abiotic and biotic factors. Estuaries 27: 426–436.
17. Hanisko, D.S., Lyczkowski Shultz, J., (2003). Occurrence and pelagic habitat of reef fish larvae in the Gulf of Mexico. pp. 205–224, *In* Stanley, D., Scarborough-Bull, A.,

(eds). Fisheries, Reefs, and Offshore Development, Am. Fish. Soc. Symposium 36, Bethesda, Maryland.
18. Robins, C.R., Ray, G.C., (1986). A Field Guide to Atlantic Coast Fishes, North America. Houghton Mifflin, New York, USA. p. 354
19. Richards, W., (ed) (2006) Early stages of Atlantic fishes. An identification guide for the western central north Atlantic, Volumes I and II Taylor and Francis, Boca Raton, USA.
20. Böhlke J.E., Chaplin, C.C.G., (1968). Fishes of the Bahamas and adjacent waters Livingston Publication Co Wynnewood, Pennsylvania, USA. p. 771
21. Hoese, H.D., Moore, R., (1977). Fishes of the Gulf of Mexico, Texas, Louisiana, and adjacent waters. Texas A&M Press. p. 377
22. Randall, J.E., (1996). Caribbean Reef Fishes. 3rd edition, TFH Publications Inc., Neptune City, New Jersey, USA. p. 368
23. Carpenter, K.E., (2002). The Living Marine Resources of the Western Central Atlantic, Vol 1 to 3, FAO Species Identification Guide for Fishery Purposes, and American Society of Icthyologists and Herpetologists Special Publication Number 5. Food and Agriculture Organization of the United Nations. Rome, Italy.
24. Kells V, Carpenter K., (2011). A Field Guide to Coastal Fishes from Maine to Texas. The Johns Hopkins University Press, Baltimore, Maryland, USA. p. 448
25. SEDAR 4 (2004) Stock Assessment of the Deepwater Snapper-Grouper Complex in the South Atlantic. Southeast Data and Assessment Review SEDAR4-SAR1, Charleston, SC, p. 594
26. Kingsley, M.C.S. (ed.). 2004. The goliath grouper in southern Florida: assessment review and advisory report. Report prepared for the South Atlantic Fishery Management Council, the Gulf of Mexico Fishery Management Council, and the National Marine Fisheries Service. Data and Assessment Review. Vii + p. 17 (See http://www.gulfcouncil.org/downloads/goliathfinalSEDAR2004.pdf).
27. SEDAR 33, (2014). Gulf of Mexico Gag Stock Assessment Report. Southeast Data and Assessment Review, North Charlston, SC p. 609. Available online at: http://www.sefsc.noaa.gov/sedar/Sedar_Workshops.jsp?WorkshopNum=33
28. Porch, C.E., Eklund, A.M., Scott, G.P., (2006). A catch-free stock assessment model with application to goliath grouper (*Epinephelus itajara*) off southern Florida. Fish. Bull., US 104: 89–101.
29. Lindberg, W.J., Frazer, T.K., Portier, K.M., Vose, F., Loftin, J., Murie, D.J., Mason, DM, Nagy, B., Hart, M.K., (2006). Density-dependent habitat selection and performance by a large mobile reef fish. Ecol. Appl. 16: 731–746
30. Lindberg, W.J., Loftin, J., (1998). Effects of Artificial Reef Characteristics and Fishing Mortality on Gag (*Mycteroperca microlepis*) Productivity and Reef Fish Community Structure. Gainesville, FL, Florida Dept. of Environmental Protection. p. 47
31. Westmeyer, M.P., Wilson, C.A., Nieland, D.L., (2007). Fidelity of red snapper to petroleum platforms in the northern Gulf of Mexico. pp. 105–122 *In* Patterson, W.F., Cowan, J.H. Jr., Fitzhugh, G.R., Nieland, D.L. (eds.) Red Snapper Ecology and Fisheries in the US Gulf of Mexico. Am. Fish. Soc., Symposium 60, Bethesda, Maryland.
32. Strelcheck, A.J., Cowan, J.H. Jr., Shah. A., (2005). Influence of Reef Location on Artifical-Reef Fish Assemblages in the Northcentral Gulf of Mexico. Bull. Mar. Sci 77: 425–440.
33. Strelcheck, A.J., Cowan, J.H. Jr., Patterson, W.F. III. (2007) Site fidelity, movement, and growth of red snapper; implications for artificial reef management. pp. 147–162 *In* Patterson, W.F., Cowan, J.H. Jr., Fitzhugh, G.R., Nieland, D.L. (eds.) Red Snapper Ecology and Fisheries in the US Gulf of Mexico. Am. Fish. Soc., Symposium 60, Bethesda, Maryland.
34. Wells R.J.D., Cowan, J.H. Jr., Fry, B., (2008). Feeding ecology of red snapper, *Lutjanus campechanus*, in the northern Gulf of Mexico. Mar. Ecol. Prog. Ser. 361: 213–225
35. Schwartzkopf, B.D., Cowan, J.H. Jr., (in review). Nutritional condition of red snapper *Lutjanus campechanus* on natural and artificial reefs in the northwestern Gulf of Mexico. J Fish Biol.

36. National Oceanographic and Atmospheric Administration, National Marine Fisheries Service (2013). Framework action to set the annual catch limit and bag limit for vermillion snapper and yellow snapper, and modify the venting tool requirement. Gulf of Mexico Fishery Management Council Pursuant to National Oceanic and Atmospheric Administration Award No. NA10NMF4410011. p. 183
37. Hood P.B., Strelcheck, A.J., Steele, P., (2007). A history or red snapper management in the Gulf of Mexico. pp. 267–284 *In* Patterson, W.F., Cowan, J.H. Jr., Fitzhugh, G.R., Nieland, D.L. (eds.) Red Snapper Ecology and Fisheries in the US Gulf of Mexico. Am. Fish. Soc., Symposium 60, Bethesda, Maryland.
38. Goodyear, CP., (1995). Red snapper in U.S. waters of the Gulf of Mexico. National Marine Fisheries Service, Southeast Fisheries Science Center, Miami Laboratory, Miami MIA-95/96-05. p. 103
39. Porch, C.E., Turner, S.C., Schirripa, M.J., (2007). Reconstructing the commercial landings of red snapper in the Gulf of Mexico from 1872–1963. pp. 337–354 *In* Patterson, W.F., Cowan, J.H. Jr., Fitzhugh, G.R., Nieland, D.L. (eds.) Red Snapper Ecology and Fisheries in the US Gulf of Mexico. Am. Fish. Soc., Symposium 60, Bethesda, Maryland.
40. Hilborn, R., (2006). Faith-based fisheries. Fisheries 32(11): 554–555.
41. Shipp, R.L., (1999). The artificial reef debate: Are we asking the wrong questions? Gulf Mex. Sci. 17: 51–55.
42. Shipp, R.L., Bortone, S.A., (2009). A Prospective of the importance of artificial habitat on the management of red snapper in the Gulf of Mexico. Rev. Fish. Sci. 17: 41–47.
43. Carpenter, J.S., (1965). A review of the Gulf of Mexico Red Snapper Fishery. US Dept. Interior., US Fish. Wildl. Serv., Bur. Comm. Fish. Circular 208. p. 41
44. Bortone, S.A., (1998). Resolving the attraction-production dilemma in artificial reef research: some yeas and nays. Fisheries 23 (3): 6–10.
45. Claudet, J., Pelletier, D., (2004). Marine protected areas and artificial reefs: A review of the interactions between management and scientific studies. Aquat. Living Resour. 17: 129–138.
46. Page, H.M., Dugan, J.E., Schroeder, D.M., Nishimoto, M.M., Love, M.S., Hoesterey, J.C., (2007). Trophic links and condition of a temperate reef fish: comparisons among offshore oil platform and natural reef habitats. Mar. Ecol. Prog. Ser. 344:245–256
47. Pitcher, T.J., Seaman, W., (2000). Petrarch's Principle: how protected human-made reefs can help the reconstruction of fisheries and marine ecosystems. Fish and Fisheries 1:73–81.
48. Cowan, J.H. Jr., Grimes, C.B., Patterson, W.P. III, Walters, C.J., Jones, A.C., Lindberg, W.L., Sheehy, D., Pine, W.E. III, Powers, J.D., Campbell, M.D., Lindeman, K.C., Diamond, S.L., Hilborn, R., Gibson, H.T., Rose, K.A., (2011). Red snapper management in the Gulf of Mexico: science- or faith-based? Rev. Fish. Biol. Fisheries 21: 187–204.

CHAPTER 15

Anti-Fatigue Activity of a Mixture of Seahorse (Hippocampus Abdominalis) Hydrolysate and Red Ginseng

Nalae Kang[1], Seo-Young Kim[1], Sum Rho[2], Ju-Young Ko[1*†] and You-Jin Jeon[1*†]

ABSTRACT

Seahorse, a syngnathidae fish, is one of the important organisms used in Chinese traditional medicine. *Hippocampus abdominalis*, a seahorse species successfully cultured in Korea, was validated for use in food by the Ministry of Food and Drug Safety in February 2016; however. the validation was restricted to 50% of the entire composition. Therefore, to use *H. abdominalis* as a food ingredient, *H. abdominalis* has to be prepared as a mixture by adding other materials. In this study, the effect of *H. abdominalis* on muscles was investigated to scientifically verify its potential bioactivity. In addition, the anti-fatigue activity of a mixture comprising *H. abdominalis* and red ginseng (RG) was evaluated to commercially utilize *H. abdominalis* in food industry. *H. abdominalis* was hydrolyzed using Alcalase, a protease, and the effect of *H. abdominalis* hydrolysate (HH) on the muscles was assessed in C2C12 myoblasts by measuring cell proliferation and glycogen content. In addition, the mixtures comprising HH and RG were prepared at different percentages of RG to HH (20, 30, 40, 50, 60, 70, and 80% RG), and the anti-fatigue activity of these mixtures against oxidative stress was assessed in C2C12 myoblasts. In C2C12 myoblasts, H_2O_2-induced oxidative stress caused a decrease in viability and physical fatigue-related biomarkers such as glycogen and ATP contents. However, treatment with RG and HH mixtures increased cell viability and the content of fatigue-related biomarkers. In particular, the 80% RG mixture showed an optimum effect on cell viability and ATP synthesis activity. In this study, all results indicated that HH had anti-fatigue activity at concentrations approved for use in food by the law in Korea. Especially, an 80% RG to HH mixture can be used in food for ameliorating fatigue.

Keywords: *Hippocampus abdominalis* Anti-fatigue activity C2C12 myoblast Mixture of *Hippocampus abdominalis* hydrolysate and red ginseng

BACKGROUND

Seahorse is a well-known ingredient in traditional Chinese medicine and is used as an invigorant for the treatment of erectile dysfunction, impotence, wheezing, and nocturnal enuresis. Modern scientific research has proven the pharmaceutical effects of seahorse. *Hippocampus kuda* has various bioactivities such as anti-tumor, anti-aging, and anti-fatigue as well as Ca^{2+} channel blocking properties (Kumaravel et al. 2010). A peptide derived from *H. kuda* has been shown to be effective in chondrocytes and inflammatory arthritis (Kumaravel et al. 2012). In addition, seahorses have a putative free radical scavenging effect in controlling aging process (Kumaravel et al. 2012). However, the natural source of seahorse has dramatically reduced owing to overfishing, unsustainable trade, and habitat destruction (Qian et al. 2012). Therefore, seahorses became the first commercially valuable marine genus to be protected and included in Appendix II of the Convention on International Trade in Endangered Species (CITES) in 2004 (Segade et al. 2015).

Hippocampus abdominalis is one of the largest seahorse species growing up to 35 cm in length (Perera et al. 2016). It was validated for use as a food ingredient by the Ministry of Food and Drug Safety in February 2016. However, the validation was restricted to 50% of the entire composition. For use in food, we should try to prepare a mixture of *H. abdominalis* by adding other materials. In addition, biological activities of *H. abdominalis* have rarely been reported so far.

Fatigue is a common distressing condition accompanied by a feeling of extreme physical or mental tiredness that often results in diverse disorders such as anemia, thyroid dysfunction, premature aging, and depression. It could also have adverse effects on work efficiency, physical activities, life quality, and social relationships (Huang et al. 2011). Fatigue is caused by sleep deprivation, inadequate rest, low mood, stress, nutritional imbalance, insufficient exercise, as well as side effects of medications. Chronic fatigue is a persistent unexplainable fatigue lasting for more than 6 months, and it is considered a complex symptom of various neurological, psychiatric, and systemic diseases (Huang et al. 2014). Recently, many researchers have presented the results on the anti-fatigue activity of natural products (Yu et al. 2008; Zhang et al. 2006). Especially, red ginseng has been mainly focused on its anti-fatigue activity with the ability mitigating exercise-related muscle damage, maintaining homeostasis of the body and enhancing vital energy (Kim et al. 2013; Kim et al. 2016). On the other hand, anti-fatigue activity of seahorse has not been scientifically proven although seahorse is a well-known traditional Chinese medicine.

Oxidative stress is caused by an imbalance between reactive oxygen species (ROS) and antioxidant molecules. Excess accumulation of ROS causes oxidative damage by reacting with biomolecules including DNA, membrane lipids, cellular proteins, and diverse pathological states (Kang et al. 2013). Oxidative stress and ROS are the most important causes of exercise-induced disturbances (Fan et al. 2016). In particular, an oxidative imbalance in the skeletal muscle results in increased muscle fatigability. Thus, antioxidants can be used to alleviate fatigue by counteracting the oxidative stress (Nam et al. 2016).

In this study, the effect of *H. abdominalis* on muscles was investigated to scientifically verify its potential bioactivity. Also, the anti-fatigue activity of a mixture comprising *H. abdominalis* and red ginseng was investigated to evaluate the synergy effect and to utilize *H. abdominalis* in the food market. The anti-fatigue activity of *H. abdominalis* and a mixture was evaluated by measuring the levels of physical fatigue-related biomarkers such as serum glycogen and ATP contents.

METHODS

Materials

H. abdominalis was kindly donated by Corea Center of Ornamental Reef & Aquariums CCORA (Jeju, Korea) and lyophilized at −70 °C using a freeze dryer. The lyophilized *H. abdominalis* powder was stored at −80 °C until use. Red ginseng extract containing 30% saponin was purchased from ILHWA Co., LTD. (Gyeonggi, Korea) and lyophilized at −70 °C using a freeze dryer. The lyophilized red ginseng powder was stored at −80 °C until use. Alcalase, a commercial food-grade protease, was purchased from Novozyme Co. (Novozyme Nordisk, Bagsvaerd, Denmark). The other chemicals and reagents used were of analytical grade.

Preparation of *H. abdominalis* hydrolysate (HH)

The enzymatic hydrolysis of *H. abdominalis* was performed using Alcalase under optimal conditions (50 °C and pH 8). The dried *H. abdominalis* powder was homogenized in distilled water and hydrolyzed using the enzyme at an enzyme/substrate (E/S) ratio of 1:100 for enzymatic reactions. The optimal pH of the homogenates was adjusted before enzymatic hydrolysis. The mixture was incubated for 24 h at the optimal temperature for each homogenate, with stirring, and then boiled for 10 min at 100 °C to inactivate the enzyme. After filtration, all hydrolysates were stored at −70 °C for further experiments.

Preparation of the *H. abdominalis* mixture

The *H. abdominalis* mixtures were prepared by adding different concentrations of red ginseng (RG). The lyophilized *H. abdominalis* powder and RG powder were mixed as indicated in Table 1, and these seven mixtures were labeled as % of RG.

Table 1 Percentage of HH and RG to prepare the *H. abdominalis* mixtures

HH (%)	**RG (%)**	**The mixture (%)**
80	20	100
70	30	100
60	40	100
50	50	100
40	60	100
30	70	100
20	80	100

Cell culture

The C2C12 myoblasts obtained from American Type Culture Collection (ATCC, Manassas, VA, USA) were cultured in Dulbecco's Modified Eagle Medium (DMEM) supplemented with 10% heat-inactivated fetal bovine serum (FBS), streptomycin (100 mg/mL), and penicillin (100 u/mL) at 37 °C in a 5% CO_2 humidified incubator. To induce differentiation, 80% confluent cultures were switched to DMEM containing 2% horse serum (HS) for 6 days with medium changes every other day.

Cell viability

The cytotoxicity of the samples on C2C12 myoblasts was determined by colorimetric MTT assays. The cells were seeded at 5×10^4 cells per well into 48-well plates. After the induction of differentiation, the cells were treated with various concentrations of the sample and incubated for an additional 24 h at 37 °C. MTT stock solution (100 μL; 2 mg/mL in PBS) was then added to each well. After incubating for 4 h, the plate was centrifuged at 500 g for 10 min, and the supernatant was aspirated. The formazan crystals in each well were dissolved in dimethyl sulfoxide (DMSO). The amount of purple formazan was determined by measuring the absorbance at 540 nm.

Cell proliferation assay

The cell proliferation effect of the samples on C2C12 myoblasts was determined by using 5-bromo-2′-deoxyuridine (BrdU) assay (Millipore, Billerica, MA, USA). The cells were seeded at 1×10^4 cells per well into 48-well plates. After the induction of differentiation by switching media, the cells were treated with various concentrations of the sample and incubated for an additional 72 h at 37 °C. Then, the cell proliferation was determined by BrdU reagent following manufacture protocol. In briefly, 10 μL of BrdU reagent was added to each well and the cells were incubated for 2 h. After incubation, the cells were fixed using 100 μL fixing solution. Then, the cells were washed using wash buffer and 50 μL of anti-BrdU monoclonal was added to each well and the cells were incubated for 1 h at RT. The cells were washed using wash buffer and 50 μL of goat anti-mouse IgG was added to each well, and the cells were incubated for 30 min at RT. Also, 50 μL of TMB substrate was added to each well and then, 50 μL of stop solution was added to each well. Finally, the cell proliferation was calculated by comparison with the absorbance at 450 nm of standard solutions of BrdU in the non-treated cells.

Anti-fatigue activity in oxidative stress-induced C2C12 myoblasts

The anti-fatigue activity was determined by measuring cell proliferation as well as the glycogen, ATP contents in H_2O_2-treated C2C12 myoblasts. The cells were seeded into 48-well plates. Then, they were treated with various concentrations of the sample during the differentiation period. After differentiation, fatigue was induced by adding H_2O_2 to each well at a concentration of 100 μM; then, the cells were incubated for an additional 24 h at 37 °C.

Measurement of fatigue-related biochemical parameters

To investigate the effect of the samples on muscle growth, we determined several factors such as glycogen and ATP contents in C2C12 myoblasts. For analysis of the effects of the sample on glycogen accumulation, the glycogen content in the cells was measured via glycogen assay (Abcam, Cambridge, MA, USA). Glucoamylase hydrolyzes glycogen to glucose, which was then specifically oxidized to form an intermediate product that reacts with OxiRed probe to generate color. The color was detected by measuring the absorbance at 450 nm. To determine ATP contents, the cell lysates were deproteinized with 4 M perchloric acid (PCA) and 2 M KOH, and the supernatant was assessed using ATP assay kits (Abcam, Cambridge, MA, USA).

Statistical analysis

All measurements were made in triplicate, and all values were represented as means ± SE. The results were subjected to analysis of variance using Tukey's test to analyze the differences. $p < 0.05$ and $p < 0.01$ were considered significant.

RESULTS AND DISCUSSION

Cytotoxicity of HH

Cell viability was estimated using the MTT assay, which is a test of metabolic competence predicated upon the assessment of mitochondrial performance. It is a colorimetric assay, which is dependent on the conversion of yellow tetrazolium bromide to its purple formazan derivative by mitochondrial succinate dehydrogenase in the viable cells (Kang et al. 2012). The viabilities of C2C12 myoblasts treated with different concentrations of HH (50, 100, 150, and 200 µg/mL) were expressed to represent 100% viability (the viability of control cells; Fig. 1). In a preliminary experiment, HH concentrations up to 200 µg/mL showed no significant cytotoxicity for 24 h.

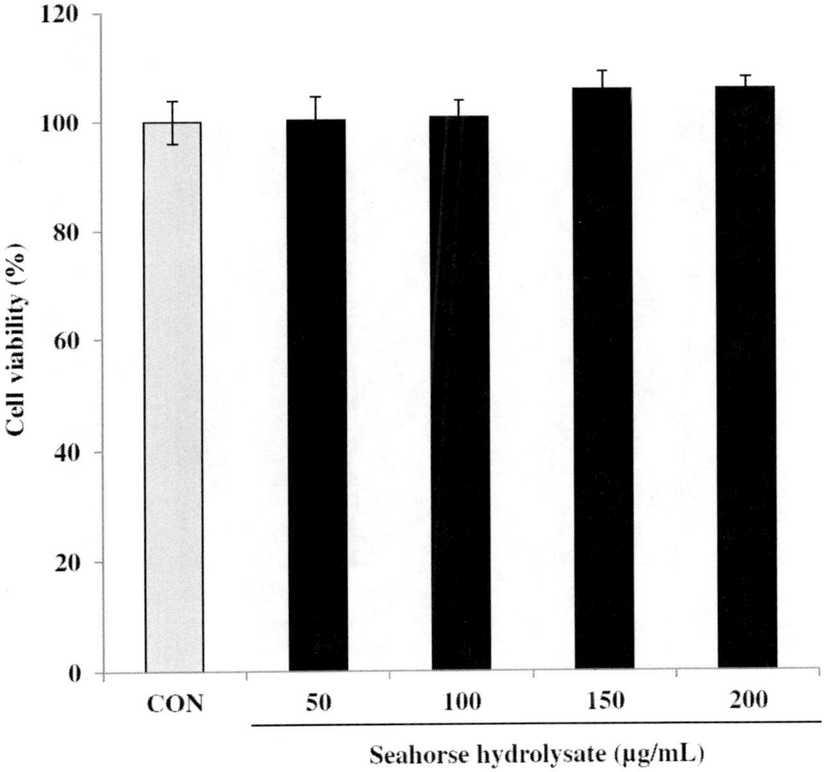

Figure. 1 Cytotoxicity of seahorse hydrolysate on C2C12 myoblasts. The cell was treated with various concentrations of seahorse hydrolysate (50, 100, 150, and 200 µg/mL) and incubated for 24 h. The cytotoxicity was assessed by MTT assay. The experiment was performed in triplicate. Each value indicates the mean ± standard error from three independent experiments

Effect of HH on cell proliferation

HH significantly enhanced cell proliferation in C2C12 myoblasts compared with the control cells (Fig. 2). In particular, HH treatment induced cell proliferation in a concentration-dependent manner in the range of 100–

200 μg/mL. The cell numbers increased approximately 1.8-fold by HH treatment at 200 μg/mL concentration (Fig. 2).

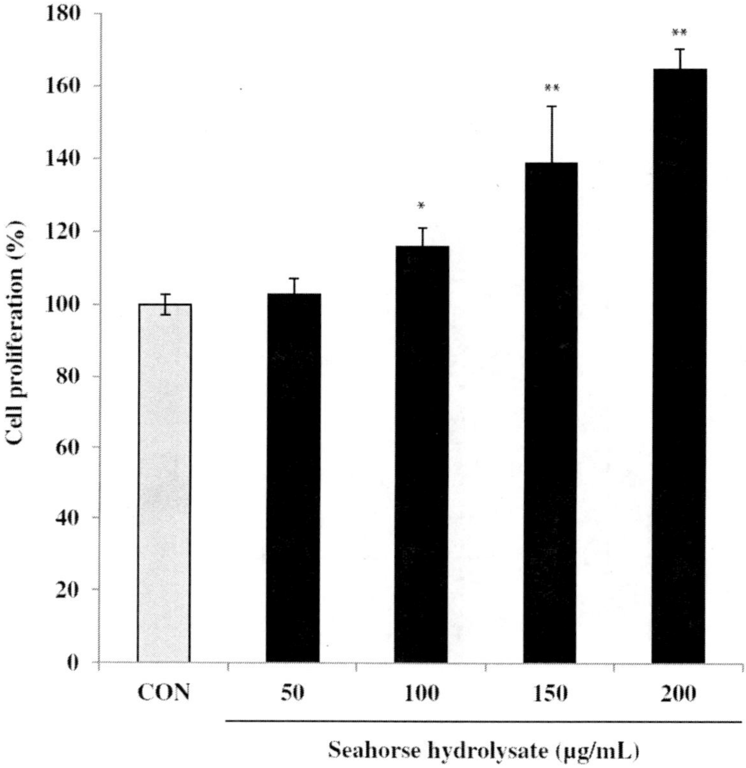

Figure. 2 Cell proliferation of seahorse hydrolysate on C2C12 myoblast. The cell was treated with various concentrations of seahorse hydrolysate (50, 100, 150, and 200 μg/mL) and incubated for 72 h. The cell proliferation was assessed by BrdU assay. The experiment was performed in triplicate. Each value indicates the mean ± standard error from three independent experiments. $^{*}p < 0.05$, $^{**}p < 0.01$

Effect of HH on glycogen contents

The skeletal muscles are the major site of glycogen storage in the body (Deshmukh et al. 2015). The glycogen content in C2C12 myoblasts was increased by HH treatment at concentrations of 50 and 100 μg/mL (Fig. 3). In C2C12 myoblasts, HH (100 μg/mL) increased glycogen content by 1.5-fold compared with that in the control cells. However, HH treatment at higher concentrations (150 and 200 μg/mL) decreased glycogen content.

Thus, it can be suggested that high concentrations of HH suppressed glycogen content.

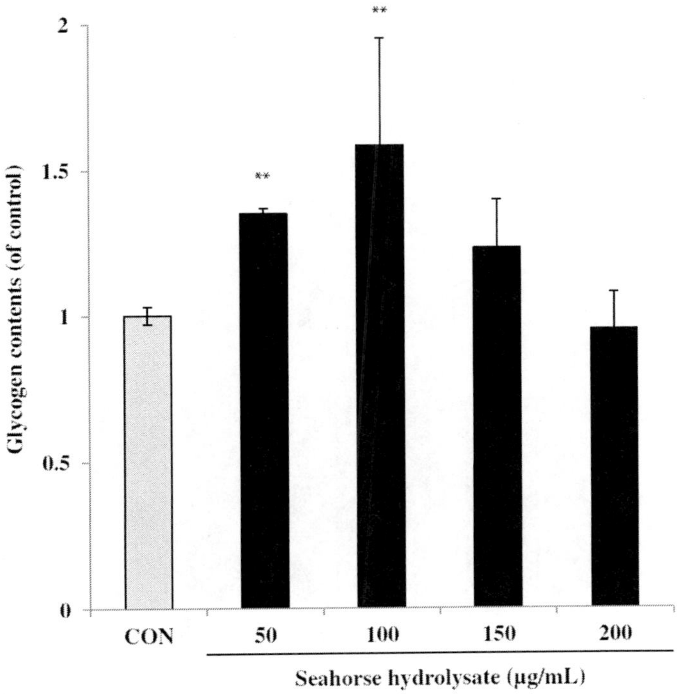

Figure. 3 Effect of seahorse hydrolysate on glycogen content in C2C12 myoblasts. The cells were incubated with various concentrations of seahorse hydrolysate (50, 100, 150, and 200 μg/mL) for 24 h. The glycogen contents were assessed. The experiment was performed in triplicate. Each value indicates the mean ± standard error from three independent experiments. $^{*}p < 0.05$, $^{**}p < 0.01$

Cytotoxicity of the mixture of HH and RG

Although *H. abdominalis* was validated for use in food by the Ministry of Food and Drug Safety in February 2016, the validation was restricted to 20% of the entire composition. To use *H. abdominalis* as a food ingredient, we should prepare an *H. abdominalis* mixture by adding other materials.

RG has been frequently used in traditional Asian medicine to treat many disorders, such as debility, aging, stress, diabetes, and insomnia (Tang et al. 2008). Especially, RG has been mainly focused on its anti-fatigue activity with the ability mitigating exercise-related muscle damage, maintaining homeostasis of the body, and enhancing vital energy (Kim et al. 2013; Kim et al. 2016). Thus, *H. abdominalis* mixtures were prepared by adding different concentrations of RG (20, 30, 40, 50, 60, 70, and 80% of RG) to investigate the synergy effect between *H. abdominalis* and RG on anti-fatigue activity.

Effect of these mixtures on cell viability was estimated using the MTT assay. The viabilities of C2C12 myoblasts treated with the mixtures at different concentrations (50, 100, 200, 250, and 500 µg/mL) were expressed to represent over 90% viability, which was similar to that of the control cells (Fig. 4). Thus, mixtures up to 500 µg/mL concentration did not show any significant cytotoxicity for 24 h.

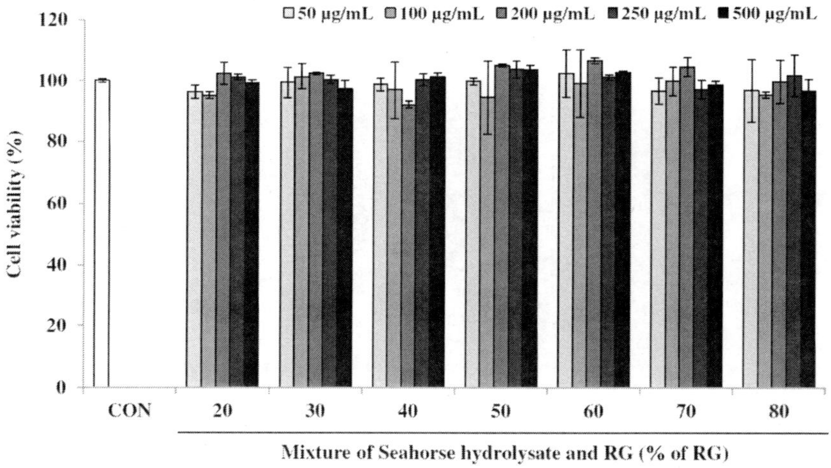

Figure. 4 Cytotoxicity of the six mixtures of seahorse hydrolysate and RG on C2C12 myoblasts. The cell was treated with various concentrations of the six mixtures (50, 100, 150, 200, 250, and 500 µg/mL) and incubated for 24 h. The cytotoxicity was assessed by MTT assay. The experiment was performed in triplicate. Each value indicates the mean ± standard error from three independent experiments

Effect of the mixture of HH and RG on cell proliferation

To assess the effect of HH (0% of RG) and RG mixture on muscle growth, cell proliferation was measured as shown in Fig. 5. HH and RG mixtures showed the significant effect on cell proliferation. Especially, at the low concentrations (200 and 250 µg/mL), HH significantly enhanced proliferation of C2C12 myoblasts compared with that of mixtures as well as that of the control cells. However, at the high concentrations (400 and

500 µg/mL), HH and RG mixtures did not show any significant effects on cell proliferation to each other.

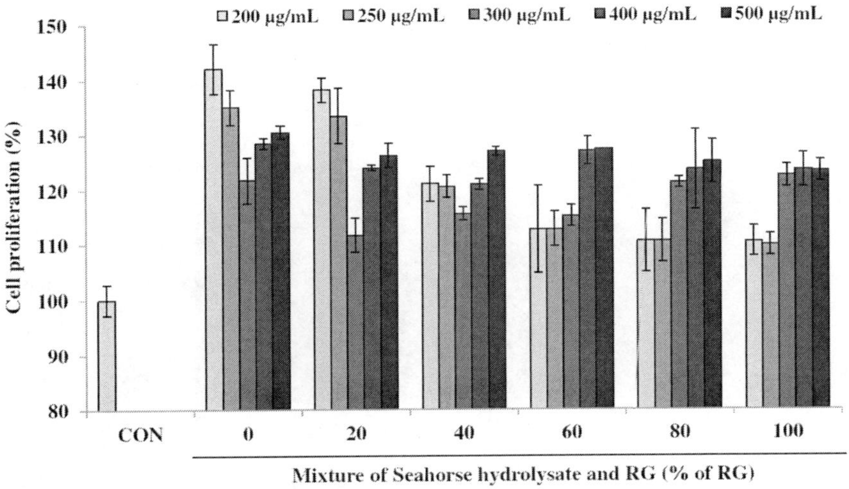

Figure. 5 Cell proliferation of the six mixtures of seahorse hydrolysate and RG on C2C12 myoblast. The cell was treated with various concentrations of the six mixtures (50, 100, 150, and 200 µg/mL) and incubated for 72 h. The cell proliferation was assessed by BrdU assay. Effect of different mixtures of seahorse hydrolysate and RG on C2C12 myoblast proliferation. The experiment was performed in triplicate. Each value indicates the mean ± standard error from three independent experiments

Anti-fatigue activity of the mixtures of HH and RG

Several biomarkers such as lactate dehydrogenase (LDH), glycogen, aspartate transaminase (AST), and alanine transaminase (ALT) have been used to investigate muscle injury during exhaustive exercise (Huang et al. 2015). Also, fatigue is related to mitochondrial dysfunction and diminished ATP levels (Singh and Singh 2014). The anti-fatigue activity of the mixtures of HH and RG was assessed in H_2O_2-treated C2C12 myoblasts by measuring cell proliferation as well as the glycogen and ATP contents. Severe and continuous exercise may elevate the formation of ROS, thereby increasing oxidative stress. A sustained elevated oxidative stress can hamper mitochondrial function resulting in low ATP synthesis and increased lactic acid in the muscles, consequently decreasing the physical efficiency. These observations suggest that improving the antioxidant status may enhance the overall physical performance by maintaining the pro-oxidant/antioxidant balance (Swamy et al. 2011). To induce oxidative stress in C2C12 myoblasts, the cells were incubated with H_2O_2 at a concentration of 100 μM. After H_2O_2 treatment, the viability of C2C12 myoblasts decreased to less than 60% compared to that in the control cells (Fig. 6). However, C2C12 myoblasts treated with HH and RG mixtures showed increased viability compared with that reported for the control cells. Notably, at a sample concentration of 500 μg/mL, cell viability increased steadily with increasing percentage of RG except at 100% RG, where only RG was present in the mixture.

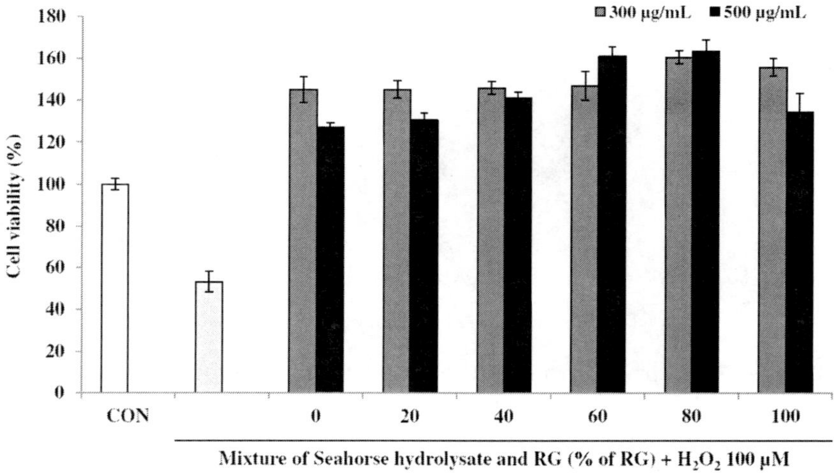

Figure. 6 Protective effect of the six mixtures of seahorse hydrolysate and RG against H_2O_2-treated C2C12 myoblasts. The treatment of H_2O_2 induced a decrease in cell viability. All mixtures showed protective effect on H_2O_2-induced oxidative stress in C2C12 myoblasts. The experiment was performed in triplicate. Each value indicates the mean ± standard error from three independent experiments

Glycogen contents

Energy expenditure during exercise leads to physical fatigue, which is mainly caused by energy consumption and deficiency. Catabolized fat and carbohydrates are considered the main sources of energy in the skeletal muscles during exercise, and glycogen is the predominant source of glycolysis for energy production. Therefore, glycogen storage directly affects exercise ability (Wu et al. 2013). The glycogen content of the H_2O_2-treated cells was lower than that of the control cells. However, in C2C12 myoblasts, the treatment with HH and RG mixtures increased the glycogen content to more than double as compared to the values reported for the control cells. In particular, RG60, RG80, and RG100 showed increased glycogen content at a sample concentration of 300 and 500 μg/mL (Fig. 7).

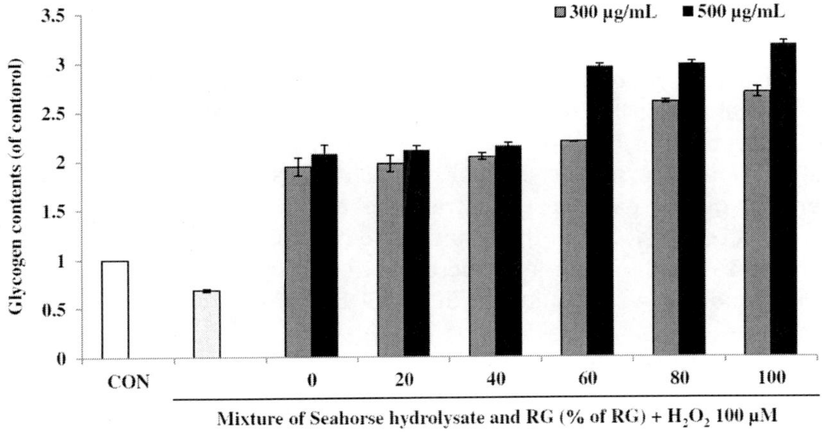

Figure. 7 Effect of the six mixtures of seahorse hydrolysate and RG on glycogen content in H_2O_2-treated C2C12 myoblasts. H_2O_2 treatment induced a decrease in glycogen contents. All mixtures showed protective effect on H_2O_2-induced oxidative stress in C2C12 myoblasts. The experiment was performed in triplicate. Each value indicates the mean ± standard error from three independent experiments

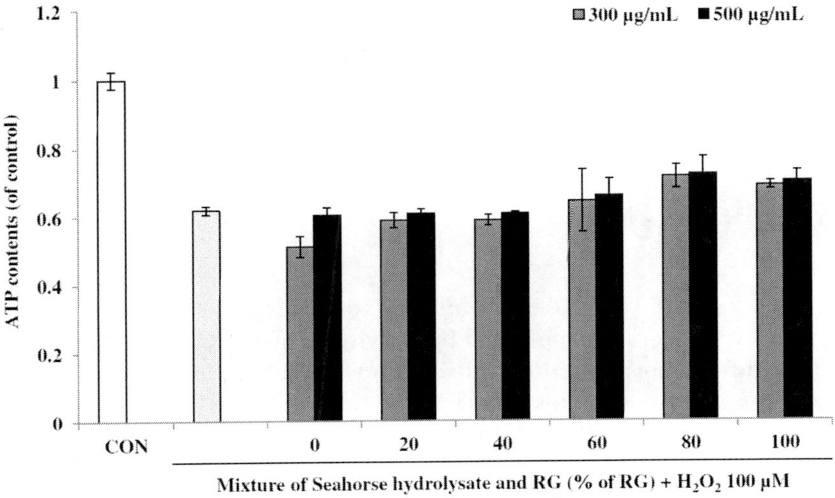

Figure. 8 Effect of the six mixtures of seahorse hydrolysate and RG on ATP synthesis in H_2O_2-treated C2C12 myoblasts. The treatment of H_2O_2 induced a decrease in ATP contents. All mixtures showed a protective effect against H_2O_2-induced oxidative stress in C2C12 myoblasts. The experiment was performed in triplicate. Each value indicates the mean ± standard error from three independent experiments

ATP contents

Muscular exercise causes rapid ATP consumption, and energy deficiency is a critical reason for physical fatigue. Therefore, compounds that promote ATP production could be candidates for alleviating physical fatigue. The skeletal muscle mainly catabolizes fat and carbohydrates as sources of energy during exercise (Nozawa et al. 2009). ATP content in the H_2O_2-treated cells was lower than that in the control cells (Fig. 8). Although HH and RG mixtures did not increase the ATP content, RG80 relatively increased the ATP content at 300 and 500 μg/mL concentrations of the mixture.

Exercise-induced oxidative stress can cause the increased muscle fatigability. Thus, antioxidants can decrease the oxidative stress and improve the physiological condition (You et al. 2011). Some reports showed that a loach peptide has not only antioxidant activities but also an anti-fatigue effect in mice (You et al. 2011). Actually, the peptide showing in vitro antioxidant activity possesses the in vivo anti-fatigue activity. The peptide acts as the scavenger for DPPH and hydroxyl radicals. Also, the anthocyanins of mulberry fruit have been assessed in vitro antioxidant activity and in vivo anti-fatigue activity (Jiang et al. 2013). These studies showed values of in vitro study to evaluate the potential anti-fatigue activity through in vivo study. In the present study, the mixtures of HH and RG acted as the antioxidant for hydrogen peroxide and showed the anti-fatigue activity on C2C12 myoblast. Furthermore, the mixtures have valuable needs to be investigated through in vivo animal study.

CONCLUSIONS

In this study, the effect of *H. abdominalis* on the muscles was investigated to scientifically verify its potential bioactivity. Also, the anti-fatigue activity of a mixture comprising HH and RG was evaluated to commercially utilize *H. abdominalis* in food industry. The treatment of HH to C2C12 myoblast induced the cell proliferation and glycogen contents. These results indicated that *H. abdominalis* had anti-fatigue activity on C2C12 myoblast. Moreover, the treatment of the mixture comprising HH and RG increased cell viability and the content of fatigue-related biomarkers such as glycogen and ATP contents. In particular, the 80% RG mixture showed an optimum effect on cell viability and ATP synthesis activity. These results indicated that HH had anti-fatigue activity at concentrations approved for use in food by the law in Korea. Especially, an 80% RG to HH mixture has the potential to ameliorate fatigue condition induced by oxidative stress by increasing the fatigue-related biochemical parameters such as glycogen and ATP contents in C2C12 myoblasts. Therefore, 80% RG to HH mixture can be used in food for ameliorating fatigue in Korea.

ABBREVIATIONS

HH: *Hippocampus abdominalis*

RG: Red ginseng

ACKNOWLEDGEMENTS

This research was financially supported by the Ministry of Trade, Industry, and Energy (MOTIE), Korea, under the "Regional Specialized Industry Development Program" supervised by the Korea Institute for Advancement of Technology (KIAT).

REFERENCES

1. Deshmukh AS, Murgia M, Nagaraj N, Treebak JT, Cox J, Mann M. Deep proteomics of mouse skeletal muscle enables quantitation of protein isoforms, metabolic pathways, and transcription factors. Mol Cell Proteomics. 2015;14:841–53.
2. Fan H, Tan Z, Hua Y, Huang X, Gao Y, Wu Y, Liu B, Zhou Y. Deep sea water improves exercise and inhibits oxidative stress in a physical fatigue mouse model. Biomed Rep. 2016;4:751–7.
3. Huang L-Z, Huang B-K, Ye Q, Qin L-P. Bioactivity-guided fractionation for anti-fatigue property of Acanthopanax senticosus. J Ethnopharmacol. 2011;133:213–9.
4. Huang W-C, Lin C-I, Chiu C-C, Lin Y-T, Huang W-K, Huang H-Y, Huang C-C. Chicken essence improves exercise performance and ameliorates physical fatigue. Nutrients. 2014;6:2681–96.
5. Huang W-C, Chiu W-C, Chuang H-L, Tang D-W, Lee Z-M, Wei L, Chen F-A, Huang C-C. Effect of curcumin supplementation on physiological fatigue and physical performance in mice. Nutrients. 2015;7:905–21.
6. Jiang D-Q, Guo Y, Xu D-H, Huang Y-S, Yuan K, Lv Z-Q. Antioxidant and anti-fatigue effects on anthocyanins of mulberry juice purification (MJP) and Mulberry marc purification (MMP) from different varieties mulberry fruit in China. Food Chem Toxicol. 2013;59:1–7.
7. Kang S-M, Heo S-J, Kim K-N, Lee S-H, Jeon Y-J. Isolation and identification of new compound, 2, 7″-phloroglucinol-6, 6′-bieckol from brown algae, Ecklonia cava and its antioxidant effect. J Funct Foods. 2012;4:158–66.
8. Kang N, Ko S-C, Samarakoon K, Kim E-A, Kang M-C, Lee S-C, Kim J, Kim Y-T, Kim J-S, Kim H. Purification of antioxidative peptide from peptic hydrolysates of Mideodeok (Styela clava) flesh tissue. Food Sci Biotechnol. 2013;22:541–7.
9. Kim H-G, Cho J-H, Yoo S-R, Lee J-S, Han J-M, Lee N-H, Ahn Y-C, Son C-G. Antifatigue effects of Panax ginseng CA Meyer: a randomised, double-blind, placebo-controlled trial. PLoS One. 2013;8:e61271.

10. Kim S, Kim J, Lee Y, Seo MK, Sung DJ. Anti-fatigue effects of acute red ginseng intake in recovery from repetitive anaerobic exercise. Iran J Public Health. 2016;45:387–9.
11. Kumaravel K, Ravichandran S, Balasubramanian T, Siva Subramanian K, Bilal A. Antimicrobial effect of five seahorse species from Indian coast. Br J Pharmacol Toxicol. 2010;1:62–6.
12. Kumaravel K, Ravichandran S, Balasubramanian T, Sonneschein L. Seahorses—a source of traditional medicine. Nat Prod Res. 2012;26:2330–4.
13. Nam S-Y, Kim H-M, Jeong H-J. Anti-fatigue effect by active dipeptides of fermented porcine placenta through inhibiting the inflammatory and oxidative reactions. Biomed Pharmacother. 2016;84:51–9.
14. Nozawa Y, Yamada K, Okabe Y, Ishizaki T, Kuroda M. The anti-fatigue effects of the low-molecular-weight fraction of bonito extract in mice. Biol Pharm Bull. 2009;32:468–74.
15. Perera N, Godahewa G, Lee J. Copper-zinc-superoxide dismutase (CuZnSOD), an antioxidant gene from seahorse (Hippocampus abdominalis); molecular cloning, sequence characterization, antioxidant activity and potential peroxidation function of its recombinant protein. Fish Shellfish Immunol. 2016;57:386–99.
16. Qian Z-J, Kang K-H, Kim S-K. Isolation and antioxidant activity evaluation of two new phthalate derivatives from seahorse, Hippocampus kuda Bleeler. Biotechnol Bioprocess Eng. 2012;17:1031–40.
17. Segade Á, Robaina L, Otero-Ferrer F, García Romero J, Molina Domínguez L. Effects of the diet on seahorse (Hippocampus hippocampus) growth, body colour and biochemical composition. Aquacult Nutr. 2015;21:807–13.
18. Singh T, Singh K. Mitochondrial dysfunction and chronic fatigue syndromes: issues in clinical care (modified version). IOSR J Dental Med Sci. 2014;13:30–3.
19. Swamy M, Naveen S, Singsit D, Naika M, Khanum F. Anti-fatigue effects of polyphenols extracted from pomegranate peel. Int J Integr Biol. 2011;11:69–72.
20. Tang W, Zhang Y, Gao J, Ding X, Gao S. The anti-fatigue effect of 20 (R)-ginsenoside Rg3 in mice by intranasally administration. Biol Pharm Bull. 2008;31:2024–7.
21. Wu R-E, Huang W-C, Liao C-C, Chang Y-K, Kan N-W, Huang C-C. Resveratrol protects against physical fatigue and improves exercise performance in mice. Molecules. 2013;18:4689–702.
22. You L, Zhao M, Regenstein J-M, Ren J. In vitro antioxidant activity and in vivo anti-fatigue effect of loach (*Misgurnus anguillicaudatus*) peptides prepared by papain digestion. Food Chem. 2011;124:188–94.
23. Yu B, Lu Z-X, Bie X-M, Lu F-X, Huang X-Q. Scavenging and anti-fatigue activity of fermented defatted soybean peptides. Eur Food Res Technol. 2008;226:415–21.
24. Zhang Y, Yao X, Bao B, Zhang Y. Anti-fatigue activity of a triterpenoid-rich extract from Chinese bamboo shavings (Caulis bamfusae in taeniam). Phytother Res. 2006;20:872–6.

CHAPTER 16

Effect of a New Phosphorus Source, Magnesium Hydrogen Phosphate (MHP) on Growth, Utilization of Phosphorus, and Physiological Responses in Carp *Cyprinus Carpio*

Tae-Hyun Yoon[1], Seunggun Won[2], Dong-Hoon Lee[3], Jung-Woo Choi[1], Changsix Ra[1] and Jeong-Dae Kim[1*]

[1]College of Animal Life Sciences, Kangwon National University, Chuncheon 24341, Korea. [2]Department of Animal Resources, Daegu University, Gyeongsan 38453, Korea.

[3]Gyenoggi Province Maritime and Fisheries Research Institute, 23-2 Sangkwang-Gil, Yangpyeong-Gun 12513, Korea.

ABSTRACT

Magnesium hydrogen phosphate (MHP, $MgHPO_4$) recovered from swine manure was prepared as an alternative phosphorus (P) source. Conventional P additives, monocalcium phosphate (MCP), dicalcium phosphate (DCP), and tricalcium phosphate (TCP) were compared with the MHP in terms of growth and P availability by juvenile carp *Cyprinus carpio*. A basal diet as a negative control was prepared using practical feed ingredients without P supplementation to which four supplemental P sources were added at the level of 2%. Five groups of 450 fish having mean body weight of 6.5 g following 24 h fasting after 2 weeks of adaptation period were randomly distributed into each of 15 tanks (30 fish/tank). Fish were hand-fed to apparent satiety twice a day for 9 weeks. Fish fed the MHP had weight gain (WG), feed conversion ratio (FCR), protein efficiency ratio (PER), and specific growth rate (SGR) comparable to those fed the MCP. Those values of both the MHP and MCP groups were significantly different ($p < 0.05$) from the other groups. Fish groups fed control and the TCP showed the lowest WG, PER, and SGR and the highest FCR among treatments. No fish were died among treatments during the experimental period. Fish fed control and the TCP showed hematocrit and hemoglobin significantly lower ($p < 0.05$) than fish fed the MHP. The lowest inorganic P (Pi) in plasma was found in the control group. Even though Pi was not significantly different ($p > 0.05$) from other phosphate groups, fish fed the MCP and MHP retained higher P in whole body than the other groups. P availability was determined to be 93.2, 62.4, 6.1, and 98.0% for MCP, DCP, TCP, and MHP, respectively. The present results suggested that the MHP recovered from wastewater stream could be used as an alternative P source in carp diet.

Keywords: *Cyprinus carpio* Weight gain Feed conversion ratio Phosphorus availability Alternative phosphorus Magnesium hydrogen phosphate

BACKGROUND

Phosphorus (P) is a component of organic phosphate in diverse forms, such as nucleotides, phospholipids, coenzymes, deoxyribonucleic acid, and ribonucleic acid. Inorganic phosphates also serve as important buffers to maintain normal pH of intra- and extra-cellular fluids (Zubay 1983). Fish have the dietary requirement of P like other animals for growth. The requirement was reported to range from 0.3 to 0.6% for channel catfish (Wilson et al. 1982), rainbow trout (Rodehutscord and Pfeffer 1995), striped bass (Brown et al. 1993; Dougall et al. 1996), and white fish (Vilema et al. 2002). However, somewhat higher requirement value of 0.7 to 0.9% was estimated for haddock (Roy and Lall 2003), red tilapia (Phromkunthong and Udom 2008), African catfish (Nwanna et al. 2009), red drum (Davis and Robinson 1987), yellow croaker (Ma et al. 2006), and carp (Kim et al. 1998). On the other hand, P discharged into water stream as a major nutrient for eutrophication is known to play a vital role in promoting algal growth (Beveridge 1984; Auer et al. 1986).

Theoretically, diet for fish should contain all essential nutrients including P above the needs and maintain total P levels as low as its available requirements in order to achieve both maximal growth and minimum P discharge. However, the practical diet contains an excessive P originating from animal and plant sources, which is excreted to the cultural water body due to the low availability by stomachless species like carp (Kim and Ahn 1993). Therefore, soluble P sources like monocalcium phosphate (MCP) and dicalcium phosphate (DCP) are being supplemented to the diet to meet the requirement of P for maximum growth. Based on this point of views, some significant discharge of P is inevitable from fish farming. Kim et al. (1998) reported dietary available P of 0.7% with 2% MCP exerted both maximal growth and minimum P loss in juvenile carp. An adequate combination of low-P protein meals supplemented with 0.5% MCP significantly reduced the P loading on rainbow trout without compromising the growth (Satoh et al. 2003; Hernandez et al. 2004, 2005).

Such an important source for living creatures as one of non-renewable elements in the nature is mostly produced from phosphate rock of which production in the world has constantly increased from 198 million metric tons in 2011 to 210 million metric tons in 2012 according to U.S. Geological Survey (USGS 2013). Shu et al. (2006) expected all reserves of phosphate rock would be depleted by 2090, assuming an annual increase of 1.5% in its demand. Considering this estimation, it is necessary to recover phosphorous from diverse downstream of agricultural and industrial fields. One of resources to retrieve phosphorous could be swine manure which contains high levels of phosphorous and nitrogen. Swine manure becomes a source of pollution on surface waters and induces eutrophication near the site when it is under improper treatment. Thus, the control of wastewater stream must be achieved and struvite precipitation could be an effective way to control phosphorus from wastewater stream with the addition of magnesium (Liu et al. 2011). If this alternative phosphate recovered from swine manure could be effectively employed as an available P source for animals including fish, it would not only substitute for import of several phosphates but also protect our environment through recycling of the waste source.

Cyprinid is the species the most cultured worldwide. In 2012, China produced more than 90% of the world's carp which relies on formulated feed with the exception of filter-feeding species (Cao et al. 2015). Considering the low availability of P in both animal and plant feedstuffs for carp (NRC 1983), the amount of dietary P supplemented through its additive would be substantial. Nevertheless, the potential use of dietary alternative phosphate as an available P source for growth of the fish was not examined until now. This study was therefore carried out to investigate the supplemental effect of various conventional phosphate sources and the alternative (MHP) on growth, feed utilization, plasma inorganic P, whole body P, and the availability of juvenile carp.

METHODS

The protocol for the present experiment was approved by the Institutional Animal Care and Use Committee of Kangwon National University, Chuncheon, Republic of Korea. Commercially selling phosphate additives (MCP, BIOFOS®, USA; DCP, SICHUAN MIANJHUSANJIA FEED Co., China; tricalcium phosphate (TCP), FOODCHEM, Shandong, China) were obtained from fish feed companies in Korea, and the testing P additive, magnesium hydrogen phosphate (MHP, $MgHPO_4$), was prepared from P-recovery process in swine farm operated by Kangwon National University.

A P-recovery process was a pilot scale with the effective volume of 400 L, and the operational condition was set with hydraulic retention time of 3 h and pH 8 to 9 controlled by aeration (33 L air/$m^3{}_{reactor}$.min) resulting in CO_2 stripping. Such a condition leads to precipitate crystal called struvite which is composed of magnesium, ammonium, and phosphate in equimolar ratio. In order to provide magnesium source that is a deficient source in swine manure, magnesium chloride ($MgCl_2$) was added to meet Mg to P ratio of approximately 1.2.

The precipitates collected was struvite, which was identified by X-ray diffractometer (XRD) (Rigaku, Model D/Max-2500 V, Japan). The MHP was obtained by removing ammonium-N through incineration of the recovered struvite at 550C for 30 min. It was finely ground to use as phosphate additive.

Preparation of diets

The diet was prepared with MCP, DCP, TCP, and MHP at the level of 2% but the control diet did not contain phosphate source of 2% in lieu of cellulose. The major ingredients were constituted with fish meal (25%), soybean meal (40%), wheat flour (27%), fish oil (2%), and soy oil (2%) as a basal diet containing 42.5% protein and 6.5% lipid (Table 1). Prior to diet formulation, chemical composition of fish meal, soybean meal, wheat flour, and four phosphate sources were determined. All the ingredients were weighed following the formula and ground to 100-mesh size by a hammer mill and thoroughly mixed for 10 min using a V-mixer (Hangjin Co., Korea) in order to make a mixture of 500 kg per diet. Then, the mixture was transferred to a twin extruder (Model ATX-2, Fesco Precision Co., Korea) and manufactured to the sinking pellets with two sizes of 1.5 and 2.5 mm, respectively, for feeding during the first 6- and second 3-week growth trials, respectively. Operating conditions of extrusion were as follows: feeder speed, 16~18 rpm; conditioner temperature, 80~90C; main screw speed, 250~320 rpm; temperature of the second and the third barrel compartment, 105~135C; steam heater pressure, 4~6 kgf/cm^2 and temperature of the fourth barrel compartment, 80~90C. Extruded pellets were dried at 60C for 6 h using a drying oven which resulted in the moderate moisture content of 5 to 8%.

Table 1 Ingredient and chemical composition of the experimental diets[a]

	Diet				
Ingredient (%)	**Control**	**MCP**	**DCP**	**TCP**	**MHP**
Fish meal	25.00	25.00	25.00	25.00	25.00
Soybean meal	40.00	40.00	40.00	40.00	40.00
Wheat flour	27.18	27.18	27.18	27.18	27.18
Soya oil	2.00	2.00	2.00	2.00	2.00
Fish oil	2.00	2.00	2.00	2.00	2.00
Vit. mix[a]	0.70	0.70	0.70	0.70	0.70
Min. mix[b]	0.30	0.30	0.30	0.30	0.30
Lysine-HCl	0.30	0.30	0.30	0.30	0.30
DL-methionine	0.20	0.20	0.20	0.20	0.20
Choline-HCl	0.30	0.30	0.30	0.30	0.30
Antioxidant	0.02	0.02	0.02	0.02	0.02
P source	–	2.00	2.00	2.00	2.00
Cellulose	2.00	–	–	–	–
Total	100.00	100.00	100.00	100.00	100.00
Composition (%, DM)					
C. protein	42.27	42.85	42.20	42.70	42.52
C. fat	6.90	6.67	6.62	6.65	6.65
C. ash	8.41	9.38	9.77	10.33	9.63
Ca	1.74	1.93	2.16	2.30	1.75
P	1.20	1.57	1.53	1.51	1.53
Av. P [c]	0.32	0.66	0.52	0.34	0.64

MCP, monocalcium phosphate, *DCP* dicalcium phosphate, *TCP* tricalcium phosphate, *MHP* magnesium hydrogen phosphate

[a]Vitamin added to supply the following (per kg diet): vitamin A, 4000 IU; vitamin D_3, 800 IU; vitamin E, 150 IU; vitamin K_3, 20 mg; thiamine HCl, 25 mg; riboflavin, 50 mg; D-Ca pantothenate, 100 mg; biotin, 1 mg; folic acid, 20 mg; vitamin B_{12}, 0.2 mg; niacin, 200 mg; pyridoxine HCl, 20 mg; ascorbic acid, 500 mg; inositol, 200 mg; BHT, 15 mg; BHA, 15 mg

[b]Mineral added to supply the following (per kg diet): copper sulfate (25.4% Cu), 30.5 mg; zinc sulfate (22.7% Zn), 230 mg; manganous sulfate (32.5% Mn), 100 mg; cobalt chloride (24.8% Co), 20 mg; potassium iodide (76.4% I), 6.5 mg; sodium selenite (45.6% Se), 2.2 mg; sodium fluoride (45.2% F), 8 mg

[c]Available P calculated based on apparent digestibility coefficient values

For the measurement of P digestibility, chromic oxide of 1.0% was mixed with an aliquot of 10 kg of each extruded diet fully ground. Then, each diet mixture was added with 20% distilled water, and the mixture was pelletized using a meat chopper and dried for 12 h in a ventilated oven at 60C. The diets were stored in a freezer at −20C for P digestibility measurement until the growth trials.

Growth trial

Carp fry of 5000 with around body weight of 5 g were kindly provided from Kyeongnam freshwater fish institute and acclimated to the experimental conditions for 3 weeks with a control diet. Following a 24-h fasting, five groups (triplicates/group) of 450 fish of a mean body weight of 6.6 g were randomly allotted to each of 15 tanks (0.4 × 0.6 × 0.36 cm, water volume of 66 L). The feeding experiment lasted 9 weeks during which each diet was hand-fed to apparent satiety twice a day (08:30 and 17:30) at the 4% of body weight in every 6 days per week. A recirculation freshwater system was employed where water temperatures for first, second, and third weeks were maintained at 26 ± 1.2, 22 ± 1.2, and 18 ± 1.5C, respectively, and dissolved oxygen 5.5~6.4 mg O_2/L for the whole experimental periods. The flow rate of 5 L/min was constantly set. The extruded pellet of 2.5 mm size was fed for the last feeding of 3 weeks. Fish were bulk-weighed at the beginning of the experiment in every 3 weeks. Daily feed intake (DFI, %/av. body weight/d), weight gain (WG, %), feed conversion ratio (FCR), protein efficiency ratio (PER), specific growth rate (SGR, %), and survival rate (SR, %) were calculated as follows:

DFI (%/av. Body wt/d)=feed intake (g, DM)/((initial wt+final wt)/2)/experimental days x 100,

WG (%)=(final weight (g) initial weight (g))×100/initial weight (g),

$$FCR = \text{feed intake (g, DM)}/\text{wet weight gain (g)},$$

$$PER = \text{wet weight gain (g)}/\text{protein intake},$$

$$SGR\ (\%) = (\text{Ln final weight (g)} - \text{Ln initial weight (g)})/\text{experimental days} \times 100,$$

and

$$SR\ (\%) = \text{final fish number}/\text{initial fish number} \times 100.$$

Digestibility trial

At the end of the growth trial, digestibility measurement was conducted to calculate the available P of the experimental diets. Following a 24-h fasting, 300 fish (mean body weight, 22.2 g) were randomly distributed into each of five 130 L capacity tanks (60 fish/tank) with a fecal collection column. Following 1 week of feeding, fecal collections were made for three consecutive weeks as described by Kim et al. (2006). Each diet was fed by hand to apparent satiety twice a day (08:30 and 16:30). One hour after final feeding of the day, the drain pipes and fecal collection columns were thoroughly cleaned with a brush to remove feed residues and feces from the system. The settled feces and surrounding water were carefully collected into 250-ml centrifuge bottles each morning (08:00). Apparent digestibility coefficient (ADC) of P in experimental diets was calculated according to Maynard and Loosli (1969). P availability of phosphate additives was calculated by dividing the difference of total P into that of available P between control and each phosphate containing diet:

$$ADC\ (\%) = \left(1 - \frac{ID \times PF}{IF \times PD}\right) \times 100$$

where ID is % indicator in the diet, PF represents % P in the feces, IF indicates % indicator in the feces, and PD is % P in the diet.

P availability of the phosphorus additives was calculated according to the following equation:

$$P\ \text{availability}\ (\%) = \frac{APDP\text{-}APCD}{TPDP\text{-}TPCD} \times 100$$

where APDP indicates % available P in the diet containing P source, APCD is % available P in control diet, TPDP shows % total P in the diet containing P source, and TPCD is % total P in control diet.

Sample collection and analysis

At the end of the experimental period, fish were anesthetized with AQUI-S (New Zealand Ltd., Lower Hutt, NZ) and bulk-weighed and counted for calculation of WG, FCR, SGR, PER, and SR. Blood samples were obtained from the caudal vessels with a heparinized syringe from two fish of each tank after fish were starved for 24 h and anesthetized with AQUI-S. Feces collected in the same bottle from each tank for 6 days a week were used as one replicate for the treatment. After collection of three replicate samples of each diet during 3 weeks, fecal samples were lyophilized, finely ground, and frozen at −20C until analysis.

Chemical analyses of feed ingredient, diets, and feces were performed by the standard procedure of AOAC (1990) for moisture, crude protein, crude fat, and crude ash. Moisture content was obtained after drying in an oven at 105C for 24 h. Crude protein ($N \times 6.25$) was determined by Kjeldahl method after acid digestion. Crude fat was determined by the soxhlet extraction method by using Soxtec system 1046 (Foss, Hoganas, Sweden) and crude ash from incineration in a muffle furnace at 550C for 12 h. Chromium in diets and feces for P digestibility measurement was analyzed using a spectrophotometer (Shimadzu, UV-120-12) at a wavelength of 440 nm after perchloric acid digestion (Bolin et al. 1952). Ca in diets and P in diets and the whole body of the final fish (five fish per replicate) were measured using inductively coupled plasma mass spectrometer (ICP-MS) (Perkin-Elmer, NexION 300D, Waltham, MA, USA) after the pretreatment of test materials following the method from US Environmental protection agency (USEPA 1996). Hematocrit (PCV, %) and hemoglobin (Hb, g/dL) were measured with the same fish (two fish per replicate) by the microhematocrit method (Brown 1980) and the cyanmethemoglobin procedure using Drabkin's solution, respectively. Hb standard prepared from human blood (Sigma Chemical, St. Louis, MO) was employed. Blood plasma from two fish per replicate was obtained after blood centrifugation (3500×*g*, 5 min, 4C) and stored at −80C until inorganic P (Pi) was analyzed. The plasma Pi was measured using a blood chemical analyzer (HITACHI 7600-210, Hitachi High-Technologies Co., Ltd., Japan) with commercial clinical investigation reagent (Clinimate IP, Sekisui medical Co., Ltd., Tokyo, Japan).

Statistical analysis

Data of growth trial (initial and final fish, DFI, WG, FE, PER, SGR, and SR), P availability among both replications and treatments, whole body P, and hematological and serological parameters were analyzed using one-way analysis of variance (ANOVA), and significant differences among treatment means were compared using Duncan's multiple range test (Duncan 1955). Prior to the analysis, homogeneity of variance of all data was verified using Cochran's test (Sokal and Rohlf 1994). All statistical analyses were carried out using the SPSS Version 10 (SPSS 1999). Statistical significance of the differences was determined by a significant level of 5% ($p < 0.05$).

RESULTS

Growth performance, whole body P, and P digestibility

As shown in Table 1, crude protein level of dry diets was determined to be 42.2 (DCP) to 42.9% (MCP) and crude fat to be 6.6 (DCP) to 6.9% (control). At the end of the 9-week growth trial, daily feed intake (DFI, %) per averaged weight of fish ranged from 1.88 (MHP) to 2.17 (control and TCP). Fish fed the MHP showed the highest WG of 278%, which was not significantly different ($p > 0.05$) from that (270%) of fish fed the MCP, while fish fed the control and TCP showed lowest WG among the treatment groups ($p < 0.05$). FCR ranged from 0.99 (MHP) to 1.29 (control). The highest value of PER and SGR were found in fish fed the MCP and MHP, while fish fed the control and TCP showed the lowest level (Table 2). Relative weight gain (RWG) was significantly higher in DCP, MCP, and MHP compared to that in control and TCP (Fig. 1). Fish fed the control showed the lowest P content (0.36%) in whole body, while those fed the MHP the highest (0.46%) after 9-week feeding trial. (Fig. 3). The significant difference in P availability of repetition was not shown from manure analyses every week. Apparent availability of P in the experimental diets varied from 24.2 to 41.9% for TCP and MHP, respectively. The values of P were found to be 93.2, 62.4, 6.1, and 97.8% for MCP, DCP, TCP, and MHP, respectively (Table 3). The available P level in diets (Table 1) was calculated based on determined P digestibility, which ranged from 0.32 (control) to 0.66% (MCP).

Table 2 Growth performance of carp fed diets containing various phosphorus sources for 9 weeks

Parameters	Diet				
	CON	MCP	DCP	TCP	MHP
Initial wt (g/fish)	6.40 ± 0.05[ns]	6.63 ± 0.07	6.63 ± 0.08	6.60 ± 0.08	6.61 ± 0.12
Final wt (g/fish)	19.19 ± 0.23[d]	24.53 ± 0.30[a]	22.41 ± 0.28[b]	20.14 ± 0.41[c]	24.94 ± 0.39[a]

DFI (%/av. wt/day)[1]	2.17 ± 0.02^{a}	1.91 ± 0.02^{c}	2.01 ± 0.03^{b}	2.17 ± 0.03^{a}	1.88 ± 0.03^{c}
WG (%)[2]	199.87 ± 5.68^{c}	269.85 ± 2.44^{a}	237.87 ± 2.76^{b}	205.19 ± 2.61^{c}	277.54 ± 12.59^{a}
FCR[3]	1.29 ± 0.03^{a}	1.00 ± 0.01^{c}	1.09 ± 0.01^{b}	1.28 ± 0.03^{a}	0.99 ± 0.03^{c}
PER[4]	1.83 ± 0.04^{c}	2.34 ± 0.03^{a}	2.17 ± 0.03^{b}	1.83 ± 0.04^{c}	2.39 ± 0.08^{a}
SGR(%)[5]	1.74 ± 0.03^{c}	2.08 ± 0.01^{a}	1.93 ± 0.01^{b}	1.77 ± 0.01^{c}	2.11 ± 0.05^{a}
SR(%)[6]	100 ± 0.00^{ns}	100 ± 0.00	100 ± 0.00	100 ± 0.00	100 ± 0.00

Values (means ± SE of triplicates) with different superscript letters in the same row are significantly different ($p < 0.05$)

ns nonsignificant

[1]Daily feed intake (%/av. wt/day) = dry feed intake (g/fish)/((initial wt + final wt)/2)/experimental days × 100

[2]Weight gain (%) = (final weight (g) – initial weight (g)) × 100/initial weight (g)

[3]Feed conversion ratio = dry feed intake/wet weight gain

[4]Protein efficiency ratio = wet weight gain (g)/protein intake

[5]Specific growth rate (%) = (Ln final weight (g) – Ln initial weight (g))/experimental days × 100

[6]Survival rate (%) = final fish number/initial fish number × 100

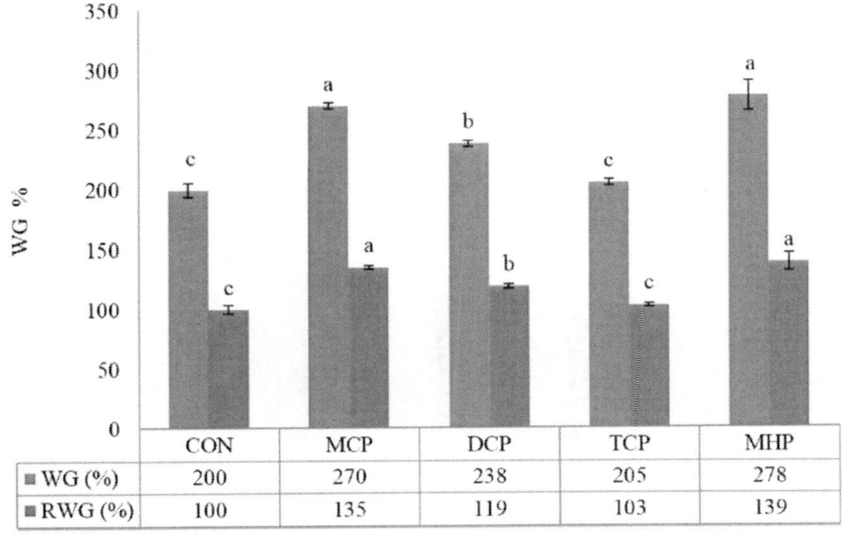

Figure. 1 Weight gain (WG) and relative WG (RWG) of juvenile carp fed various phosphorus sources for 9 weeks

Table 3 Apparent availability of phosphorus in the experimental diets and various phosphate sources

	P availability				
	CON	**MCP**	**DCP**	**TCP**	**MHP**
Diet	26.57 ± 1.07^{c}	42.27 ± 0.47^{a}	34.30 ± 0.46^{b}	22.37 ± 0.74^{d}	41.97 ± 0.68^{a}
Phosphorus sources		93.19 ± 1.96^{b}	62.42 ± 3.37^{c}	6.11 ± 2.02^{d}	97.97 ± 2.06^{a}

Means ± S.D of three replicates of each group with different superscript letters in the same row are significantly different ($p < 0.05$)

Hematological parameters and plasma inorganic P

Hematological parameters and inorganic P in plasma of fish fed the experimental diets are shown in Figs. 2and 3, respectively. PCV (%) of fish fed TCP (24.2) and control (25.5) showed a significant difference ($p < 0.05$) from that of fish fed MHP (28.8). Hb (g/dL) of fish ranged from 7.5 (TCP) to

8.8 (MHP), which were significantly different ($p < 0.05$). Significantly lower P (mg/dL) in plasma was found in fish fed the control (4.3), while the other groups did not show any significant difference ($p > 0.05$) each other.

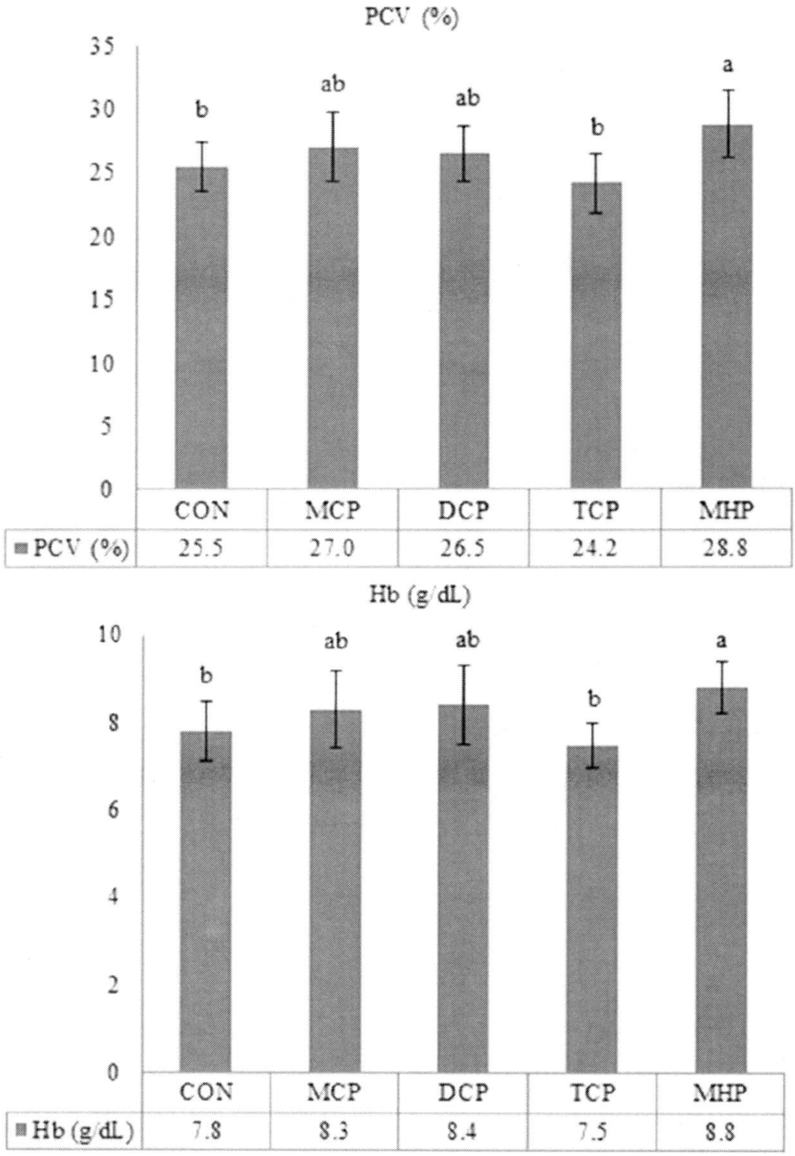

Figure. 2 Hematocrit (PCV) and hemoglobin (Hb) levels of juvenile carp fed the experimental diets for 9 weeks

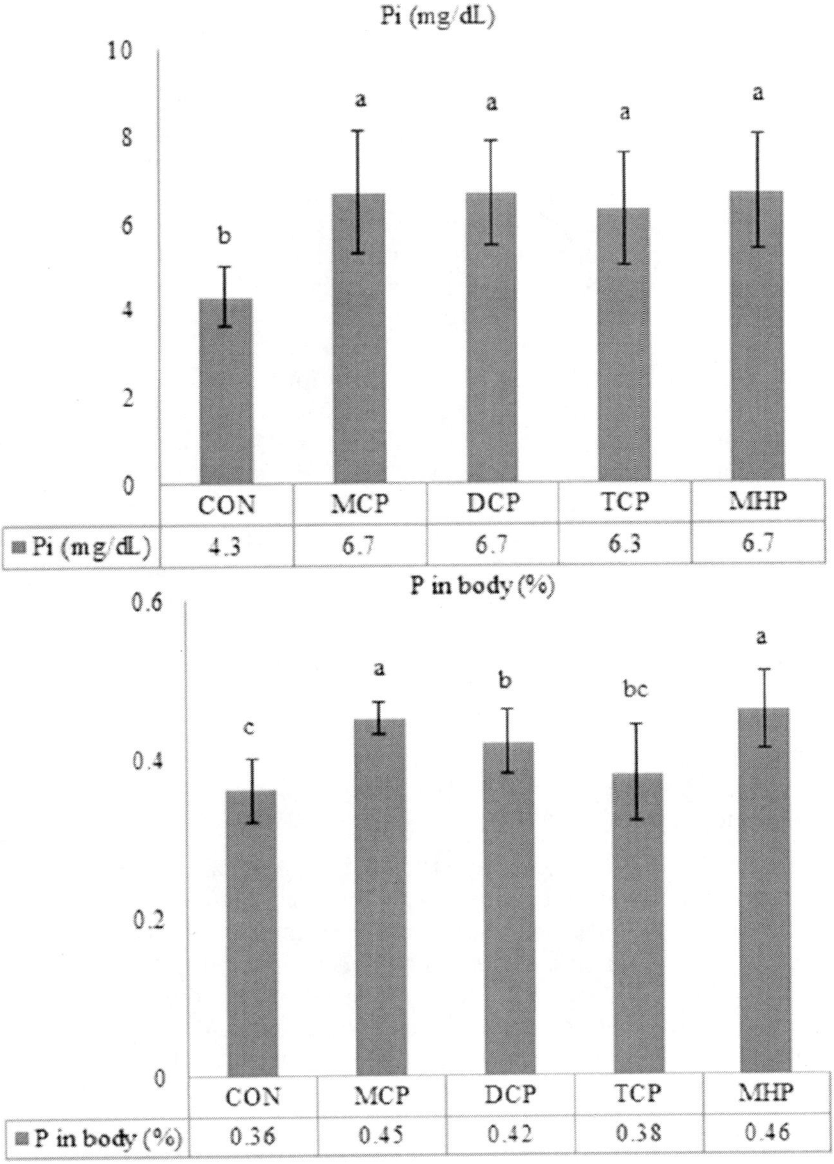

Figure. 3 Plasma inorganic phosphorus (Pi) and P in whole body of juvenile carp

DISCUSSION

As the environmental risks has been increased by releasing animal fecal wastes including fisheries, the P-recovery from swine manure has been introduced in the form of struvite and widely studied (Yoon et al. 2015). Its application has been only limited as a fertilizer but it is much desirable to apply P recovered to new applications after appropriate manufacturing and securing safety. Magnesium hydrogen phosphate (MHP) was newly manufactured from struvite which was to recycle P from swine manure. It is first attempt to investigate the effect of the MHP as dietary P source on growth and feed utilization of carp. Based on the earlier study (Kim et al. 1998), who found that 2% of MCP in diet for carp improved growth and decreased P loads, the same level of various P sources was incorporated into the experimental diets (Table 1). Fish fed the MHP showed the best WG at the end of growth trial, although it was not significantly different ($p > 0.05$) from that of fish fed the MCP. Same tendencies were found in FCR, PER, and SGR (Table 2). The result suggests that juvenile carp could utilize P from MHP as effectively as MCP. From the results, it was evident that MHP was a good P source competitive with MCP in terms of WG and FCR in juvenile carp. As shown in Fig. 1, RWG of MCP, DCP, TCP, and MHP to control at the level of 2% in each diet were 135, 119, 103, and 139% on WG, respectively. Such relative differences among various P sources might be due to the difference in availability of P, by which the requirement could be met or not. Available P requirement is known to be 0.6~0.7% (Ogino and Takeda 1976; Kim et al. 1998) for carp. The present study confirmed such a requirement to be optimal.

Ogino et al. (1979) extensively studied the P availability of inorganic P sources and various feed ingredients by fish. They reported that the availability of MCP, DCP, and TCP was 94, 46, and 13 by carp, respectively, using egg albumin-based diet. Though somewhat higher values were obtained for DCP and TCP in the present study, the value for MCP corresponds well with their result. On the other hand, MHP had the higher P availability than MCP, suggesting its potential use for carp. Similar results were reported in the previous study conducted using juvenile catfish (Yoon et al. 2014). They found that MHP (90.9%) had P availability comparable to MCP (88.1%) at 2% incorporation level in diet. Sarker et al. (2009) determined P availability of different P sources by yellowtail using albumin-based diet. They incorporated MCP (2.7%), DCP (3.7%), and TCP (3.6%) into the respective diets and obtained P availabilities of 92.4, 59.2, and 48.8% for MCP, DCP, and TCP, respectively. Lovell (1978) reported that the availability was found to be 94 and 65% for reagent grade MCP and DCP, respectively, in channel catfish. On the other hand, Eya and Lovell (1997) determined the net absorption of P from various P sources using all-plant basal diet in channel catfish. They obtained the values of 81.2, 74.8, and 54.8% for MCP, DCP, and TCP, respectively.

To our knowledge, this is the first time P availability of feed grade MCP, DCP, and TCP as well as MHP was determined for juvenile carp. In the present study, P availabilities of both MCP and DCP were found to be high while that of TCP was very low (Table 3). P availability of MCP seems to be comparable to those reported by Ogino et al. (1979), Lovell (1978), Kim et al. (1997), and Sarker et al. (2009). However, it was higher than those reported by Pimentel-Rodrigues and Oliva-Teles (2007) and Eya and Lovell (1997). The availability of DCP was comparable to those obtained in yellowtail (Sarker et al. 2009) and European sea bass (Pimentel-Rodrigues and Oliva-Teles 2007), while it was lower than those obtained in channel catfish (Lovell 1978), rainbow trout (Ogino et al. 1979), and far eastern channel catfish (Yoon et al. 2014). On the other hand, reported P availability of TCP by stomach fish ranged from 42% (Pimentel-Rodrigues and Oliva-Teles 2007) to 64% (Ogino et al. 1979). In contrast, the value from the present study was as low as that obtained by carp, stomachless species (Ogino et al. 1979). Bioavailability of dietary P is influenced by several factors including chemical form, digestibility of diet, particle size, interaction with other nutrients, feed processing, and water chemistry (Lall 1991). It remains to be explained whether such discrepancies in P availability of various P sources, especially TCP, are due to the differences in species and diet composition employed for the digestibility measurement as well as methodological approach in fecal collection (Kim et al. 1996). On the other hand, P availability of MHP was higher than those of MCP obtained from carp and rainbow trout by Ogino et al. (1979) and from yellowtail by Sarker et al. (2009), suggesting that MHP could be a potential P source for fish.

Hematological and serological parameters are useful in monitoring the physiological status of fish and as indicators of the health of the aquatic environment, although they are not routinely used in fish disease diagnosis (El-Sayed et al. 2007). Hematocrit (PCV, %) provides an indirect measurement of the body's oxygen carrying ability, while Hb (g/dL) a direct measurement of the oxygen carrying capacity of the blood (McClatchey 2002). It was reported that hematological parameters could be influenced by nutritional status (Spannhof et al. 1979), infectious disease (Barham et al. 1980; Iwama et al. 1986), environmental changes (Giles et al. 1984), and stress (Ellsaesser and Clem 1986). The normal ranges of healthy adult Atlantic salmon were reported 44 to 49 and 8.9 to 10.4 for PCV and Hb, respectively (Sandnes et al. 1988). Somewhat lower values for cichlid fish were reported by V′azquez and Guerrero (2007), which were 22.5 to 39.2 for PCV and 5.2 to 8.3 for Hb. Recently, Rahimnejad and Lee (2013) reported 30.7 to 34.3 for PCV and 4.4 to 5.4 for Hb of red sea bream fed various dietary valine levels. Our findings for the parameters are in good agreement with those obtained from tilapia (Hrubec et al. 2000) and striped bass (Hrubec et al. 2001), although there is no available information on the effect of dietary available P on the parameters of fish. In the present study (Fig. 2), PCV and Hb of fish groups fed low available P in diet (control and TCP) were significantly lower than the MHP group ($p < 0.05$). Differently from the present results, Yoon et al. (2014) observed that dietary low P resulted in PCV significantly lower but did not affect Hb. An increase in available P in diet resulted in an increase in plasma Pi (Vielma and Lall 1998; Bureau and Cho 1999; Avila et al. 2000). A clear evidence was observed by Yoon et al. (2014) that plasma Pi increased in fish fed MCP and MHP with higher available P, resulting in the significant improvement in WG by meeting dietary P need. However, any significant difference in plasma Pi was not found in fish groups fed diets with various P sources ($p > 0.05$) in the present study. Nevertheless, P in whole body was greatly affected by dietary available P levels (Fig. 3), suggesting successive P intake through diet maintains plasma Pi constant then deposition in skeletal tissues occurs when P requirement is met. More researches remain to be clarify the difference in plasma Pi level observed in previous (Yoon et al. 2014) and present studies.

CONCLUSIONS

In conclusion, the potential use of MHP recovered from swine manure was proven sufficiently to replace MCP as an alternative P source with respect to WG and FCR as well as P availability. "Such a re-use of P from swine manure could positively influence on the development of other useful sources from livestock manure."

ABBREVIATIONS

DCP: Dicalcium phosphate; **DF:**Daily feed intake; **FCR:**Feed conversion ratio; **Hb:** Hemoglobin; **MCP:** Monocalcium phosphate; **MHP:** Magnesium hydrogen phosphate; **P:** Phosphorus; **PCV:** Hematocrit**PER:** Protein efficiency ratio;**Pi:** Inorganic P; **RWG:** Relative weight gain; **SGR:** Specific growth rate; **SR:** Survival rate; **TCP:** Tricalcium phosphate; **USGS:** U.S. Geological Survey; **WG:** Weight gain; **XRD:** X-ray diffractometer

ACKNOWLEDGEMENTS

We are grateful to the funding by Rural Development Administration, Korea. Also, this study was partially supported by 2015 Research Grant from Kangwon National University.

Funding

This study was supported by the grant from Rural Development Administration, Korea (PJ008390) for the design of study, experimental procedures, and interpretation of data derived from this study. In addition, this study was partially supported by 2015 Research Grant from Kangwon National University (No. 520150143) for the experiments in this manuscript.

REFERENCES

1. AOAC. Official Methods of Analysis. Association of Official Analytical Chemists:Arlington, VA, USA;1990. p. 1298.
2. Auer MT, Kiesser MS, Canale RP. Identification of critical nutrient levels through field verification of models for phosphorus and phytoplankton growth. Can J Fish Aquat Sci. 1986;43:379–88.
3. Avila EM, Tu H, Basantes S, Ferraris RP. Dietary phosphate regulates intestinal transport and plasma concentrations of phosphate in rainbow trout. J Comp Physiol. 2000;170B:210–09.
4. Barham WT, Smit GL, Schoonbee HJ. The haematological assessment of bacterial infection in rainbow trout, *Salmo gairdneri* Richardson. J Fish Biol. 1980;17:275–81.
5. Beveridge MCM. Cage and pen fish farming: carrying capacity models and environmental impact. FAO Fish Tech Pap. 1984;255:131.
6. Bolin DW, King RP, Klosrerman EW. A simplified method for the determination of chromic oxide (Cr_2O_3) when used as an inert substance. Science. 1952;116(3023):634–5.
7. Brown BA. Routine hematology procedures, Hematology: Principles and Procedures. Philadelphia: Lea and Febiger; 1980. p. 71–112.
8. Brown ML, Jaramillo Jr F, Gatlin III DM. Dietary phosphorus requirement of juvenile sunshine bass, *Morone chrysops* × *M. saxatilis*. Aquaculture. 1993;113:355–63.
9. Bureau DP, Cho CY. Phosphorus utilization by rainbow trout (*Oncorhynchus mykiss*): estimation of dissolved phosphorus waste output. Aquaculture. 1999;179:127–40.
10. Cao L, Naylor R, Henriksson P, Leadbitter D, Metian M, Troell M, Zhang W. China's aquaculture and the world's wild fisheries. Science. 2015;347:133–5.
11. Davis DA, Robinson EH. Dietary phosphorus requirement of juvenile red drum *Sciaenops ocellatus*. J World Aquacult Soc. 1987;18:129–36.
12. Dougall DS, Wood III LC, Douglass LW, Soares JH. Dietary phosphorus requirement of juvenile striped bass *Morone saxatilis*. J World Aquacult Soc. 1996;27:82–91.
13. Duncan DB. Multiple range and multiple 'F' tests. Biometrics. 1955;11:1–42.
14. Ellsaesser CF, Clem LW. Haematological and immunological changes in channel catfish stressed by handling and transport. J Fish Biol. 1986;28:511–21.

15. El-Sayed YS, Saad TT, El-Bahr SM. Acute intoxication of deltamethrin in monosex Nile tilapia, *Oreochromis niloticus* with special reference to the clinical, biochemical and haematological effects. Environ Toxicol Phamacol. 2007;24:212–7.
16. Eya JC, Lovell RT. Net absorption of dietary phosphorus from various inorganic sources and effect of fungal phytase on net absorption of plant phosphorus by channel catfish *Ictalurus punctatus*. J World Aquacult Soc. 1997;28:386–91.
17. Giles MA, Majewski HS, Hobden B. Osmoregulatory and haematological responses of rainbow trout (*Salmo gairdneri*) to extended environmental acidification. Can J Fish Aquat Sci. 1984;41:1686–94.
18. Hernandez A, Satoh S, Kiron V, Watanabe T. Phosphorus retention efficiency in rainbow trout fed diets with low fish meal and alternative protein ingredients. Fish Sci. 2004;70:580–6.
19. Hernandez A, Satoh S, Kiron V. Effect of monocalcium phosphate supplementation in a low fish meal diet for rainbow trout based on growth, feed utilization, and total phosphorus loading. Fish Sci. 2005;71:817–22.
20. Hrubec TC, Cardinale JL, Smith SA. Hematology and plasma chemistry reference intervals for cultured tilapia (*Oreochromis* hybrid). Vet Clin Pathol. 2000;29:7–12.
21. Hrubec TC, Smith SA, Robertson JL. Age related changes in haematology and chemistry values of hybrid striped bass chrysops *Morone saxatilis*. Vet Clin Pathol. 2001;30:8–15.
22. Iwama GK, Greer GL, Randall DJ. Changes in selected haematological parameters in juvenile Chinook salmon subjected to a bacterial challenge and a toxicant. J Fish Biol. 1986;28:563–72.
23. Kim JD, Ahn KH. Effects of monocalcium phosphate supplementation on phosphorus discharge and growth of carp (*Cyprinus carpio*) grower. Asian Australas J Anim Sci. 1993;6:521–6.
24. Kim JD, Kim KS, Song JS, Lee JY, Jeong KS. Optimum level of dietary monocalcium phosphate based on growth and phosphorus excretion of mirror carp, *Cyprinus carpio*. Aquaculture. 1998;161:337–44.
25. Kim JD, Kim KS, Song JS, Jeong KS, Woo YB, Choi NJ, Lee JY. Comparison of feces collection methods for determining apparent phosphorus digestibility of feed ingredients in growing mirror carp (*Cyprinus carpio*). Korean J Anim Nutr Feed. 1996;20:201–6.
26. Kim JD, Kim KS, Lee SB, Jeong KS. Nutrient and energy digestibilities of various feedstuffs fed to Israeli strain of growing common carp (*Cyprinus carpio*). J Aquacult. 1997;10:327–34.
27. Kim JD, Tibbets SM, Milley JE, Lall SP. Effect of the incorporation level of herring meal into test diet on apparent digestibility coefficients for protein and energy by juvenile haddock, *Melanogrammus aeglefinus* L. Aquaculture. 2006;258:479–86.
28. Lall SP. Digestibility, metabolism and excretion of dietary phosphorus in fish. In: Cowey CB, Cho CY, editors. Nutritional Strategies and Aquaculture Wastes. Ontario: University of Guelph; 1991. p. 21–36.
29. Liu YH, Kumar S, Kwag JH, Kim JH, Kim JD, Ra CS. Recycle of electronically dissolved struvite as an alternative to enhance phosphate and nitrogen recovery from swine wastewater. J Hazard Mater. 2011;195:175–81.
30. Lovell RT. Dietary phosphorus requirements of channel catfish. Trans Am Fish Soc. 1978;107:617–21.
31. Ma K, Zhang C, Ai Q, Duan Q, Xu W, Zhang L, Luifu Z, Tan B. Dietary phosphorus requirement of large yellow croaker, *Pseudosciaena crocea* R. Aquaculture. 2006;251:346–53.
32. Maynard LA, Loosli JK. Animal nutrition. 6th ed. New York: McGraw-Hill; 1969. p. pp 613.
33. McClatchey KD. Clinical laboratory medicine. Baltimore: Williams & Wilkins; 2002. p. 833–9.
34. NRC. Nutrient requirements of warmwater fishes and shellfishes. Washington: National Academy Press; 1983.

35. Nwanna LC, Adebayo IA, Omitoyin BO. Phosphorus requirements of African catfish, *Clarias gariepinus*, based on broken-line regression analysis methods. Science Asia. 2009;35:227–33.
36. Ogino C, Takeda H. Mineral requirements in fish-III calcium and phosphorus requirements in carp. Bull Jap Soc Sci Fish. 1976;42:793–9.
37. Ogino C, Takeuchi L, Takeda H, Watanabe T. Availability of dietary phosphorus in carp and rainbow trout. Bull Jap Soc Sci Fish. 1979;45:1527–32.
38. Phromkunthong W, Udom U. Available phosphorus requirement of sex-reversed red tilapia fed all-plant diets. Songklanakarin J Sci Technol. 2008;30:7–16.
39. Pimentel-Rodrigues A, Oliva-Teles A. Phosphorus availability of inorganic phosphates and fish meal in European sea bass (*Dicentrarchus labrax* L.) juveniles. Aquaculture. 2007;267:300–7.
40. Rahimnejad S, Lee KJ. Dietary valine requirement of juvenile red sea bream *Pagrus major*. Aquaculture. 2013;416–417:212–8.
41. Rodehutscord M, Pfeffer E. Requirement for phosphorus in rainbow trout (*Oncorhynchus mykiss*) growing from 50 to 200 g. Water Sci Technol. 1995;31:137–41.
42. Roy PK, Lall SP. Dietary phosphorus requirement of juvenile haddock (*Melanog rammus aeglefinus* L.). Aquaculture. 2003;221:451–68.
43. Sandnes K, Lie O, Waagbo R. Normal ranges of some blood chemistry parameters in adult farmed Atlantic salmon, *Salmo salar*. J Fish Biol. 1988;32:129–36.
44. Sarker PK, Fukada H, Masumoto T. Phosphorus availability from inorganic phosphorus sources in yellowtail (*Seriola quinqueradiata* Termminck and Schlegel). Aquaculture. 2009;289:113–7.
45. Satoh S, Hernandez A, Tokoro T, Morishita Y, Kiron V, Watanabe T. Comparison of phosphorus retention efficiency between rainbow trout (*Oncorhynchus mykiss*) fed a commercial diet and a low fish meal based diet. Aquaculture. 2003;224:271–82.
46. Shu L, Schneider P, Jegatheesan V, Johson J. An economical evaluation of phosphorus recovery as struvite from digester supernatant. Bioresour Technol. 2006;97:2211–6.
47. Sokal RR, Rohlf FJ. Nested analysis of variance, Biometry. New York: Freeman W. H; 1994. p. 272–342.
48. Spannhof L, Nasev D, Kreutzmann HL. Early recognition of metabolic disturbances in trout (*Salmo gairdneri* Rich.) stocks. Aquaculture. 1979;18:317–23.
49. SPSS Inc. SPSS Base 10.0 for Windows User's Guide. Chicago: SPSS Inc; 1999.
50. USEPA (US Environmental protection agency). 1996. Acid digestion of sediments, sludge, and solids, Method 3050B, Revision 2 (December 1996). Available at: https://www.epa.gov/sites/production/files/2015-06/documents/3050b.pdf.
51. USGS (US Geological survey). 2013. Mineral commodity summaries. Digital Data Series. Available at: http://minerals.usgs.gov/minerals /pubs/mcs/ 2013/mcs 2013.pdf.
52. V′azquez GR, Guerrero GA. Characterization of blood cells and hematological parameters in *Cichlasoma dimerus*(Teleostei, Perciformes). Tissue Cell. 2007;39:151–60.
53. Vilema J, Koskela J, Ruohonen K. Growth, bone mineralization, and heat and low oxygen tolerance in European whitefish (*Coregonnus lavaretus* L.) fed with graded levels of phosphorus. Aquaculture. 2002;212:321–33.
54. Vielma J, Lall SP. Control of phosphorus homeostasis of Atlantic salmon (*Salmo sala*) in fresh water. Fish Physiol Biochem. 1998;19:83–93.
55. Wilson RP, Robinson EH, Gatlin III DM, Poe WE. Dietary phosphorus requirement of channel catfish. J Nutr. 1982;112:1197–202.
56. Yoon TH, Lee DH, Won SG, Ra CS, Kim JD. Effects of dietary supplementation of magnesium hydrogen phosphate ($MgHPO_4$) as an alternative phosphorus source on growth and feed utilization of juvenile far eastern catfish (*Silurus asotus*). Asian Australas J Anim Sci. 2014;27:1141–9.

57. Yoon TH, Lee DH, Won SG, Ra CS, Kim JD. Optimal incorporation level of dietary alternative phosphate ($MgHPO_4$) and requirement for phosphorus in Juvenile far eastern Catfish (*Silurus asotus*). Asian Australas J Anim Sci. 2015;28:111–9.
58. Zubay GL. Biochemistry. Reading: MA, Addison-Wesley Publishing Company, Inc.; 1983. p. pp 388.

Index